Mechanical Properties and Characterization of Additively Manufactured Materials

The book highlights mechanical, thermal, electrical, and magnetic properties, and characterization of additive manufactured products in a single volume. It will serve as an ideal reference text for graduate students and academic researchers in diverse engineering fields including industrial, manufacturing, and materials science.

This text:

- Explains mechanical properties like hardness, tensile strength, impact strength, and flexural strength of additive manufactured components.
- Discusses characterization of components fabricated by different additive manufacturing processes including fusion deposition modelling and selective laser sintering.
- Highlights corrosion behaviour of additive manufactured polymers, metals, and composites.
- Covers thermal, electrical, and magnetic properties of additive manufactured materials.
- Illustrates intrinsic features and their influence on mechanical properties of additive manufactured products.

This text discusses properties, wear behaviour, and characterization of components produced by additive manufacturing technology. These products find applications in diverse fields including design, manufacturing and tooling, aerospace, automotive industry, and biomedical industry. It will further help the readers in understanding the parameters that influence the mechanical behaviour and characterization of components manufactured by additive manufacturing processes. It will serve as an ideal reference text for graduate students and academic researchers in the fields of industrial engineering, manufacturing engineering, automotive engineering, aerospace engineering, and materials science.

Mechanical Properties and Characterization of Additively Manufactured Materials

Edited by
Dr. K. Ravi Kumar,
Dr. S.C. Vettivel, and Dr. R. Subramanian

CRC Press
Taylor & Francis Group
Boca Raton London New York

CRC Press is an imprint of the
Taylor & Francis Group, an **informa** business

First edition published 2024
by CRC Press
2385 NW Executive Center Drive, Suite 320, Boca Raton, FL 33431

and by CRC Press
4 Park Square, Milton Park, Abingdon, Oxon, OX14 4RN

CRC Press is an imprint of Taylor & Francis Group, LLC

ISBN: 9781032392776 (hbk)
ISBN: 9781032553368 (pbk)
ISBN: 9781003430186 (ebk)

DOI: 10.1201/9781003430186

Typeset in Sabon
by codeMantra

Contents

Preface

The development of industries relies on innovative cutting-edge technologies related with the materials, manufacturing processes and product design. Currently, manufacturing industries are connected to complex product requirements, shorter life cycles, customized products and shorter delivery times. Therefore, there is always a need for design and development of new manufacturing processes. Additive manufacturing has the ability to convert any virtual shape design developed by CAD software into a real-time product. Additive manufacturing (AM) is a technique of merging materials layer by layer due to fusion, binding or solidification. The commonly used materials are resins, plastics and metals that may be in the form of powders, liquids or solids. Additive manufacturing develops components layer by layer using the three-dimensional CAD model. Terminologies such as Direct Digital Manufacturing (DDM), 3D printing (3DP), Solid Freeform Fabrication (SFF), and Rapid Manufacturing (RM) are used to describe additive manufacturing processes. AM is suitable for high design complexity, low production volumes and areas that require frequent changes in design. Additive manufacturing technology has exploited many areas like aerospace, automobile, transportation, health care, military and so on. Additive manufacturing will be the future for production in engineering and manufacturing processes.

In the current scenario, research work is going on in the area of additive manufacturing components for various applications like automobile, aeronautical, electronics, electrical, medical, engineering, etc. This book primarily focuses on the mechanical properties such as tensile strength, impact strength, flexural strength, hardness, wear, surface roughness, and material characterization of the components fabricated by different additive manufacturing technologies. Tremendous research is also taking place in the medical industry using additive manufacturing. This book also focuses on the properties and characterization of components used for engineering and medical applications. We believe that the book will serve as a reference for all the researchers in the domain of additive manufacturing.

Dr. K. Ravi Kumar
Dr. S. C. Vettivel
Dr. R. Subramanian

Editors

Dr. K. Ravikumar is currently the Professor and Associate Head of the Department of Mechanical Engineering at KPR Institute of Engineering and Technology, Coimbatore, India. He completed his B.E. Mechanical Engineering in 2001 from Bharathiyar University and M.E. Production Engineering in 2002 from Annamalai University. He received his Ph.D. degree from Anna University in the area of Composite Materials in 2013. He is into the teaching profession since 2003 and has handled engineering education in various capacities. His areas of research include composite materials, unconventional machining, optimization, characterization and additive manufacturing. He has received grants from agencies like ICMR, SERB, Anna University for conducting seminars, workshops and FDPs. He has also published five patents to his credit. He has published more than 30 research articles and two book chapters in various reputed publications. His publications have widely been cited by researchers. He has also presented and published more than 20 papers in various international and national conferences. He has authored a book on "CNC Technology" and also authored a textbook on Additive Manufacturing which is under production. He is an active reviewer and editorial board member for various international journals that include Elsevier, Taylor & Francis, SAGE, and De Gruyter publications. Presently, seven scholars are pursuing their Ph.D. programmes under his guidance. He is an active member of professional societies namely IE (I), ISTE and SAE.

Dr. Vettivel Singaravel Chidambaranathan is currently working in the Department of Mechanical Engineering, Chandigarh College of Engineering and Technology (Degree Wing), Sector 26, Chandigarh, India. He has completed his B.E. Mechanical Engineering in 2001 from Manonmaniam Sundaranar University, Tamil Nadu and M.E. Production Engineering in 2002 from Annamalai University, Tamil Nadu, India. He received his Ph.D. degree from Anna University, Chennai, Tamil Nadu, India, in the area of Nano Composite Materials through Powder Metallurgy in 2013. He has 20 years of teaching experience and has handled engineering education in various capacities. His areas of research include composite materials, unconventional machining, optimization, welding, characterization and additive manufacturing. He has been granted with one international patent and also published five patents to his credit. He has published four books, 70+ research articles and one book chapter in various reputed publications. Presently, two scholars are pursuing their Ph.D. programmes under his guidance. He is an active member of ISTE.

Prof. R. Subramanian is currently working as Professor, Department of Metallurgical Engineering, PSG College of Technology, Coimbatore. He completed his M.Tech. in Metallurgy from IIT Kanpur and Ph.D. from Bharathiar University. He has about 35 years of work experience in teaching and research. He has received funds worth nearly 100 lakhs from DRDO, DST and TATA STEEL Ltd to carry out research work. He has published more than 75 articles in reputed national and international journals. He has co-authored: two books (*Heat Treatment* and *Non-Destructive Testing and Powder Metallurgy*) and a chapter in a monograph on boron carbide. Twenty scholars have received Ph.D. under his guidance.

List of Contributors

Mahaboob Basha Adam Khan
Department of Mechanical Engineering
Kalasalingam Academy of Research & Education
Virudhunagar, India

Babu Raj Anushraj,
Department of Mechatronics Engineering
Chennai Institute of Technology
Chennai, India

Rajesh Kannan Arasappan
Post-doctoral Researcher, Mechanical Engineering
National Institute of Technology
Tiruchirappalli, India

Dhanabal A
Department of Mechanical Engineering
M/s. Fabforge Innovations Pvt. Ltd.
Coimbatore, India

A. Ganapathi
Periyar University
Salem, India

Senthilmurugan Arumugam
Department of Mechanical Engineering
Coimbatore Institute of Technology
Coimbatore, India

Kumaragurubaran Balasubramanian
University College of Engineering
Anna University
Trichy, India

Abhi Bansal
DPG Institute of Technology and Management
Gurugram, India

John Baruch
Department of Aeronautical Engineering
MVJ College of Engineering
Bangalore, India

Pushpendera S. Bharti
Guru Gobind Singh Indraprastha University
Delhi, India

Emmy Prema Chellam
Bethlahem Institute of Engineering
Karungal, India

Deepak Chhabra
Department of Mechanical Engineering
Maharshi Dayanand University
Rohtak, India

Vettivel Singaravel Chidambaranathan
Department of Mechanical Engineering
Chandigarh College of Engineering and Technology (Degree Wing)
Chandigarh, India

Jenson Joseph Earnest
SCMS School of Engineering and Technology
Ernakulam, India

Sivakumar Ganesan
Department of Mechanical Engineering
Coimbatore Institute of Technology
Coimbatore, India

Ramesh Kumar Garg
Deenbandhu Chhotu Ram University of Science and Technology
Sonipat, India

Gibin George
SCMS School of Engineering and Technology
Ernakulam, India

G. Paulraj
SRM TRP Engineering College
Tiruchirappalli, India

Alphonse Bovas Herbert Bejaxhin
Saveetha School of Engineering, SIMATS
Chennai, India

Nagarajan K. Jawaharlal
Thiagarajar College of Engineering
Madurai, India

Devasahayam Soosai Jenaris
PSN Engineering College
Tirunelveli, India

Arunprasath Kanagaraj
Department of Mechanical Engineering
PSN College of Engineering and Technology
Tirunelveli, India

K. Prakash
Periyar University
Salem, India

K. A. Ramesh Kumar
Periyar University
Salem, India

Pitchandi K
Department of Mechanical Engineering
Sri Venkateswara College of Engineering
Sriperumbudur, India

Ashish Kaushik
Deenbandhu Chhotu Ram University of Science and Technology
Sonipat, India

Francis Luther King
Department of Mechanical Engineering
Swarnandhra College of Engineering and Technology
Narsapur, India

Ravikumar K
Department of Mechanical Engineering
KPR Institute of Engineering and Technology
Coimbatore, India

Haiter Lenin
Kombolcha Institute of Technology
WOLLO University
Kombolcha, Ethiopia

Jatinder Madan
Department of Mechanical Engineering
Chandigarh College of Engineering and Technology (Degree Wing)
Chandigarh, India

M. Thilak
SRM TRP Engineering College
Tiruchirappalli, India

Sowrirajan Maruthasalam
Department of Mechanical Engineering
Coimbatore Institute of Engineering and Technology
Coimbatore, India

Kumar M
Department of Mechanical Engineering
KPR Institute of Engineering and Technology
Coimbatore, India

Vijayakumar Murugesan
Department of Mechanical Engineering
PSN College of Engineering and Technology
Tirunelveli, India

Mohandass Muthukrishnan
Department of Mechanical Engineering
Sri Venkateswara College of Engineering
Sriperumbudur, India

Navaneetha Krishnan Muthu Nadar
Amrita College of Engineering and Technology
Nagercoil, India

Aezhisai Vallavi Muthusamy Subramanian
Government College of Technology
Coimbatore, India

Sridhar Nagarajan
Government College of Technology
Coimbatore, India

Easwaran Naveen
Sri Sai Ram Engineering College
Chennai, India

Bahadur Singh Pabla
Department of Mechanical Engineering
National Institute of Technical Teachers Training and Research (NITTTR)
Chandigarh, India

P. Maadeswaran
Periyar University
Salem, India

Piychapillai Periyaswamy
St Peter's Institute of Higher Education and Research
Chennai, India

Arunkumar P
Department of Mechanical Engineering
KPR Institute of Engineering and Technology
Coimbatore, India

Selvakumar Ponnusamy
Department of Mechanical Engineering
PSN College of Engineering and Technology
Tirunelveli, India

Stanlykochappa Premila Jani
Department of Mechanical Engineering
Marri Laxman Reddy Institute of Technology and Management
Hyderabad, India

Upender Punia
Deenbandhu Chhotu Ram University of Science and Technology
Sonipat, India

Mohan Pushparaj
Department of Mechanical Engineering
Coimbatore Institute of Technology
Coimbatore, India

Raghav Gurumoorthy Raaja
SCMS School of Engineering and Technology
Ernakulam, India

Prashanna Rangan R
Department of Mechanical Engineering
Sri Venkateswara College of Engineering
Sriperumbudur, India

Pavendhan Rajangam
Mohamed Sathak Engineering College
Kilakarai, Tamilnadu, India.

R. Muraliraja
Vels Institute of Science, Technology and Advanced Studies
Chennai, India

Ashok Kumar Rajendran
SRM Madurai College for Engineering and Technology
Ernakulam, India

Ashokraj Rajendran
J.J. College of Engineering and Technology
Trichy, India

Nandagopan Ramanan
Sri Jayaram Institute of Engineering and Technology
Chennai, India

Rajesh Ranganathan
Coimbatore Institute of Technology
Coimbatore, India

S. Padmanabhan
Vel Tech Rangarajan Dr. Sagunthala R&D Institute of Science and Technology
Chennai, India

V.S. Shaisundaram
Vels Institute of Science, Technology and Advanced Studies
Chennai, India

S. Balasubramanian
Kumaraguru College of Technology
Coimbatore, India

Nallathambi Siva Shanmugam
Department of Mechanical Engineering
National Institute of Technology
Tiruchirappalli, India

Anmol Sharma
Guru Gobind Singh Indraprastha University
Delhi, India

Madhuvanthi Sigamani
Government College of Technology
Coimbatore, India

Suresh Sundarraj
University College of Engineering
Nagercoil, India

Vijay Ananth Suyamburajan
Vels Institute of Science, Technology and Advanced Studies
Chennai, India

Mugilan Thanigachalam
Government College of Technology
Coimbatore, India

Sridharan Vadivel
Government College of Technology
Coimbatore, India

Prashant Veer
Department of Mechanical Engineering
National Institute of Technical Teachers Training and Research (NITTTR)
Chandigarh, India

Gurusamy Visvanathan
Department of Mechanical Engineering
Sri Venkateswara College of Engineering
Sriperumbudur, India

Jebbas Thangiah Winowlin Jappes
Department of Mechanical Engineering & Centre for Surface Engineering
Kalasalingam Academy of Research & Education
Virudhunagar, India

Mohit Yadav
Department of Applied Science
Maharshi Dayanand University
Rohtak, India

Y. Brucely
SRM TRP Engineering College
Tiruchirappalli, India

Chapter 1

4D Printing technology for health care applications

G. Paulraj, Y. Brucely, and M. Thilak
SRM TRP Engineering College

CONTENTS

1.1 INTRODUCTION

Nearly three decades have passed since the invention of 3D printing technology. While the additive manufacturing industry advances new applications, materials, and 3D printers, another technology is gaining traction. It is called 4D printing, which is a futuristic technology. 4D printing is the process by which the structure of a 3D printed object changes in response to an external energy source such as heat, light, or other environmental stimuli. Consider applying 4D printing on a very small scale, for example, in sectors such as medicine. 4D printed proteins may be an excellent application because they are self-configuring. Self-folding protein is another

DOI: 10.1201/9781003430186-1

novel material on which researchers are currently working. Another possible application of 4D printing in medicine is in the design of sentinel organs. Stents could be programmed to travel through the human body and automatically open upon arrival. This chapter discusses the numerous health care applications of 4D printing technology.

1.2 FOUR-DIMENSIONAL PRINTING'S ADVANTAGES OVER THREE-DIMENSIONAL PRINTING

4D printing enables the creation of products whose characteristics and properties change over time in response to environmental changes such as temperature (1–3). Among the numerous benefits of 4D printing are the following:

Capabilities of printed intelligent products: Four-dimensional printing enables users to create products using smart material, which has a wide range of applications in engineering, medicine, dentistry, and material science (Figures 1.1 and 1.2).

Modification of the product's shape to meet the specifications: The shape of a 3D printed model changes over time as a result of environmental parameters such as temperature, humidity, and so on.

1.2.1 Size changing

The most apparent benefit of 4D printing is that it enables items that are bigger than printers to be produced as just one component. This is accomplished by computational folding. Because the items that are printed in 4D

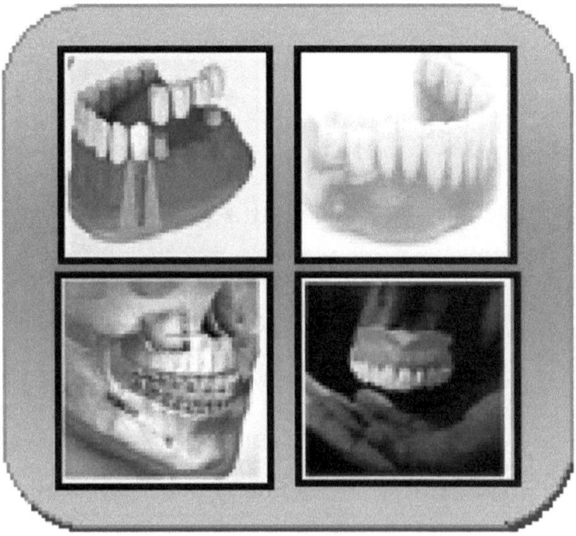

Figure 1.1 Some of the medical applications of 4D technology.

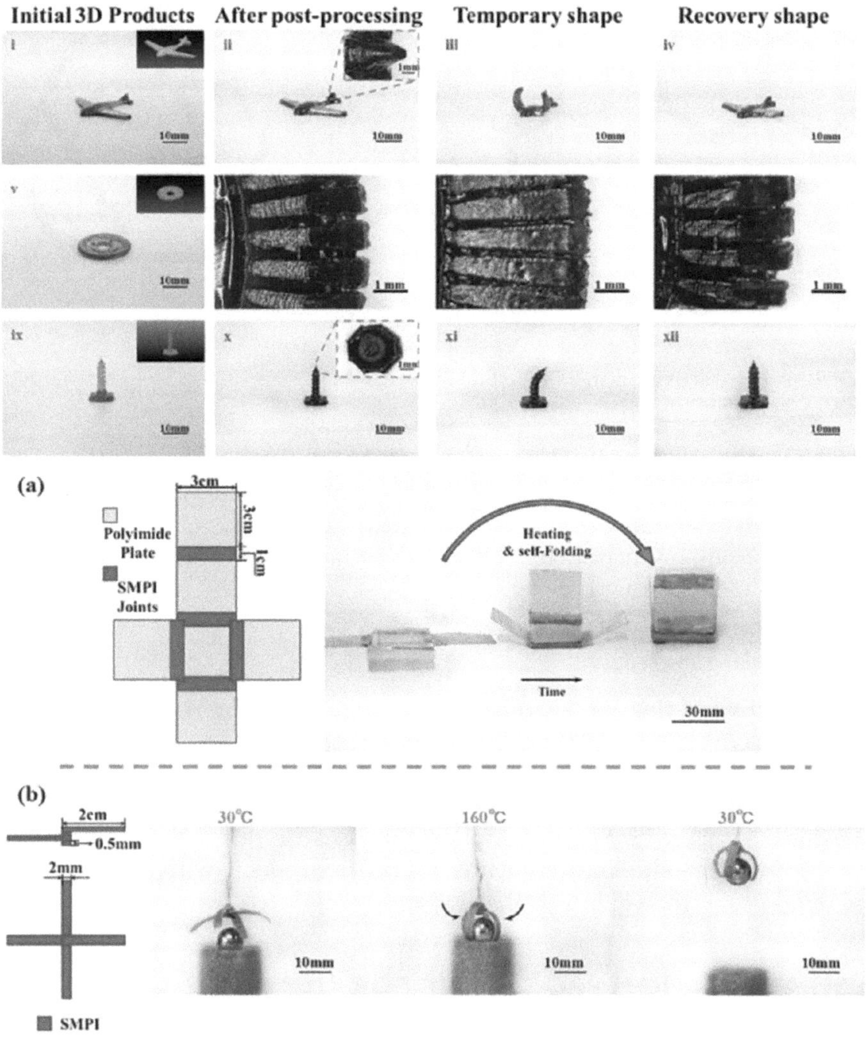

Figure 1.2 Diagrammatic representation of (a) shape memory behaviour of printed structures like aeroplane (top), ultrasonic motor, and pagoda using SMPI. (b) Application of SMPI to print self-folding box (top) and gripper (bottom). Copyright (2020), Elsevier.

may change shape, they can shrink and unfold, which allows for objects that are too huge to fit in a printer to be compressed into its secondary form so that it can be produced in 3D.

1.2.2 Innovate

This technology advances the product's design and development phases. Additionally, research is successfully conducted using the technology. The most important results of adopting 4Dinnovations are the chances afforded

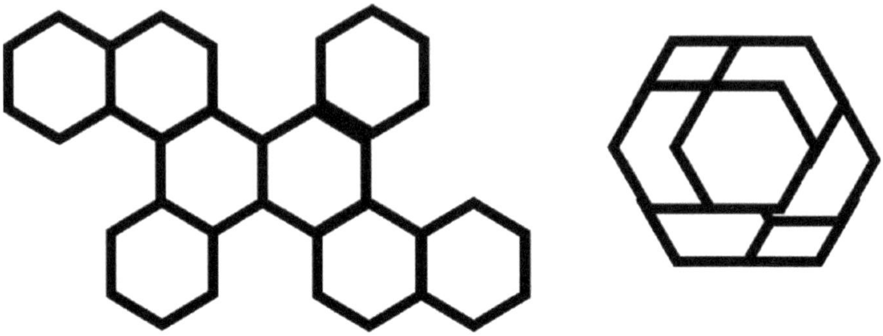

Figure 1.3 Four-dimensional printing of octagonal shapes developed at MIT's self-assembly laboratory.

by the fostering of feedback loops that reduces the transactional distance. The possible consequences increases throughout the planning effort due to the obvious resultant of the added interactions with members of the construction group or client team. This tends to increase the quality and truthfulness of the plan that is generated.

1.2.3 Self-assembly

The objective of the self-assembly and configurable material technology is to rethink construction, manufacturing, the assembly of goods, and performance. Printing in four dimensions is used in a variety of fields, including medical, engineering, and others. Proteins that have been printed in four dimensions might be a useful application. Because this technology makes use of smart materials, it frequently results in products with self-assembly material properties (Figure 1.3).

1.3 FABRICATION PROCEDURE FOR 4D PRINTING

By layering material atop a digital CAD model, the process of "4D printing" produces a three-dimensional physical replica of a smart material. The important stages in the process of using 4D printing in the medical field are listed in the subsequent sections.

1.3.1 Scanning and imaging

A medical image is created using a variety of scanning technologies, including computed tomography (CT), magnetic resonance imaging (MRI), and scanners. Additionally, these technologies are used to analyse the defect and broken bone of the patient (4–14). Multi-dimensional scanning methods, such as 4D CT and 4D MRI, enable the detection of even the smallest deviations from the norm. These scanning methods create a video replay of the scan pictures. Thus, they aid in the observation, tracking, and analysis of internal body motions. Scanners like this are the next stage in imaging, and they're very precise and speedy. Smart material is used to build 3D physical products/implants by layering it on top of each other, receiving input from computer-aided design (CAD) data. When temperature, humidity, or pressure varies, the geometry of the smart material changes with it.

1.3.2 Transformation of data into a three-dimensional digital model

To convert scan data to a three-dimensional model, software such as OsiriX, 3D slicer, Mimics, Magics, 3D doctor, and In Vesalius must be used. It is critical to convert data to a 3D model before printing or modifying a design.

1.3.3 Design modification and simulation

During the design modification stage of the product design process, various modifications are made to the designs of medical implants and devices. These modifications are done in line with the philosophy of the product design process. We are able to employ mechanical simulation to investigate the particular component of the body that is designed to carry out the essential task.

1.3.4 Printing in four dimensions

This process involves layer-by-layer printing of medical implants in the desired shape and size using smart materials. This technology fabricates complex-shaped medical devices efficiently and models them according to each patient's unique anatomy. It enables greater flexibility in the fabrication of an internal structure.

1.4 IMPLEMENTATION OF PRINTED MODEL IN THE MEDICAL FIELD

The goal of 4D printing is to be of benefit to medical practitioners, particularly in areas that are not covered by technology associated with 3D printing. 4D printing is a method that assists in the production of a 3D physical object by building it up layer by layer using intelligent material with computer-aided design. The technique of 4D printing is especially utilized to make implants, which are then implanted in the patients. Because of this procedure, surgery is simplified, and its success rate and dependability are both increased. Following surgical intervention, the patient may experience improved comfort as a result of the placement of a smart material implant produced using 4D printing technology (Figure 1.4).

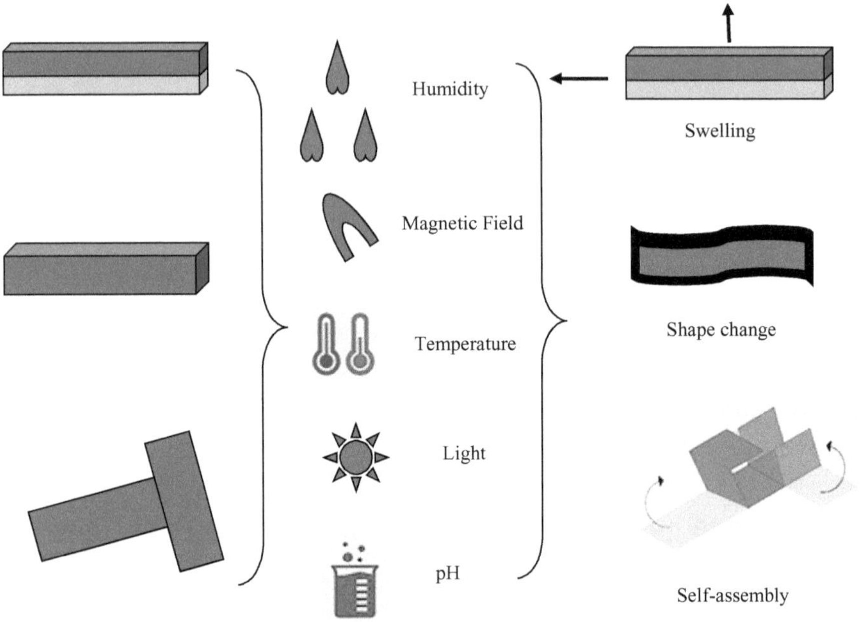

Figure 1.4 Types of stimuli and responses observed in smart materials.

1.5 THE STATE OF THE ART IN 4D PRINTING RESEARCH

According to Scopus data, the number of research publications on 4D printing is increasing. We identified 171 articles on this technology up until September 2018. The first research article on 4D printing was published in 2007, wherein only one article on the subject was published. In 2008, 2009, 2010, and 2011, there was not a single article published. In 2012 and 2013, one article each was published, while 2014 saw an increase to six articles. In 2015, 13 articles on 4D printing were published; in 2016, 26 articles were published; in 2017, 66 articles were published; and, through September 2018, 55 articles on 4D printing were published. As a result, we observed that research increased significantly in 2016, 2017, and 2018. Numerous journals and other publications have featured articles on four-dimensional printing. With nine publications, the *International Journal of Precision Engineering and Green Manufacturing Technology* is the most prolific. The same eight articles were published in both the *Smart Materials and Structures* and the *Virtual and Physical Prototyping* journals. Six publications appear in *ACS Applied Materials & Interfaces*, while *Assembly Automation* and *Materials and Design* each have five. Three papers were published in each journal, *3D Printing* and *Additive Manufacturing and Materials.* Three research articles have been published in the scientific reports journal. Additionally, the remaining journals and sources have published articles on 4D printing.

1.6 THE STATE OF RESEARCH IN THE FIELD OF 4D PRINTING IN MEDICINE

There is a dearth of research in the field of 4D printing in medicine. We found only 13 articles published between September 2017 and September 2018 that contained the keywords "4D Printing" and "medical." The first article was published in 2012; no publications were reported in 2013 or 2014. One article was published in 2015; two articles were published in 2016, three articles were published in 2017, and six articles have been published in this field as of September 2018.

Materials science makes a maximum contribution of 33% to the field of "4D printing in the medical field." Engineering journals contribute 15%; computer science, medicine, physics, and astronomy journals each contribute 11%; chemical engineering journals contribute 8%; chemistry journals contribute 7%; and mathematics journals contribute 4% via 4D printing technology.

1.7 MEDICAL APPLICATIONS OF 4D PRINTING

The use of 4D printing might result in the creation of stents that, when exposed to the heat of the patient's body, can expand and assume the desired form. The most recent remedy may be used to immediately save a patient's life during a challenging operation. Time and temperature both have an effect on the form that something takes. This burgeoning technology may be used to build intricate three-dimensional organs and for printing organs on which the patient's own cells may be placed in order to save the patient's life. It is an encouraging approach to the problem of organ scarcity. Because of this, the possibility of the material that can change its form will also be created. The possible use of this technology might include medical implants or tissue engineering to create structures that can alter their form while they are within the body; They can also be used for the regeneration of tissues such as muscle, bone, and cartilage, which are characterized by dynamic changes in their mechanical properties while the body is in motion. Babies who are having difficulty breathing may have their lives saved because of technological advancements. It swiftly produces a medical implant that is capable of changing form over time to accommodate growing infants and to assist those newborns in maintaining their breathing.

By making use of the smart material, 4D printing will one day be able to reproduce internal organs such as the heart, kidneys, and liver. It has the ability to print these components, which have a high degree of elasticity as well as a precise fit and a perfect genetic match. It offers a wide range of opportunities to print skin grafts in a colour that is similar to the patients' natural skin tone. It is also helpful for patients who have suffered burns since the tissue is simple to implant in the patient's body and continues to develop normally (15–17). With the help of this technology, it is possible to produce 3D printed versions of complicated, intelligent medical equipment that have superior functional features, adapted to the needs of the surgical procedure with regard to timing. It is possible that it might build a haptic model that represents not only the look but also the movement of the body. In the not-too-distant future, it may find use in more complex surgical procedures that are now beyond the capabilities of conventional manufacturing technologies. Through the use of a smart material and with the assistance of CT and MRI scans, it is possible to construct a model that can exactly imitate the movement of the hand or another body part (18–21). It is able to depict anatomical features in a way that is both exact and accurate. Printed medical models may alter form and function over time thanks to a new 4D bio-printing process that includes time-of-use into 3D bio-printing. The model's ramifications for biological scaffolds and other key applications are also noteworthy (22–27) (Figure 1.5).

In medicine, precise anatomical detail is required to improve clinical outcomes, reduce effective duration, minimize donor site morbidity, and minimize overall operative complications (28,29). 4D printing has a lot of potential in biomedical applications such as self-bending stents and other self-shrinking/tightening staples. In orthopaedics, the use of 4D printing technology has the potential to fundamentally alter the surgeon's approach to diagnostic and treatment planning (30).

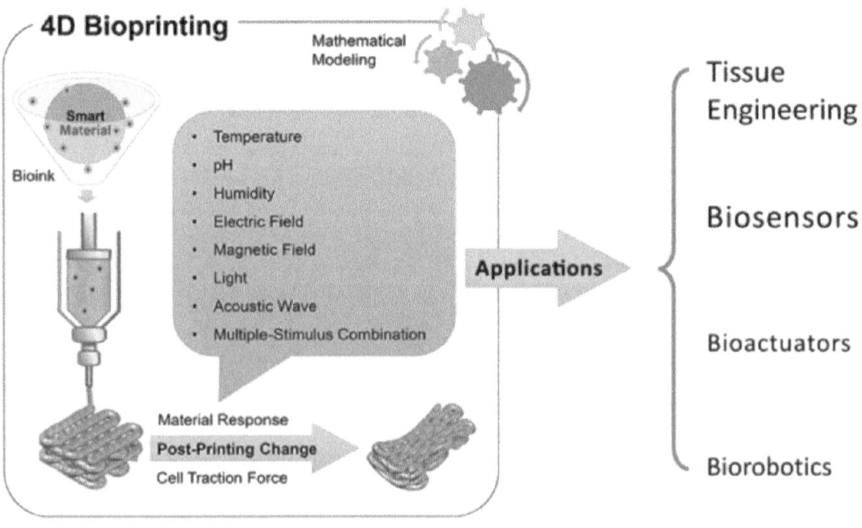

Figure 1.5 Future perspective in bio-printing.

1.8 CONCLUSIONS

4D printing is an advanced version of 3D printing, a cutting-edge technique for creating smart medical implants and devices that adapt to the patient's growth using a smart material. CT, MRI, and other scanning technologies are used to collect patient data, which is then printed layer by layer using four-dimensional printing technology. The implant's shape may change over time or in response to the growth of the human body. It is uniquely capable of fabricating stents and organs that can adapt their shape in response to changing conditions. This technology has the potential to significantly improve patient outcomes and provide a more viable alternative to severe organ shortages. It has enormous potential in the field of tissue engineering and intelligent implants. 4D printing has been investigated for its potential application in medicine, with the goal of customizing intelligent multi-material printing and clearly demonstrating multiple body parts in a 3D printed model. It resolves infants' breathing problems by combining a precise fit and perfect genetic match of the heart, kidney, and liver with the increased flexibility of the printed material. 4D printing is an excellent method for printing a variety of intelligent medical tools and devices, as well as natural colour skin grafts. It is used for extremely complex surgery, has a high success rate, and has the potential to save a patient's life.

REFERENCES

1. Bakarich SE, Gorkin R, Panhuis M, Spinks GM. 4D printing with mechanically robust, thermally actuating hydrogels. *Macromol Rapid Commun.* 2015;36:1211–1217.
2. Baker AB, Wass DF, Trask RS. 4D sequential actuation: combining ionoprinting and redox chemistry in hydrogels. *Smart Mater Struct.* 2016;25(10):1–9.
3. Tibbits S. 4D printing: multi-material shape change. *Architect Des.* 2014;84:116–121.
4. Chae MP, Hunter-Smith DJ, De-Silva I, Tham S, Spychal RT, Rozen WM. Four-dimensional (4D) printing: a new evolution in computed tomography-guided stereolithographic modeling. Principles and application. *J Reconstr Microsurg.* 2015;31(6):458–463.
5. Ge Q, Sakhaei AH, Lee H, Dunn CK, Fang NX, Dunn ML. Multi-material 4D printing with tailorable shape memory polymers. *Sci Rep.* 2016;6:31110. doi: 10.1038/srep31110.
6. Miao S, Zhu W, Castro NJ, et al. 4D printing smart biomedical scaffolds with novel soybean oil epoxidized acrylate. *Sci Rep.* 2016;6:27226. https://doi.org/10.1038/srep27226.
7. Javaid M, Haleem A. Additive manufacturing applications in orthopaedics: a review. *J Clin Orthop Trauma.* 2018;9(3):202–206. doi: 10.1016/j.jcot.2018.04.008.
8. Gao B, Yang Q, Zhao X, Jin G, Ma Y, Xu F. 4D bioprinting for biomedical applications. *Trends Biotechnol.* 2016;34(9):746–756.

9. Zhou Y, Huang WM, Kang SF, Wu XL, Lu HB, Fu J, Cui H. From 3D to 4D printing: approaches and typical applications. *J Mech Sci Technol.* 29(10):4281–4288.
10. Kwok JKS, Lau RWHL, Zhao ZR, et al. Multi-dimensional printing in thoracic surgery: current and future applications. *J Thorac Dis.* 2018;10(6):756–763.
11. Bakarich SE, Gorkin R, Panhuis M, Spinks GM. 4D printing with mechanically robust, thermally actuating hydrogels. *Macromol Rapid Commun.* 2015;36:1211–1217.
12. Baker AB, Wass DF, Trask RS. 4D sequential actuation: combining ionoprinting and redox chemistry in hydrogels. *Smart Mater Struct.* 2016;25(10):1–9.
13. Tibbits S. 4D printing: multi-material shape change. *Architect Des.* 2014;84:116–121.
14. Herodotus, *Histories*, trans. A. D. Godley. Cambridge: Harvard University Press, 1920.
15. Zarek M, Mansour N, Shapira S, Cohn D. 4D Printing of shape memory-based personalized endoluminal medical devices. *Macromol Rapid Commun.* 2017;38(2). doi: 10.1002/marc.201600628.
16. Ge Q, Sakhaei AH, Lee H, Dunn CK, Fang NX, Dunn ML. Multi-material 4D printing with tailorable shape memory polymers. *Sci Rep.* 2016;6:31110. https://doi.org/10.1038/srep31110.
17. Haq IU. 4D Printed Implant Saved Babies with Breathing Problems, 2015; http://computingcage.com/4d-implant-saves-babies-with-breathing-problems/.
18. Hendrikson WJ, Rouwkema J, Clementi F, Blitterswijk CA, Fare S, Moroni L. Towards 4D printed scaffolds for tissue engineering: exploiting 3D shape memory polymers to deliver time-controlled stimulus on cultured cells. *Biofabrication.* 2017;9(3):031001.
19. Mandon CA, Blum LJ, Marquette CA. 3D–4D Printed objects: new bioactive material opportunities. Micromachines. 2017; doi: 10.3390/mi8040102.
20. Gosnell J, Pietila T, Samuel BP, Kurup HKN, Haw MP, Vettukattil JJ. Integration of computed tomography and three-dimensional echocardiography for hybrid three dimensional printing in congenital heart disease. *J Digit Imaging.* 2016;29(6):665–669.
21. Khoo ZX, Teoh JEM, Liu Y, Chua CK, Yang S, An J, Leong K, Yeong WY. 3D printing of smart materials: a review on recent progress in 4D printing. *Virtual Phys Prototyp.* 2015;10(3):103–122.
22. He P, Zhao J, Zhang J, Li B, Gou Z, Gou M, Li X. Bioprinting of skin constructs for wound healing. *Burns Trauma.* 2018;6(5):1–10.
23. Pei E, Loh GH. Technological considerations for 4D printing: an overview. *Prog Addit Manuf.* 2018;3(1–2):95–107.
24. Castro NJ, Meinert C, Levett P, Hutmacher DW. Current developments in multifunctional smart materials for 3D/4D bioprinting. *Curr Opin Biomed Eng.* 2017;2:67–75.
25. Zhao T, Yu R, Li X, Cheng B, Zhang Y, Yang X, Zhao X, Zhao Y, Huang W. 4D printing of shape memory polyurethane via stereolithography. *Eur Polym J.* 2018;101:120–126. https://doi.org/10.1016/j.eurpolymj.2018.02.021.
26. Lee AY, An J, Chua CK. Two-way 4D printing: a review on the reversibility of 3D printed shape memory materials. *Engineering.* 2017;3(5):663–674.

27. Javaid M, Haleem A, Kumar L. Current status and applications of 3D scanning in dentistry. *Clin Epidemiol Global Health.* 2019;7(2):228–233. doi: 10.1016/j.cegh.2018.07.005.
28. Awasthi S, Pandey N. Rural background and low parental literacy associated with discharge against medical advice from a tertiary care government hospital in India. *Clin Epidemiol Global Health.* 2015;3:24–28.
29. Awasthi S, Verma T, Agarwal M, Singh JV, Srivastava NM, Nichterd M. Developing effective health communication messages for community-acquired pneumonia in children under five years of age: a rural North Indian qualitative study. *Clin Epidemiol Global Health.* 2017;5:107–116.
30. Haleem A, Javaid M, Vaishya R. 4D printing and its applications in orthopaedics. *J Clin Orthop Trauma.* 2018;9(3):275–276. doi: 10.1016/j.jcot.2018.08.016 https://doi.org/10.1016/j.jcot.2018.08.016.

Chapter 2

Additive manufactured components for medical applications

S. Balasubramanian
Kumaraguru College of Technology

CONTENTS

DOI: 10.1201/9781003430186-2

2.1 INTRODUCTION TO ADDITIVE MANUFACTURING

Additive manufacturing (AM) is a special manufacturing process controlled by a computer that creates objects by depositing materials layer by layer. 3D printing technology has become a common application in many industries such as aerospace and medical component manufacturing and medicine. 3D printing has many potential uses in medicine and the fabrication of tissues and organs, customized prosthetics, implants, and anatomical models, as well as pharmaceutical research regarding drug dosage forms, delivery, and discovery. The three most produced custom-made medical equipment and products are SLS, TIJ, and FDM.

A 3D CAD file, in which a geometrical model of the part is stored, is directly used for additive fabrication. Several commercial 3D modellers can be used to design the part model, which is exported in STL format [1]. Titanium alloys are suitable research materials because of their excellent properties like corrosion resistance, low modulus, high fatigue strength, low density, and good mechanical properties.

2.1.1 Advantages

AM process is fast and accurate.

There are separate core materials for components and removable material for supports.

There is no requirement for jigs, fixtures, mould, or other tools [2].

Material wastage is minimum.

Product development time is reduced considerably.

2.1.2 Limitations

The limitations of AM process include expensive hardware and expensive raw materials, which result in costly parts, and post-processing is required.

2.1.3 Materials used in additive manufacturing

2.1.3.1 Polymers and metals

Polymers and metals consist of PLA, ABS, nylon, fibre, fibre glass, HSHT fibre, bakelite, delrin, titanium, zirconium, cobalt, stainless steel, silver, aluminium composite, silver, and gold.

Biocompatibility is the foremost requirement of AM materials. Frequent materials used are cobalt–chromium alloy, titanium alloy and zirconia, gold, silver, stainless steel, and titanium-based biomaterials to fabricate mini-implants.

The structure of a normal bone is distinctly different from osteoporotic bone having lesser density. The performance of the implant depends on the right combination of mechanical and physical properties. In conventional arthroplasty surgery, stainless steel was considered a viable implant material mainly because of its availability and processing ease. The manufacturing of an artificial bone with different densities at different portions called functionally graded material is possible with AM process.

2.1.4 Classification of additive manufacturing

The AM process is classified according to fusion as melting, sintering, and photopolymerization. The types are as follows:

1. Material extrusion (FDM)
2. Binder jet process (no need to heat the raw material)
3. Powder bed fusion (LM, DMLS, EBM, SLS, SLA)
4. Sheet lamination (LOM, UAM, DED)
5. Wire arc additive manufacturing (WAAM)

2.2 COMPONENTS FOR MEDICAL APPLICATIONS

2.2.1 Equipment and instruments

2.2.1.1 Glaucoma screener

To measure eye pressure and monitor the condition of glaucoma, a small tip is gently rubbed on the surface of the eye. When measuring eye pressure, a fixed area of the cornea is gently flattened to estimate the force required. A small tip gently touches the surface of the eye. The sensitivity of the components in the glaucoma screener is important when designing for the sleek elegant human-centric screener. The portable glaucoma screener weighs around 5 kg and takes 2 minutes for the visual field screening test. The body of the screener, like a robot is three axes, adjustable to suit different structured patients and to hold various types of lenses in line screen (Figure 2.1). Materials used to produce the body and parts of the glaucoma screener are aluminium and zinc extrusion, ABS in the SLA, SLS, and DMLS processes.

2.2.1.2 Endotracheal block detector

An endotracheal block detector is a flexible plastic tube placed to deliver oxygen to the lungs through the nose or mouth into the trachea and

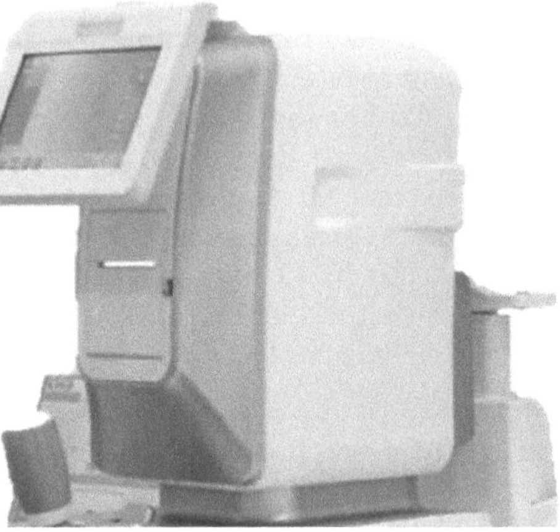

Figure 2.1 The overall image of a glaucoma screener with 3D-printed parts for monitoring the condition of glaucoma of patient.

Figure 2.2 An endotracheal block detector made by additive manufacturing processes with trachea tube, windpipe, end cap, and balloon shown at the left side and tube assembly at the right side with precise parts.

connected to a ventilator unit. Figure 2.2 shows the trachea tube, windpipe, end cap, and balloon.

If the tube moves beyond a distance set by the operator, the opt-endotracheal tube generates a visual and audible warning. For displacement prediction, it is found to have an accuracy of at least 1.0 mm. ETT tube sizes ranging from 2 to 10 mm (smaller ones) are used for babies. ETT tube with non-return valve has an indicator to facilitate the cuff inflation and deflation; the smooth cuff inflates symmetrically, eliminating any risk of hernia; smooth front end, pilot balloon, and tiny tube for pressure measurement are critical. Materials selected are optical fibre, light-emitting diode, photodetectors, polyurethane tubes, and connectors. Processes suitable for manufacturing the ETT are SLS and SLA.

2.2.1.3 *Smart stethoscope*

The instrument is used to listen to the sounds produced within the heart or lungs of patients. The challenge in robust stethoscope manufacturing is optimized for low manufacturing and maintenance costs. Reducing the number of parts movement leads to efficient performance of high sensitivity noise transduction in the instrument. In Figure 2.3 the additively manufactured stethoscope is shown with its parts.

Three major parts of a smart 3D-printed stethoscope are a chest piece with the diaphragm, stem, bell, an air-filled tube with a Y tube, and a connector tube headset consisting of an aural metal tube and ear tip. The total weight of the complete stethoscope is 160 g approximately. As the diaphragm on touching the body of the patient vibrates, an acoustic pressure wave is created inside the hollow stem. The sound is then routed through an air-filled tube to the ear tips at the end of the headset, where it reaches the listener.

The materials used are stainless steel for aura tube in headset and chest piece, ABS for ear tips, and polyethylene for diaphragm in the chest piece. Thermoplastic elastomers for tubing are used as flexible connectors. The parts are customized to the individual listener's adjustment and easiness to maintain cleaning and replacement of the diaphragm. In 3D printing, the parts produced in a modular design improve sensitivity to noise transduction. The precision parts are manufactured in FDM, SLS, and SLA processes.

2.2.1.4 *Scaffold for diabetic foot ulcer*

A scaffold is a three-dimensional porous, fibrous, permeable biomaterial intended to transport body liquids and gases, which promotes cell interaction, viability, and extracellular matrix deposition with the least amount

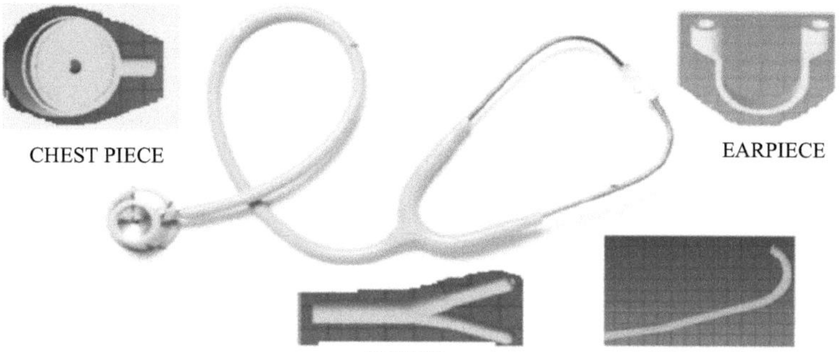

Figure 2.3 3D-printed stethoscope assembly comprising the tube, ear tip, stem, and bell (chest piece) avoids loss of sound waves and their detailed view near to the assembly.

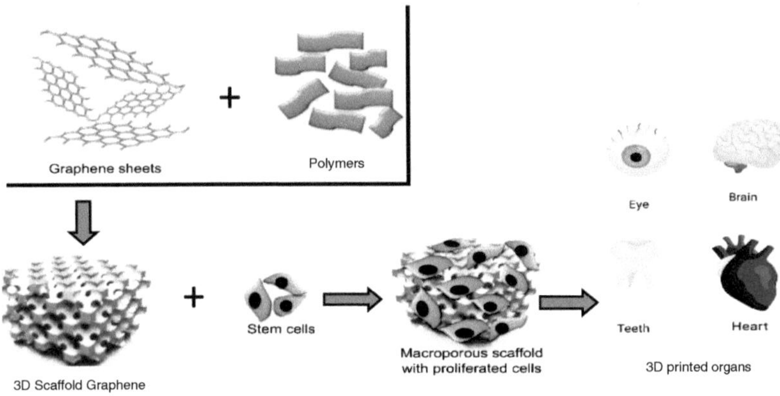

Figure 2.4 3D-printed scaffold mixed with graphene for the preparation of organ in tissue engineering shown by the added content and process by arrow marks.

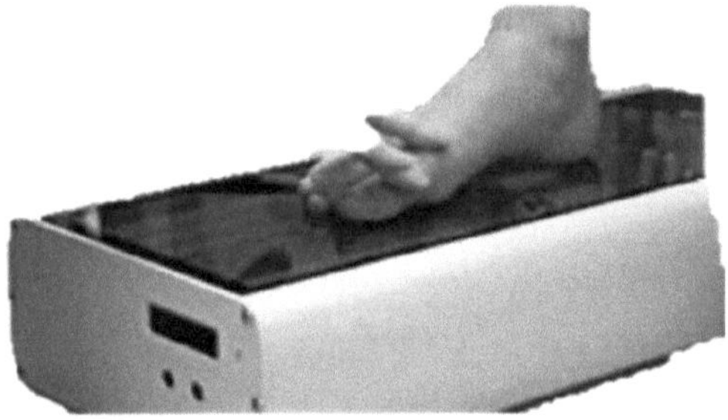

Figure 2.5 A patient's right foot is scanned for sensation using a diabetic foot screener having accurate 3D-printed sensors.

of inflammation, and dissolves at a controlled rate. Figure 2.4 shows the flow of scaffold preparation [3]. Materials for the scaffold are fine polymer, graphene fed in FDM, and SLS processes.

2.2.1.5 Screener for diabetic foot

Diabetic foot screeners are used to assess loss of protective sensation in diabetic feet. The screener helps identify patients at risk of developing diabetes [4]. Infections, injuries, and abnormalities of the bones of diabetic patients are examined. Figure 2.5 shows the leg on the diabetic screener. The monofilament 10g is a simple and objective instrument for identifying the loss of protective sensation in diabetic feet. To ensure that 10g of linear force is applied, a properly calibrated device needs to be used. The pressure will ensure an accurate measurement. Monofilament is a synthetic fibre made of

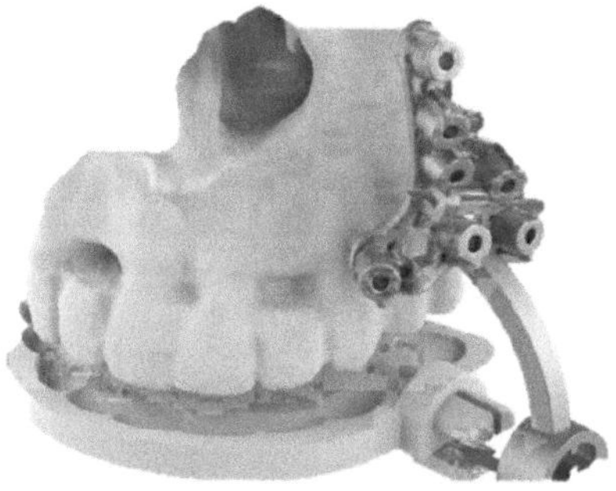

Figure 2.6 A demonstration model shows a 3D-printed customer-specific fixture with suitable support provision during skull surgery.

a single filament. Polymers are melted and mixed to create a monofilament. Polymer mixtures are then extruded through 3D printers, creating lines of various thicknesses.

2.2.1.6 Surgical guide

Surgical guides assist the surgeon in drilling implants into the bone with optimal accuracy. Just like jigs and fixtures, using sleeves, the surgical guide guides surgical instruments and implants to the correct location once placed in a patient's body. Surgical guides were fabricated from acrylic resin using the processes of 3D printing SLA. Figure 2.6 shows a skull fitted with a 3D-printed fixture.

2.2.2 Prosthetics and orthotics

Orthotics are supports that assist an existing body part, whereas prosthetics are artificial replacements for missing body parts. Orthotics is of health care technology concerned with the design, manufacture, and application of orthoses. A prosthesis is an external fitting device used to replace wholly, or in part, an absent or deficient limb segment.

2.2.2.1 Examples of orthotics

Shoe insert, knee brace, braces for teeth, neck brace, weightlifting belt, orthopaedic boot, and splint.

2.2.2.2 Examples of prosthesis

Artificial hand, foot, finger, and toe prostheses, artificial breast, hearing aids, artificial eyeballs, ear, nose, or eye socket, and artificial soft or hard palate.

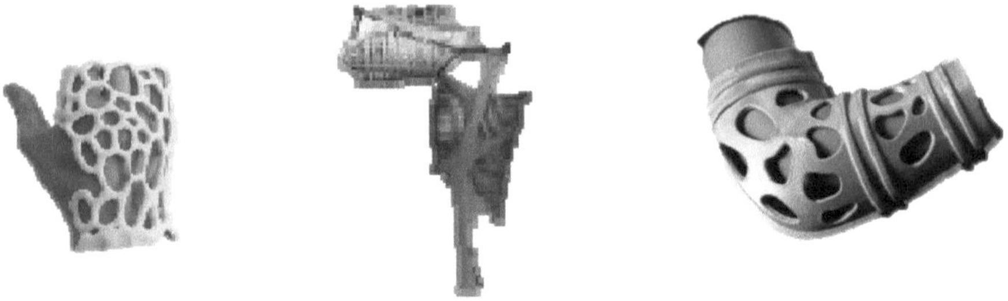

Figure 2.7 Different types of 3D-printed patient-specific splints fitted on wrists (white flexible material and black net-shaped material), lower arm with grey colour polymer support, and leg with pad-like support.

2.2.3 Orthotics

2.2.3.1 *Splints*

Splints are medical equipment used to prevent an injured body part from moving and to prevent further damage. Broken bones are commonly treated with splints; fractures, sprains, tendon injuries, and injuries awaiting orthopaedic treatment can be treated with a splint [5]. There are two types of splints: static, which prevents motion, and dynamic, which allows controlled motion. Stabilizing splints are usually made from flexible materials and consist of poles or pole-like structures.

Low-temperature thermoplastics and studies with a wide range of varying materials, from nylon to resins to epoxy photopolymers, are used in the SLS 3D printing process. Figure 2.7 shows the additively manufactured splints for the wrist, lower leg, and knee joint.

2.2.3.2 *Respiratory face masks*

The virus is efficiently prevented by 3D-printed high-end respiratory face masks [6]. Figure 2.8 shows a high-end mask's exploded view. The world now sees what it can do to save lives, and how easily it enables smartness and adaptability which push forward for manufacturing the patient-specific masks by the FDM process to improve quality of life and public safety.

2.2.3.3 *Menstrual cup*

The cup is made from TPU/TPE filament rubber or silicone and is inserted into the vagina to catch and collect period fluid. Menstrual cups are reusable feminine hygiene products. It is easy to fold this cup multiple times. Figure 2.9 is normal menstrual cup in firmness and softness. It opens with some force and is also easy to fold. Shore hardness recommended is 80–90 A. Material is processed in FDM.

Figure 2.8 3D-printed hygiene respiratory face mask fitted on a face and the parts holder, front frame, back frame, and middle filter are explored at the right side.

Figure 2.9 Two flexible menstrual cups made by additive manufacturing process, held vertically, show the fixing position.

2.2.3.4 *Insole*

An insole is the part of a shoe that runs under the foot and supports it; it is also called footbeds. Proper footwear removes pain from the heel, arch, ball foot, plantar fasciitis, and Achilles problems. Figure 2.10 shows the patient-specific pattern from the image to data acquisition.

Indigenous insoles for individuals with a different kind of spring geometry produced from the software are manufactured by 3D printing process. The pressure area of the foot gets this levelled-out responsiveness in the insole [7]. Material for the insole is TPU; it has recycled possibility prepared under ISO 18562 Standards and the AM Process is SLS.

2.2.4 Prosthesis

2.2.4.1 *Cornea of the eye*

The transparent part of the eye is called the cornea which covers the iris and the pupil and allows light to enter. The anatomy of the eye shows the

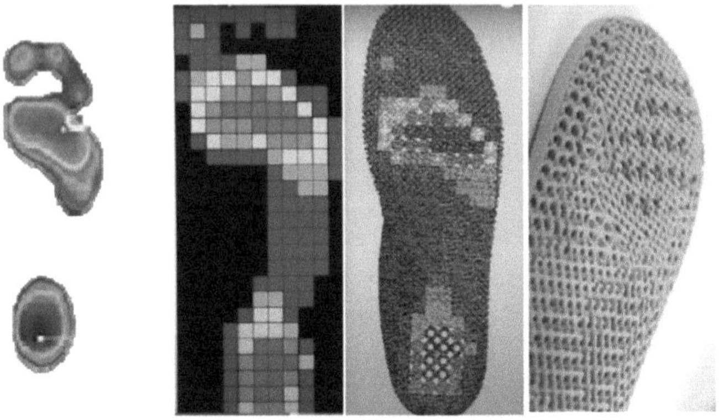

Figure 2.10 (a) Diabetic patients use specific individual insoles as their feet are scanned for body pressure. (b) Patient-specific scanned image of body pressure. (c) The scanned image is converted to programmable file as resolution data. (d) 3D-printed functionally graded pattern of sole exactly match the required foot pressure of the patient. Centre red portion meant high-pressure variety of insole structures with variable holes from 10% to 20%.

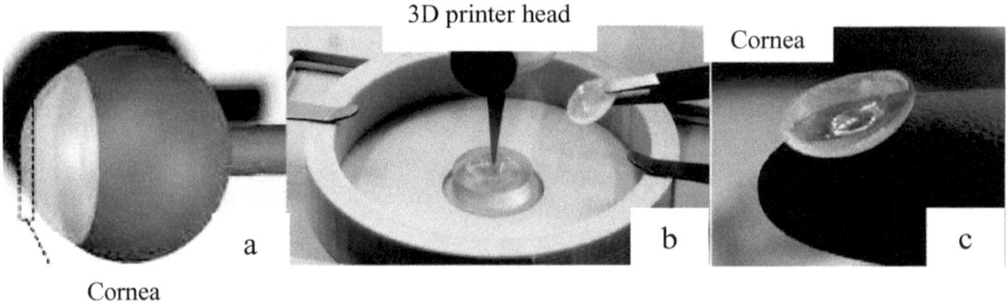

Figure 2.11 (a) Position of cornea in the eye anatomy shown in white colour front side. (b) The process showing the 3D printing of cornea at centre of the bed in FDM printer and the tools removing the finished cornea. (c) Completed cornea placed on a masked finger.

outside and inside of the eye including the eyelid, pupil, sclera, iris, cornea, lens, ciliary body, retina, choroid, vitreous humour, and optic nerve.

The replacement of the cornea involves implementing sensor function for real-time monitoring and connected feedback. Figure 2.11 shows the cornea in the eye anatomy and cornea. Recent advances in 3D printing and electronics have brought us closer to sensors with multiplex advantages, and additive manufacturing approaches offer new scope for sensor fabrication [8]. A 3D-printed artificial cornea is suitable for transplantation made with the help of an aluminium mould, an FFF 3D printer to fabricate a PVA-chitosan corneal construct with light-bending properties.

Chitosan is a biopolymer, which is an amino polysaccharide derived from the N-deacetylation of chitin. Decellularized human tissue is a natural polymer because of the presence of the degradable enzyme chitosan. Chitosan is blended with synthetic polymer PVA with formaldehyde to enhance thermal stability.

2.2.4.2 *Knee prosthetics*

The knee joins the thigh with the leg and consists of the top joint and the bottom joint. The top joint is between the femur and tibia whereas the bottom joint is between the femur and patella, and it is the largest joint in the human body too.

The knee joint comprises the lower end of the thigh bone (femur), the upper end of the lower leg (tibia), and the kneecap (patella). The ankle joint is the connection point of the lower end of the lower leg bone (tibia) and upper foot bone (talus). Normally, the ends of these bones are made up of cartilage that allows the joints to move in smooth, painless motion. Arthritis is a painful and degenerative condition of the joints [9]. The most common cause of ankle and kneecap arthritis is trauma, such as a fracture of the calcaneus (heel bone). Knee joint replacement surgery is to replace a knee joint with a man-made artificial joint. Advanced arthritis that is very painful has been treated primarily with bone fusion. This procedure obliterates the ankle joint completely. Pain is relieved but at the cost of absolute stiffness. Figure 2.12 shows the human implanted 3D-printed tibias and patella. Cobalt, chromium, and titanium porous powder are recommended as per FDA use in the DMLM production process [10].

2.2.4.3 *Ankle prosthetics*

2.2.4.3.1 Talus

The ankle joint is connected to the foot by the talus, which helps transfer weight and pressure forces across it. Joint surfaces are largely covered by

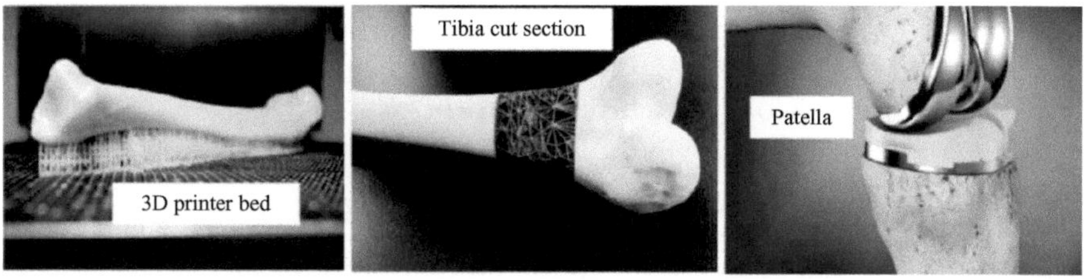

Figure 2.12 (a) 3D-printed tibias complete with supports at the bottom during additive manufacturing. (b) 3D-printed tibias with inner supports in hollow portion shown through cut section. (c) 3D-printed patella fitted on the defective bone of the lower leg and tibia.

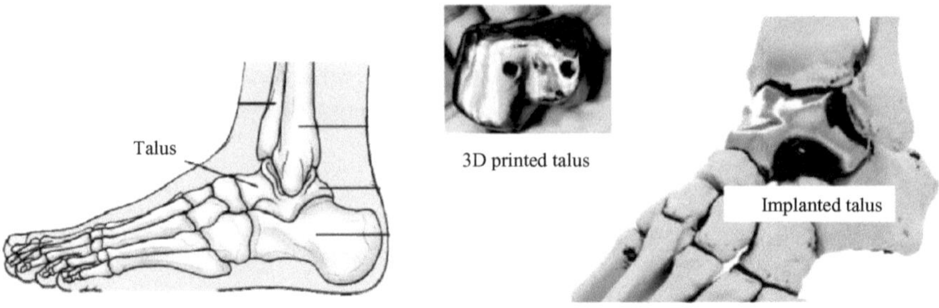

Figure 2.13 (a) Nomenclature of an ankle showing a normal talus in the ankle. (b) 3D-printed talus implant in the ankle made from Cobalt – Chromium material and shown separately at the right side with the fixing on it.

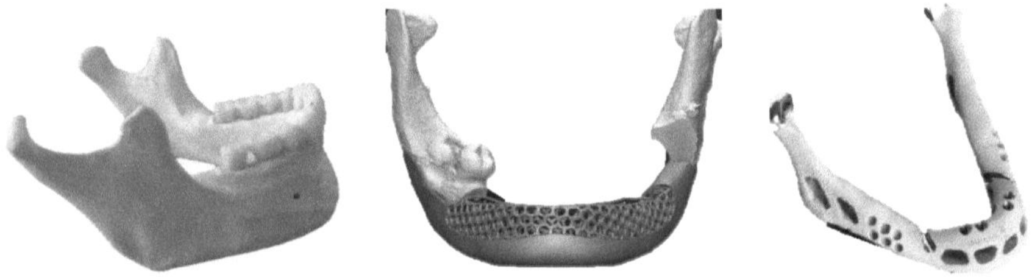

Figure 2.14 (a) Normal human mandible shape and profile with teeth on it. (b) 3D-printed patient-specific mandibles replaced the defective area with suitable 3D-printed metal portions matching the mandible shape.

articular cartilage, a slippery white substance. The cartilage between the talus and its neighbouring bones allows it to move smoothly. Figure 2.13 shows talus in the ankle and 3D-printed talus implant in the ankle. Total ankle replacement involves removal of arthritic bone and cartilage and replacement with metal implants and strong plastics [11]. These components allow the ankle joint to move smoothly, mimicking the cartilage that was lost.

CT scan images are converted into.stl data which is an input for selective laser sintering machine to develop the talus using the cobalt–chromium powder. Material – Cobalt – Chromium, process – SLS.

2.2.4.4 *Mandible*

The mandible is the largest bone in the human skull for achieving a better shape of the face. It holds the lower teeth in place, assists in mastication, and forms the lower jawline. The mandible is composed of the support frame and the ramus and is located inferior to the maxilla. The implant is remedial for hourglass facial deformity. Figure 2.14 shows the human mandible and 3D-printed patient-specific mandibles.

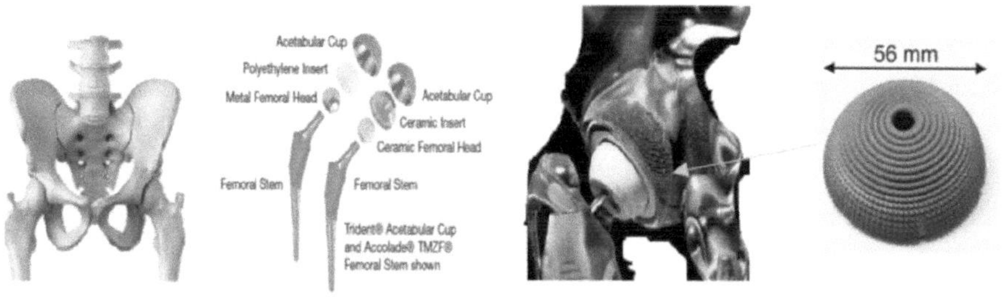

Figure 2.15 (a) Front view of a hip joint showing the thigh bones at bottom, socket area near to it, hip bone at centre, spinal discs at top. (b) Exploded views show nomenclature separately of thigh joint with its parts made from both metal and ceramic material. (c) 3D-printed hip implant assembly shows the position of femoral head, acetabular cup, and femoral stem of one side thigh joint.

The material used is titanium because of biocompatibility, strength to weight ratio, and direct structural and functional connection to bone and best suitable process is DMLS.

2.2.4.5 *Hip joint*

The hip joint is a ball-and-socket joint that allows motion and gives stability to bear body weight. The socket area (acetabulum) is inside the pelvis. The ball part of this joint femur joins with the acetabulum to form the hip joint as a ball and socket. The femoral head is the ball and the acetabulum (part of the pelvis) is the socket. These surfaces of the ball and socket are covered by articular cartilage, which is a specialized lining allowing pain-free motion at the contact surface [12]. Damage to this lining disturbs the smooth motion and results in arthritis. Figure 2.15 displays the hip joint overview, 3D-printed stem parts, 3D-printed hip implant assembly, and acetabular made from EBM and LRM. The joint is lined by a specialized synovial layer, which secretes fluid. The artificial hip has two bearing parts that carry out the whole function. These bearings can be metal-on-polyethylene, metal-on-metal, ceramic-on-polyethylene, or ceramic-on-ceramic. Using titanium powder, implants are manufactured by EBM, DMLM process with complex geometries and porous structures to encourage bone growth. Lattice structures are designed for a clinically optimized pore size of 650 μm.

2.2.4.6 *Spinal joint*

The spinal cord is the backbone of our body and carries nerves and signals from the brain to the body. It is a long tube-like structure composed of a band of tissue and bones. The damage of the structure affects nerve signals that do not help to feel sensations and move our body. The spine encloses the spinal cord and the fluid surrounding the spinal cord that allows articulation motion to facet joints.

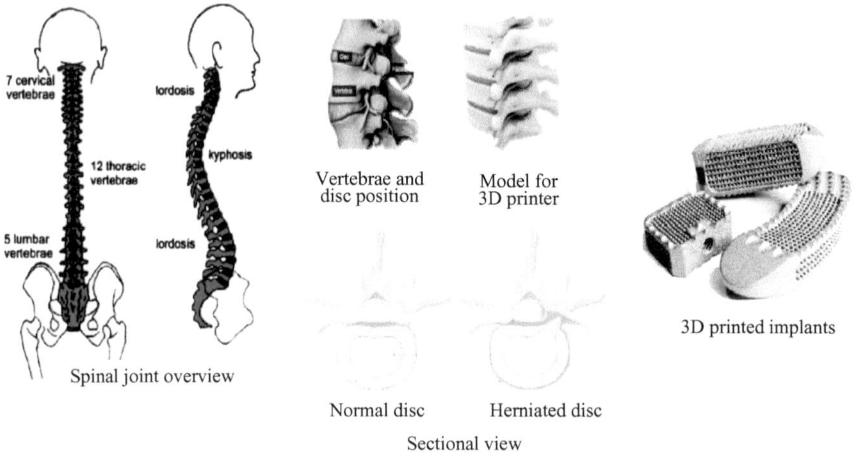

Figure 2.16 (a) Front view and side view (three disc portions of spinal) vertebrae. (b) Spinal disc normal and herniated damage at top with leakage of nucleus. (c) SLS printed titanium discs made for implant.

The spine is a flexible joint composed of 33 bones, called vertebrae, divided into five sections according to the connectivity, which is cervical, thoracic, lumbar spine sections, sacrum, and coccyx bones. The cervical section named C1–C7 of the spine is made up of the top seven vertebrae in the spine, and is connected to the base of the skull. The other sections consist of 26 bones called vertebrae; each vertebra has a strong stub by the section that supports our weight, and a hole for the spinal cord to pass through, which has three main parts, cervical (neck), thoracic (chest), lumbar (lower back). The lightweight, strong high strength weight ratio material titanium is mostly used in the production of spinal cord disc and the process is SLS 3D printing [13]. Figure 2.16 shows spinal joint overview, vertebrae and disc position and model and joint, disc herniation, and 3D-printed disc implants.

2.2.4.7 *Dental prosthesis*

Dental prostheses are artificial fittings that are used to restore a missing or damaged tooth. It is also utilized to restore an unattractive or unfunctional tooth. Dentures, bridges, and crowns are types of dental prostheses that significantly improve the patient's quality of life and health.

2.2.4.7.1 Dentures

Dentures are dental plates for artificial teeth. Generally, it is acrylic resin, chrome cobalt metal, or nylon polymer suitable to attach false teeth. By attaching it to some of the natural teeth with metal clasps or fitting it permanently, it is usually held securely in place in the mouth by clipping onto

Figure 2.17 A denture produced after 3D printing process and before removing the vertical supports between bottom of the denture and bed of the 3D printer.

some of your natural teeth. A denture is typically quite durable, and it can handle a high amount of pressure while chewing all kinds of foods. Nevertheless, they are designed to handle that pressure. The dentures are made from acrylic resin by SLA printing process. Figure 2.17 shows additive manufactured dentures with supports.

2.2.4.7.2 Dental implant

An implant is a medical device surgically implanted into the jaw to restore an individual's chewing ability and aesthetic appeal. They provide support for false teeth against biomechanical forces created during chewing action. The maximum chewing force applied on the molar tooth, at the time of chewing the food, noted by dentists is 150 N minimum and 700 N maximum. Lab simulations using computed tomography determine the position of the implants with adjacent teeth. The life of the implant is estimated as 10–15 years.

Dental implant systems consist of a dental implant body, abutment, and abutment fixation screw. The dental implant body is surgically inserted in the jawbone in place of the tooth's root and the abutment is attached to the body by the screw and extends through the gums into the mouth to support the attached artificial teeth [9]. Figure 2.18 shows the dental implant position adjacent to teeth and components of the dental implant. Titanium and ceramics are used as powders in a tub of selective laser sintering in a 3D printing machine to produce dental implants.

2.2.4.7.3 Crowns

The crown restores the tooth's shape, size, strength, and appearance. The dental crown is cemented into place on the tooth, and it covers the visible

Figure 2.18 A dental implant front of the jaw position adjacent to real teeth and components of dental implants are explored.

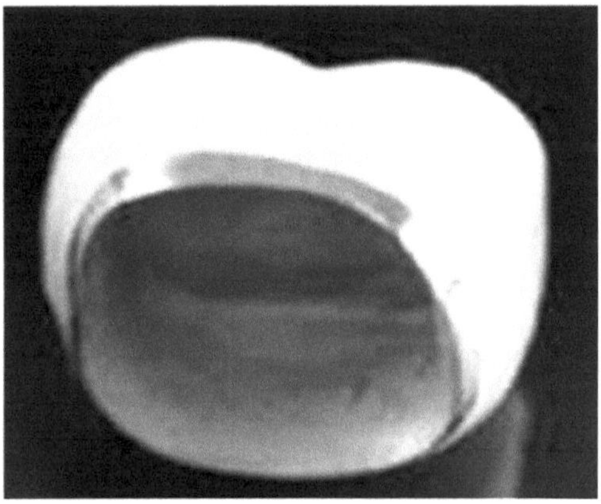

Figure 2.19 A zirconium crown shows top and metal protected bottom portion made as per 3D printer scanner profile.

portion of the tooth. The most used materials for dental crown are porcelain, porcelain fused to metal, gold, and zirconium.

The preparation of the existing defective tooth is important to halt nerve damage. Filing the tooth too thin may affect the nerve root and too abrasive crown is also destructive to the adjacent teeth. Figure 2.19 shows the zirconium crown with metal insert. The 3D printing processes SLA, SLS, and DMLS are used according to the material selected.

2.2.4.8 Aortic branches

Blood is supplied to the heart by the coronary arteries, which is the branch of the ascending aorta. Blood flows to the head, neck, and arms through the aortic arch. The ascending aorta arises from the heart and is about 50 mm long.

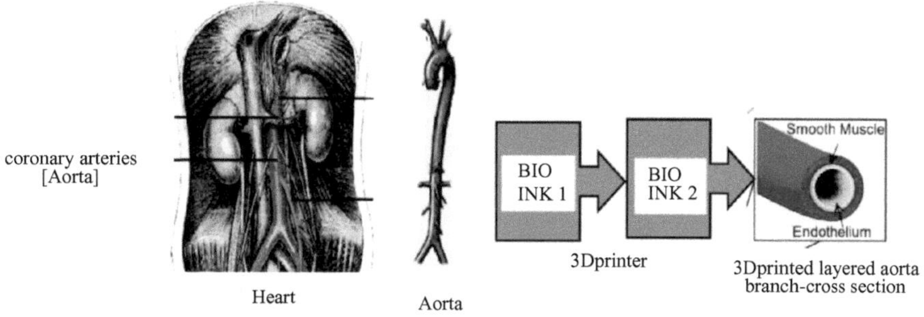

Figure 2.20 (a) The position of natural aorta and mesenteric artery shown in red colour in the heart. (b) The aortic branch is made for replacement by the 3D printing process using bioinks. (c) The anatomy and cross section of natural aortic branch zoomed separately at the right side.

The multilayered blood vessels are fabricated with TPU and bio-ink by the triple-coaxial 3D cell printing technique. The blood vessels that have the unique biomolecules needed to transform into functional blood vessels when they are implanted were formulated from smooth muscle cells from a human aorta and endothelial cells from an umbilical vein. The result is a fully functional blood vessel with a dual-layer architecture that outperforms existing engineered tissue and brings 3D-printed blood vessels several fundamental steps closer to clinical use. Figure 2.20 shows the aorta and mesenteric artery in the heart, aortic branch, and 3D-printed aortic branch.

2.2.4.9 Hearing aid

Hearing aids make sound audible to a partial hearing person to improve hearing. Sound waves are converted into electrical signals and then made louder with analogue hearing aids. The digital aid works as sound waves are converted into numerical codes like computer codes, then they are amplified and usually less expensive and have simple volume controls. The main parts of a hearing aid are the microphone that receives sound and converts it into a digital signal. An amplifier increases the strength of the digital signal. And a speaker produces the amplified sound into the ear.

The conventional method of manufacturing the parts is with injection moulding techniques. These are replaced based on a laser scan of an impression of the inner ear and on an intra-aural scan in 3D printing techniques. A hollow shell with the internal space is configured to accommodate the functional electronic hearing aid unit [3]. Figure 2.21 shows conventionally manufactured hearing aid parts and additively manufactured hearing aid. Low-viscosity photopolymer is used in the SLA 3D printing machine to create toughened, water-resistant shell part to the hearing aid.

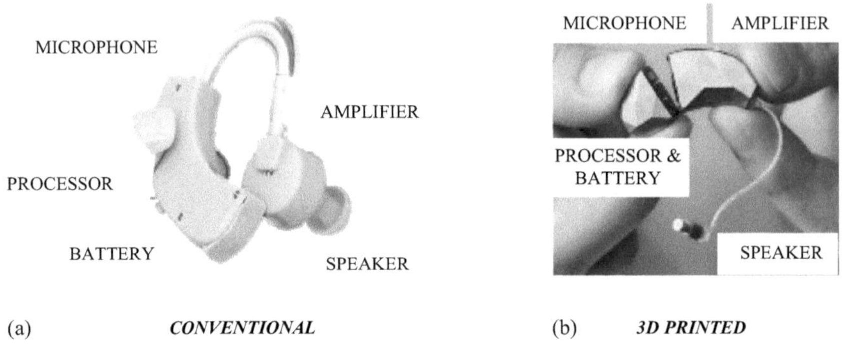

Figure 2.21 (a) Conventionally manufactured hearing aid with its functional parts more or less equal to an ear size fitted on the right side of a man. The parts necessary for a hearing aid device are named on a hearing aid device. (b) An additively manufactured smart minimized hearing aid device with all the necessary parts held in the hands.

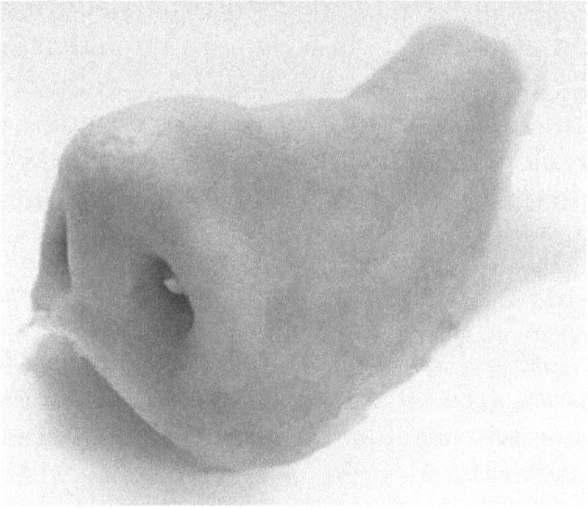

Figure 2.22 A customer-specific nose prosthetics made from silicone using 3D printer.

2.2.4.10 Nose prosthesis

Patients who have a partial nose or traumatic injury need a nose prosthesis. Figure 2.22 shows the nose made up of silicon with stereolithography 3D printing process. The nose is retained by the adhesive especially made for biocompatibility. These prostheses help humidify the nasal cavity during breathing and like a real nose protects the delicate mucosal tissue that lies below.

2.2.4.11 Myoelectric hand

Myoelectric-controlled prostheses are powered by electrical signals emitted by muscles and controlled with those signals. A myoelectric hand detects electrical signals in a stump's muscles and converts them into movement. A spinal cord injury affects both the hand and locomotion and patients with the injury are being provided with new medical technologies, including wearable devices, as well as rehabilitation treatments. A major advantage of 3D printing is that it can produce low-cost, personalized devices with scientific advancements. Using electromyography signals the researchers developed a novel 3D-printed hand prosthesis. Patients suffering from cervical injuries can use this device to assist with grasping.

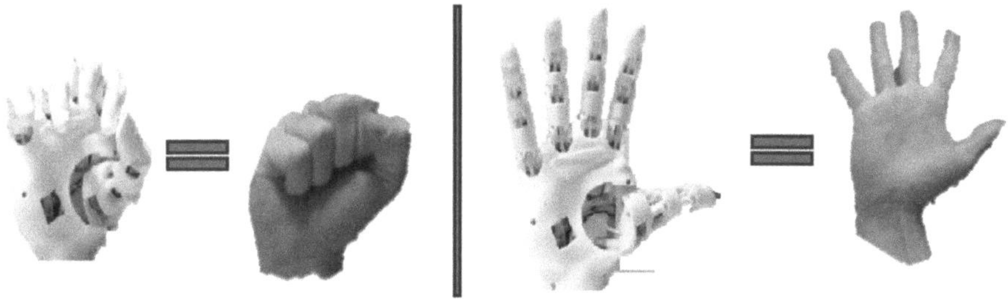

Figure 2.23 A demonstration shows the movement of natural hand's fingers in closed position at left side operates the same as the 3D printed myoelectric fingers in closed position. Similarly in open position at right side w.r.t electromyography signal from the hand.

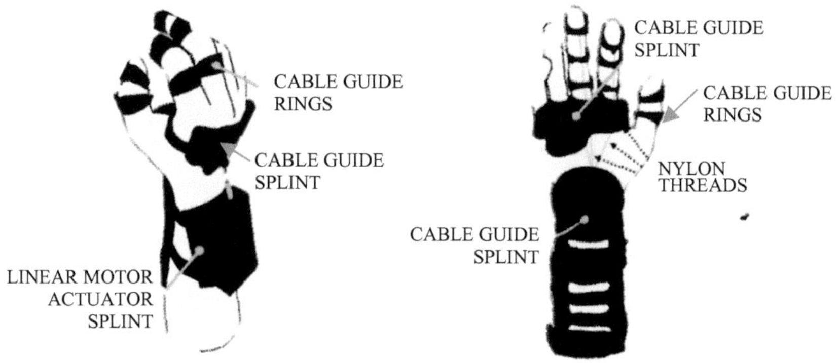

Figure 2.24 The figure shows the parts of myoelectric device made up of additive manufacturing process to be fitted on the natural hand. The detailed view of ring part, hand part, dorsal forearm splint, volar forearm splint shows the structures and the positions on the top and bottom portions of the hand.

Figure 2.23 shows the 3D-printed myoelectric fingers in closed open position synchronized to electromyography signal from hand weaned device, and Figure 2.24 shows the additive manufactured myoelectric parts and positions on the hand. The parts consisting of ring, hand part, dorsal splint, and volar splint are manufactured with polylactic acid filament in fused deposition modelling 3D printing machine.

GLOSSARY

DMLM: Direct metal laser manufacturing
DMLS: Direct metal laser sintering
EBM: Electron beam melting
ECM: Extracellular matrix
FDA: Food and Drug Administration
FDM: Fused deposition modelling
LOM: Laminated object manufacturing
LRM: Laser rapid manufacturing
SLA: Stereolithography
SLS: Selective laser sintering
TIJ: Thermal inkjet printing

REFERENCES

[1] Bailey, C, Aguilera, E, Espalin, D, Motta, J, Fernandez, A, Perez, MA, Dibiasio, C, Pryputniewicz, D, Macdonald, E & Wicker, RB 2018, 'Augmenting Computer-Aided Design Software With Multi-Functional Capabilities to Automate Multi-Process Additive Manufacturing', IEEE Access, vol. 6, pp. 1985–1994.
[2] Beyca, OF, Hancerliogullari, G & Yazici, I 2018 'Additive Manufacturing Technologies and Applications, in: Industry 4.0: Managing The Digital Transformation', Springer Series in Advanced Manufacturing, Springer, Cham, pp. 217–234.
[3] Katie Glover, Alexandros Ch. Stratakos, Aniko Varadi, Dimitrios A. Lamprou, 2021 'd3Dscaffolds in the treatment of diabetic foot ulcers: new trends vs conventional approaches' International Journal of Pharmaceutics, Vol.599, Doi- S0378517321002271.
[4] Katie Glover, Essyrose Mathew, Giulia Pitzanti, Erin Magee, Dimitrios A Lamprou, 2022, '3D bio-printed scaffolds for diabetic wound-healing applications', Springer, https://doi.org/10.1007/s13346-022-01115-8.
[5] Brian Katt, MD, Casey Imbergamo, BS, Daniel Seigerman, MD, Michael Rivlin, MD,2021, 'The Use of 3D Printed Customized Casts in Children with Upper Extremity Fractures: A Report of Two Cases', Vol.9, pp. 126–130.
[6] Bos, F, Wolfs, R, Ahmed, Z & Salet, T 2016, 'Additive manufacturing of concrete in construction: potentials and challenges of 3D concrete printing', Virtual Phys. Prototype, vol. 11, pp. 209–225.

[7] Brajlih, T, Valentan, B, Balic, J & Drstvensek, I 2011, 'Speed and accuracy evaluation of additive manufacturing machines', RapidPrototyp. J, vol. 17, pp. 64–75.
[8] Costabile, G, Fera, M, Fruggiero, F, Lambiase, A & Pham, D 2017, 'Cost models of additive manufacturing: A literature review', Int. J. Ind. Eng. Comput, vol. 8, pp. 263–283.
[9] M.Walbran, K.Turner, A.J.McDoid, 2016 'Customized 3D printed ankle-foot orthosis with adaptable carbon fibre composite spring joint', Tailor & Francis,, doi/10.1080/23311916.2016.1227022.
[10] Cartola. V, 2020'Shapeshift 3D on developing AM software for mass customization, 3D Natives, 240620204.
[11] Rich Wetzel, 2017,' Dinsmore and the Children's Hospital of Orange County Show Value of Patient-Specific 3D Printed Tibia Model'3D Print.com, 192704.
[12] Hongwei Wu, Shuo Yang, Jianfan Liu, Linqin Li, Yi Luo, Zixun Dai, Xin Wang, Xinyu Yao, Feng Zhou,2021,'3D printing guide plate for accurate hemicortical bone tumor resection in metaphysis of distal femoral: a technical note', Journal of Orthopaedic Surgery and Research, Vol. 343, Doi-10.1186/s13018-021-02374.
[13] Lucie Gaget, 2018,'3D printing in the medical industry: The 3D printed knee replacement', Sculpteo, 2018/05/04.

Chapter 3

Mechanical property of additive manufactured stainless steels

K.A. Ramesh Kumar, K. Prakash, P. Maadeswaran, and A. Ganapathi

Periyar University

CONTENTS

3.1 INTRODUCTION

Additive manufacturing techniques are used for various fields such as aerospace industry, medical, automobile and for jigs and fixtures design. So there is a need to develop awareness of AM process [1]. The fundamental principles and innovative ideas are required to know the complicated geometry for the construction of machinery. Industry and research need to collaborate with modern manufacturing techniques. The most important metal processes are selective laser melting and selective laser sintering [2].

The commercialization of additive manufactured products has predominantly increased in the past few decades. This increases the market competition of manufacturing [3]. The manufacturing process shows the transformation of technology. This gives freedom for the designers and

DOI: 10.1201/9781003430186-3

engineers to attain almost exact solution for the problems. New design methods are required to understand the transformation of modern technology. So there is a need to impart knowledge for those who are the future creators [4]. The selection of additive manufacturing techniques requires some parameters that will enhance the solution. Analytic hierarchy method is used to select the correct technology for the correct product [5].

Three-dimensional objects with simple geometry were manufactured by material extrusion process. It is purely dependent on flow rate of melted polymer. The temperature and viscosities will cause shear on flowing polymer. The constant flow with low shear can give extrusion rate in nozzle to attain optimized output of product [6]. Novel methods are used for the production of acoustic fibre incorporated with natural fibre composites. Natural fibre composite gives perfect replacement of synthetic-based absorbers. Panels designed are biodegradable after the usage [7]. Fusion Deposition Modelling technology for three-dimensional printing has major advantages over other AM methods. This method has not been preferred due to economic aspects since 2000. Atomization brings flexibility in manufacturing [8]. In the framework of paediatric medicine, 3D printing is capable of providing a measure of individualization and converting the precision medicine approach into practical implementation. Important quality features and process factors still have to be greatly enhanced for this new technology to be understood. The good news is that commercial start-ups and consortia are actively working on the requisite technological breakthroughs [9].

Bulk metallic glasses and high-entropy alloys are both important multicomponent alloys in additive manufacturing. The microstructures and exceptional properties make them promising for applications in many industries. With the introduction of metal additive manufacturing, new potential for fabricating geometrically complicated objects have emerged [10]. Increased laser power lowers porosity in printed objects, but it also causes glassy alloy devitrification. During the procedure, re-melting of powder bed layers was discovered to be critical in ensuring dense and crack-free material [11].

3.2 TYPES OF STAINLESS STEEL FOR ADDITIVE MANUFACTURED PRODUCTS

3.2.1 Austenitic stainless steels

Superior corrosion resistance, biocompatibility, and ductility of austenitic stainless steels make them a popular choice among industrial alloys. Biomedical, aerospace, defence, oil and gas, and petrochemical sectors are only few of the areas where they may be used. AM has been found to have promising qualities; however, there are still several obstacles to overcome.

3.2.1.1 Mechanical properties

In its as-built state, 304L stainless steel (304LSS) exhibits fatigue qualities comparable to those of commonly produced materials. Selective laser melting methods may be used to produce 316L cyclically loaded components. SLM-processed 316L has no substantial influence on its monotonic characteristics after stress-relieving heat treatment. In tensile testing, AM austenitic stainless steels exhibit behaviour that is distinct from that of traditionally manufactured components. When it comes to 304L stainless steel, LPBF has proved that it can create a product that is stronger and more ductile than its wrought/cast equivalent [12]. There are certain structural reliability difficulties because of a lack of knowledge of additive manufacturing process–structure–property interrelationships. The unique microstructural properties of the additively formed stainless steel alloy result in optimized fatigue resistance compared to wrought iron. Thermal behaviour of steel is affected by internal porosity. Figure 3.1 shows the ductility of LB-PBF is better to have higher yield stress [13].

The wear qualities of selective laser-melted 304LSS were examined to determine the wear characteristics with impact of the microstructure. From RT to 600°C, wear tests were performed by utilizing a ball on disc setup with an alumina counter-ball in that configuration. Applications that need wear-resistant stainless steel at high temperatures may be able to benefit from SLM316L. SLM316L SS had a smaller deformation area below the curve than typical 316 SS, which had a lower deformation zone of 27–30–40 m. Its shows the strain behaviour on reversing loads up to failure. The creation of persistent oxide glazing layers in SLM SS reduces the wear rate by two orders of magnitude compared to ordinary 316SS at 300°C, 400°C, and 600°C [14].

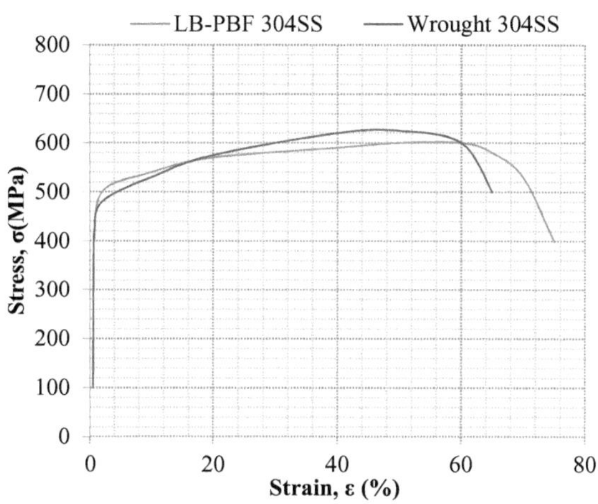

Figure 3.1 Comparison of ductility for laser beam powder bed fusion 304LSS and wrought 304LSS.

3.2.1.2 Corrosion properties

Stainless steel pitting and crevice corrosion caused by sulphide inclusions is discussed in detail. Manganese sulphides, especially those with a central oxide core, are ideal for pitting. Pit initiation becomes more difficult when manganese is replaced by chromium in the sulphide. Pitting happens at the less active inclusions when the ecosystem deteriorates. Finally, it may be stated that altering the sulphide composition can increase pitting and crevice corrosion resistance usefully but limitedly [15].

For fused coatings, fragmentation of MnS inclusions reduced metastable pit formation and stable pit transition susceptibility. Due to the tensile elastic strain in grains, the film stability was degraded, resulting in elastic strain. By using Ar shielding gas, SFC might enhance corrosion resistance by improving the overall stability of the passive film [16].

Sulphur-rich inclusions have been shown to predominate as pit nucleation sites, and the life span of unstable pits has been demonstrated to be correlated with the size of the inclusion particle. Sulphur produced from the inclusions may stabilize pit expansion, according to the researchers. The pitting resistance of type 304L stainless steel was significantly improved via laser surface melting (LSM). Pit nucleation in diluted neutral chloride solutions on type 304L stainless steel occurs mostly at sulphur-rich inclusions, according to microstructural characterization. Pitting susceptibility is determined by the size and distribution of sulphide inclusions in an alloy, not the total sulphur concentration. Quick evaluation and ranking of alloys based on pitting resistance may be done by collecting data on unstable pitting activity [17]. Austenitic stainless steel 316L is manufactured by 3D printing. Specimens were tested for corrosion resistance using potentiodynamic and potentiostatic methods. Laser power and scanning speed were used to create the different specimens. The porosity of each specimen was examined and correlated with its cyclic polarization and metastable pitting properties. These specimens are more resistant to pit formation than wrought 316L specimens. When comparing porosity of specimens, the 3D-printed product has less pit formation. In other words, in specimens with a larger porosity, repassivation following stable pitting caused by anodic polarization was less preferred. This contradicts the idea that greater corrosion resistance is associated with lower metastable pit frequency [18].

3.2.2 Maraging and precipitate hardening stainless steels

A laser beam is used to fuse small metal powder particles together through layer-by-layer scanning cross sections obtained from a 3D computer-aided design model. SLM can process a broad variety of materials, including steels, titanium, aluminium, and cobalt–chromium alloys. Materials such as maraging steel 300 (18Ni-300) are typically used in situations where high

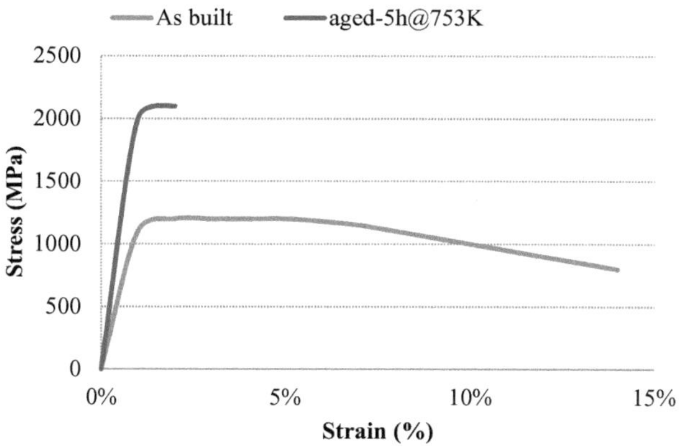

Figure 3.2 Stress–strain curve for ultrahigh strength low alloy.

fracture toughness and strength are needed or where dimensional changes must be kept to a minimum. Testing on specimens made with a combination of SLM and laser re-melt processing is conducted. The temperature and time ageing were varied to attain better hardness within a short time period. It may be inferred that dense portions of 18Ni (300) maraging steel with better strength and hardness can be produced [19].

The usage of ultrahigh strength steels in important aviation and aerospace structural applications has grown in recent years. A high-strength NiSiCrCoMo with low-alloy steel was described in this research. Tonnage scale melts have shown the reproducibility of desirable strength and toughness parameters. In the softened state, this steel has excellent forming and machining properties. After a certain time, SLM components lose their ductility and Charpy impact energy, which need more investigation as shown in Figure 3.2 [20].

As a result of its excellent printability and wide variety of applications, high-performance stainless steel has been concentrated on 17-4 PH, a single alloy that provides high strength with corrosion resistance. The casted 17-4 PH pieces are treated with a solution for removal of stress by annealing heat treatment process. Although thicker samples need longer reaction periods, this is commonly done at 1040°C for 1 hour [21].

3.2.2.1 Mechanical properties

The tensile strength, fluctuating load withstanding ability, and fracture resistance of Ti–6Al–4V and 15-5 PH steel samples were examined and compared in this study utilizing selective laser melting. The tensile characteristics of the materials were superior to those of the vertically constructed samples. Due to the variation in surface quality, selective laser-melted PH1 steels have a 20% lower endurance limit than conventional PH1 material.

The strain hardening behaviour of vertical and horizontal oriented SLM process has a 34% increase in UTS for LPBF at 593°C; however, this is offset by an almost 50% reduction in flexibility for LPBF, which is measured at a 9% extension to failure. According to the results of another investigation, horizontally constructed samples of 15-5 PH that had undergone an age hardening heat treatment had a 10% higher YS than ASM worked samples [22].

A combined experimental analysis and simulation technique was used to evaluate the room temperature and impact behaviour of 3D-printed stainless steel. It was found that the experimentally observed impact energy was 10,851.20 J/cm^2, which is equivalent to the impact energy of normally worked and non-heated stainless steel. The impact load was simulated using a Computational Fluid Dynamics based on the Johnson–Cook model with damage parameters. At room temperature, the surface indicates that the specimen was fractured [23].

3.2.2.2 *Corrosion resistance*

In light of the benefits of AM over conventional metal production processes, a number of relevant research have been conducted. Stainless steel AM-fabricated corrosion properties are the only topic of this extensive research study. The most important articles in direct laser deposition focused on the influence of heat treatments on passive films and the amount of chromium in these films. Other AM technologies, including wire arc additive manufacturing, do not have a corrosion effect. The corrosion of SS was influenced by factors such as grain size, grain porosity, chromium distribution, surface roughness, and embedded components. The corrosion investigations of SS manufactured by the AM methods other than SLM and DLD have not shown a distinct trend. New AM methods like WAAM and droplet-based 3D printing are expected to see an increase in corrosion research [24].

3.2.3 Ferritic/martensitic steels

Manufacturing technology known as additive manufacturing, or 3D printing, has the potential to revolutionize the industry. Layer-by-layer construction of intricate technical components is possible, utilizing a high-energy heating source and feed materials such as powder, wire, or sheet [25].

A martensitic 420 stainless steel with allowable stress of 1.67 GN/m^2, yield stress of 0.6 GN/m^2, and displacement of 3.5% was successfully manufactured using selective laser melting. A scanning electron microscope and X-ray diffraction revealed that the microstructure was composed of 0.5–1 μm-sized cells and submicron austenite-rich martensitic needles. Ultrahigh strength and great ductility were achieved by heat treating the material at 400°C [26]. The quality of the product is ensured by the high carbon equivalent of 4140 steel. Additive manufacturing processes using laser powder bed fusion for steel or iron powder have been investigated. An average tensile strength of 1.482 GN/m^2 and an impact toughness of 33 J at 18°C were achieved using as-welded AM constructions [27].

When it comes to additive manufacturing, the nuclear sector has been sluggish to take use of its apparent advantages. Nuclear-relevant materials, such as ferritic-martensitic steels, are being evaluated by ORNL using possible additive manufacturing processes. Before any functional components can be manufactured and deployed, there are a number of difficulties that need to be resolved. The problem is that there are no certification and qualifying processes or standards for these components. The near-net-shape creation of complicated geometries is made possible by additive manufacturing. Unique potential exists for site-specific control of chemistry and grain structure, as well. New designs for fusion reactors may be developed using additive manufacturing in order to increase efficiency by improving heat removal. Before any functional component can be created and deployed, the behaviour of additively generated steels under neutron and ion irradiation must be investigated [28].

3.2.4 High-strength low-alloy steels

Wire and arc additive manufacturing (WAAM) is an effective method for producing big and complicated components. For the microstructural development of the samples as they grew in height, the complicated thermal cycles operating on the material were used to rationalize component fabrication of high-strength low-alloy steel using WAAM. Using an infrared camera, it is discovered that heat input affects cooling rates, interlayer temperatures, and residence periods in the 800°C–500°C range [29].

3.2.5 Carbon-bearing tool steels

Layer-by-layer orientation according to a structure model is a unique production approach in additive manufacturing, particularly laser additive manufacturing. The benefits of lamination method are too quick prototyping, customization, better material consumption, and tendency to build complicated structures. Alloys, such as titanium and aluminium, are included in the list of metallic materials under research. The constraints and limits of LAM are discussed, as well as the anticipated development trend in the future.

The automobile, aircraft, architectural, biomedical, and electronics sectors all use AM to varying degrees of success. There are three main kinds of AM for metal fabrication: LAM, EBAM, and WAAM. Selective Laser Sintering is regarded as the oldest powder bed forming AM technology;

it is only suitable for alloys and cannot print with pure metal. SLM, on the other hand, can be used on almost all metallic materials because of its development from SLS [30].

3.2.6 Transformation-induced plasticity and twinning-induced plasticity steels

An inherent constraint in standard processing prevents the use of alloys with superior mechanical characteristics such as HMnS. Additive manufacturing offers a different way to use HMnS's unique features owing to drastically different processing conditions. At this time, there is no well-established way for customizing metallic alloys for additive manufacturing. Compression testing shows the alloy X30MnAl23-1 with the lowest specific energy absorption and the highest deformation behaviour. Grain boundary embrittlement suppression makes HMnS for future use in AM. Thermally induced martensite formation in X30Mn23 and the simultaneous activation of transformation-induced plasticity and twinning-induced plasticity in the microstructure were both triggered by the segregation of Mn and C. Figure 3.3 shows the stress–strain curves of the X30MnAl23 alloys in Laser Metal Deposition condition [31].

Materials with complicated geometries may be produced using selective laser melting (SLM) technology. An SLM-based method for processing a high-manganese steel was investigated to characterize the microstructure and element distribution in order to find the solidification behaviour [32].

Using selective laser melting, metastable austenitic Fe–Cr–Ni steel was produced with essentially random crystallographic microstructure. Despite the existence of porosity, tensile loading, and neutron diffraction with better ductility and a prominent strain-induced martensitic transition, mechanical

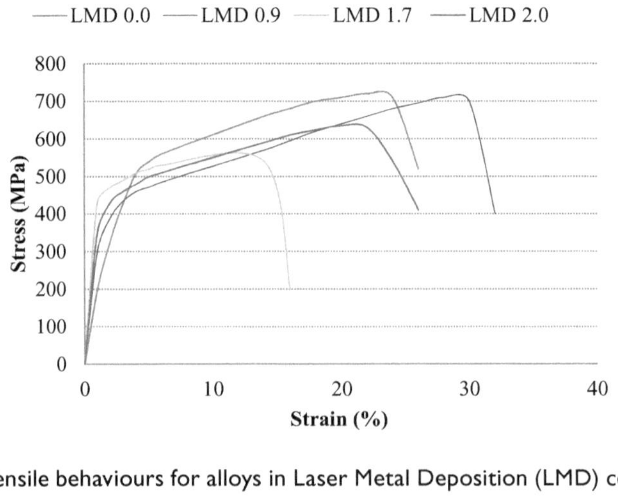

Figure 3.3 Tensile behaviours for alloys in Laser Metal Deposition (LMD) condition.

data shows a secondary hardening as a result of the martensite bearing a large load when it formed [33].

Under uniaxial deformation, strain-induced martensite formation was detected in selective laser melting-processed 304L material that was created with low laser power and checkerboard laser scanning pattern. Despite the apparent porosity, ductility and martensite production were clearly seen [34].

High yield strength and excellent ductility may be achieved despite the material's high porosity via additive manufacturing (AM) of 316L stainless steel. Using 3D printing, nitrogen gas is responsible for the majority of twinning. Deformation twinning is thought to be influenced by the presence of N solute atoms, which reduce the stacking fault energy of the steel. When developing microstructures, AM can provide for a fresh approach.

3.3 APPLICATIONS

Automotive and Aerospace Manufacturing: As a result of AM's capacity to shorten product design and development cycles, automated manufacturers have been able to get new items into the market swiftly and predictably, using this aspect of this technology. Aerospace is interested in these technologies because the titanium parts are manufactured without tooling. In the aviation and aerospace sectors, both EBM and SLS are presently in use. Lightweight automobiles and planes are the primary objective of automotive and aerospace industries; hence these AM-based technologies are particularly capable of producing lightweight components. In the aircraft business, there are several instances of the usage of AM technology. Design verification of an aircraft electrical generator, fabrication of an impeller compressor shroud casting pattern, and flight-certified production castings are examples. AM has been used to prototype gearbox housings for design verification, construct engine blocks, and create a rear wiper motor cover production tool. Manufacturing amphibious AM techniques is compared to gravity die casting, cross-drilling, electrochemical deburring, and hole blanking to create aerospace and vehicle components. The same item manufactured using AM utilized 40% less material than with the traditional method.

3.4 CONCLUSION

Austenitic stainless steels are the most commonly utilized in AM. Nano-inclusions and low angle grain boundaries restrict dislocations, making them robust and ductile. Due to AM's quick solidification, these steels are more resistant to pitting corrosion because MnS inclusions are less likely to occur. 17-4 PH stainless steel is the most extensively used PH stainless steel in AM because of its printability and versatility. AM can trigger "in situ" phase shifts like tempering and precipitation. Surface temperature and post-AM heat treatment modify ferritic/martensitic steel microstructure and properties. Because of their crack-prone nature, carbon-bearing tool steels present a particular challenge while being processed by AM. Up to 6.9% silicon steel has been successfully produced by AM and has shown interesting magnetic characteristics. With AM, the hardening stage for C-bearing tool steels might be eliminated. Compared to their wrought equivalents, AM parts have a larger density of volume defects such as holes and cavities, lamellar structures, and higher O contents. Because of AM's ability to affect local chemistry and consequently stack fault energy, the predominant deformation process is determined. In structural engineering, steel is the most widely used material because of the numerous design options it provides by systematically controlling allotropic changes during heating and cooling. It is possible to create new steel microstructures thanks to extreme temperature gradients, high cooling speeds, and inherent chemical heterogeneity in AM. Although AM steel has a promising future, more work must be done before AM can completely replace conventional steel production in additional applications.

REFERENCES

[1] M. Al-Makky, D. Mahmoud (2016) The importance of additive-manufacturing processes in industrial applications, *Proceedings of the 17th International Conference on Applied Mechanics and Mechanical Engineering (AMME), Conference*, 19–21 April, pp. 1–14.

[2] J. Go, A. J. Hart (2016) A framework for teaching the fundamentals of additive manufacturing and enabling rapid innovation, *Additive Manufacturing*, vol. 10, pp. 76–87.

[3] D. R. Eyers, A. T. Potter (2014) Industrial additive manufacturing: A manufacturing systems perspective, *Computers in Industry*, vol. 92–93, pp. 208–218.

[4] C. Beyer (2014) Strategic implications of current trends in additive manufacturing, *The Journal of Manufacturing Science and Engineering*, vol. 136, no. 6, p. 064701.

[5] E. de S. Zancul, C. G. Mançanares, J. Cavalcante da Silva, P. A. Cauchick Miguel (2015) Additive manufacturing process selection based on parts' selection criteria, *The International Journal of Advanced Manufacturing Technology*, vol. 80, pp. 1007–1014.

[6] A. Das, E. L. Gilmer, S. Biria, M. J. Bortner (2021) Importance of polymer rheology on material extrusion additive manufacturing: Correlating process physics to print properties, *ACS Applied Polymer Materials*, vol. 3, pp. 1218–1249.

[7] V. Sekar, M. H. Fouladi, S. N. Namasivayam, S. Sivanesan (2019) Additive manufacturing: A novel method for developing an acoustic panel made of natural fiber-reinforced composites with enhanced mechanical and acoustical properties, *Journal of Engineering* vol. 2019, Article ID 4546863, https://doi.org/10.1155/2019/454686.

[8] P. Becker, N. Spielbauer, A. Roennau, R. Dillmann (2020) Real-time in-situ process error detection in additive manufacturing, *International Conference on Robotic Computing*, Taichung, Taiwan, pp. 426–427. doi: 10.1109/IRC.2020.00077

[9] J. Quodbach (2022) Quality of FDM 3D Printed Medicines for Pediatrics: Considerations for Formulation Development, Filament Extrusion, Printing Process and Printer Design, *Therapeutic Innovation & Regulatory Science, vol.* 56, pp. 910–928. https://doi.org/10.1007/s43441-021-00354-0.

[10] X. Li (2018) Additive manufacturing of advanced multi-component alloys: Bulk metallic glasses and high entropy alloys advanced engineering material, pp. 1–18, 1700874. doi: 10.1002/adem.201700874

[11] J. J. Marattukalam (2020) Development of process parameters for selective laser melting of a Zr-based bulk metallic glass, *Additive Manufacturing*, vol. 33, 101124.

[12] A. Riemer, S. Leuders, M. Thöne, H. A. Richard, T. Tröster, T. Niendorf (2014) On the fatigue crack growth behavior in 316L stainless steel manufactured by selective laser melting, *Engineering Fracture Mechanics*, vol. 120, pp. 15–25.

[13] J. W. Pegues, M. D. Roach, N. Shamsaei (2020) Additive manufacturing of fatigue resistant austenitic stainless steels by understanding process-structure–property relationships, *Materials Research Letters*, vol. 8, pp. 8–15.

[14] S. Alvi, K. Saeidi, F. Akhtar (2020) High temperature tribology and wear of selective laser melted 316L stainless steel, *Wear*, vol. 448–449, p. 203228.

[15] A. J. Sedriks (1983) Role of sulphide inclusions in pitting and crevice corrosion of stainless steels, *International Metals Reviews*, vol. 28, pp. 295–307.

[16] D. Guo, C. T. Kwok, S. L. I. Chan (2019) Strengthened forced convection – A novel method for improving the pitting corrosion resistance of friction-surfaced stainless steel coating, *Materials & Design*, vol. 182, p. 108037.

[17] J. Stewart, D. E. Williams (1992) The initiation of pitting corrosion on austenitic stainless steel: On the role and importance of sulphide inclusions, *Corrosion Science*, vol. 33, no. 3, pp. 457–463.

[18] G. Sander (2017) On the corrosion and metastable pitting characteristics of 316L stainless steel produced by selective laser melting, *Journal of the Electrochemical Society*, vol. 164, no. 6, pp. C250–C257.

[19] M. Laleh, A. E. Hughes, W. Xu, I. Gibson, M. Y. Tan (2019) Unexpected erosion-corrosion behaviour of 316L stainless steel produced by selective laser melting, *Corrosion Science*, vol. 155, no. January, pp. 67–74.

[20] J. Okado (2010) Microstructure and mechanical properties of maraging steel 300 after selective laser melting. *Collections of 2010 International Solid Freeform Fabrication Symposium.* http://dx.doi.org/10.26153/tsw/15205

[21] G. Malakondaiah, M. Srinivas, P. R. Rao (1995) Basic studies leading to the development of an ultrahigh strength, high fracture toughness low-alloy steel, *Bulletin of Materials Science*, vol. 18, pp. 325–341.

[22] ASM International (1991) *Heat Treating*, Vol. 4, ASM International, Materials Park, OH, p. 860.

[23] H. K. Rafi, D. Pal, N. Patil, T. L. Starr, & B. E. Stucker. (2014) Microstructure and mechanical behavior of 174 precipitation hardenable steel processed by selective laser melting. *Journal of Materials Engineering and Performance*, vol. 23, no. 12, pp. 4421–4428.

[24] S. Sagar, Y. Zhang, L. Wu, et al. (2018) Room-temperature charpy impact property of 3D-printed 15-5 stainless steel. *Journal of Materials Engineering and Performance*, vol. 27, pp. 52–56. https://doi.org/10.1007/s11665-017-3085-9.
[25] G. Ko, W. Kim, K. Kwon, T. K. Lee (2021) The corrosion of stainless steel made by additive manufacturing: A review, *Metals*, vol. 11, no. 3, pp. 1–21.
[26] N. Haghdadi, M. Laleh, M. Moyle, S. Primig (2021) Additive manufacturing of steels: A review of achievements and challenges, *Journal of Materials Science*, vol. 56, pp. 64–107.
[27] K. Saeidi (2019) Ultra-high strength martensitic 420 stainless steel with high ductility, *Additive Manufacturing*, vol. 29, p. 100803.
[28] W. Wang, S. Kelly (2016) A metallurgical evaluation of the powder-bed laser additive manufactured 4140 steel material, *JOM*, vol. 68, no. 3, pp. 869–875.
[29] N. Sridharan, K. Field (2019) A road map for the advanced manufacturing of ferritic-martensitic steels, *Fusion Science and Technology*, vol. 75, no. 4, pp. 264–274.
[30] G. Gong (2021) Research status of laser additive manufacturing for metal: A review, *International Materials Reviews*, vol. 15, pp. 855–884.
[31] F. Kies (2018) Design of high-manganese steels for additive manufacturing applications with energy-absorption functionality, *Materials & Design*, vol. 160, pp. 1250–1264.
[32] C. Haase, J. Bültmann, J. Hof, S. Ziegler, S. Bremen, C. Hinke, A. Schwedt, U. Prahl, & W. Bleck. (2017) Exploiting process-related advantages of selective laser melting for the production of high-manganese steel. *Materials (Basel)*, vol. 10, no. 1, p. 56. doi: 10.3390/ma10010056.
[33] E. Polatidis, J. Čapek, A. Arabi-Hashemi, C. Leinenbach, M. Strobl (2020) High ductility and transformation-induced-plasticity in metastable stainless steel processed by selective laser melting with low power, *Scripta Materialia*, vol. 176, pp. 53–57.
[34] M. S. Pham, B. Dovgyy, P. A. Hooper (2017) Twinning induced plasticity in austenitic stainless steel 316L made by additive manufacturing, *Materials Science and Engineering: A*, vol. 704, pp. 102–111.

Chapter 4

Fundamentals of additive manufacturing technology and future outlooks

R. Muraliraja, VijayAnanth Suyamburajan, V.S. Shaisundaram, and V. Gowsikan
Vels Institute of Science Technology and Advanced Studies

S. Padmanabhan
Vel Tech Rangarajan Dr. Sagunthala R&D Institute of Science and Technology

CONTENTS

DOI: 10.1201/9781003430186-4

4.1 INTRODUCTION

Also known as 3D printing, additive manufacturing (AM) is a procedure that uses a computer-aided design model to generate 3D objects by adding materials one at a time. An array of items from medical gadgets to aerospace construction may be reimagined via the use of this technology, which has the potential to revolutionize the industry. To develop and fabricate unique three-dimensional devices, engineers may use a broad variety of functional materials when fabricating three-dimensional components in a layer-by-layer manner. Because of these unique capabilities, engineers may design structures that are both light and strong, as well as assemble several components into a single 3D printed item that can be functionally used. With the goal of providing an overview of existing AM methods and a fundamental scientific knowledge of this growing technology, the course is prepared. In the previous several decades, 3D printing technology has grown rapidly, enabling a wide variety of inventive projects and new production and prototype procedures [1]. To create a three-dimensional object, additive manufacturing uses material to build it from the ground up. Originally designed for fast prototyping, this manufacturing service has expanded over the years to serve a broad range of sectors and is increasingly gaining traction as a tool for mass customization of complicated end products. The evolution of 3D printing technology may be summarized as follows:

1800s – Early additive manufacturing concepts began in the 1800s, and 3D scanning concepts began to take form in their own unique manner. When François Willème created "photographic sculpture" in 1859, he used 24 cameras at various angles to create 3D representations of humans. This method of making three-dimensional topographical maps was first patented in 1892 by Joseph E. Blanther.

1990s – Technologies developed and innovations grew. Many new technologies emerged in the 1990s, including Direct Metal Laser Sintering and Binder Jetting. The development of bioprinting, which uses 3D printing to precisely arrange layers of cells and their supporting structures to generate functional tissue, also started around this time. Researchers at Wake Forest University's Institute for Regenerative Medicine developed a human bladder in 1999, making it the first organ to be grown using stem cells. As a result, they 3D printed the bladder using synthetic, biodegradable scaffolds that were covered with the patient's own cells. Medical 3D bioprinting and printing would be able to take form as a result. Machine prices began to fall in the early 2000s, and 3D printing became more widely available. With rapid prototype (RepRap), Adrian Bowyer created the first low-cost desktop printer capable of printing its own components for building a second version of itself. RepRap's "Darwin" was initially published in 2007, and there are now innumerable variations of the printer.

2000s – Increase in accessibility to 3D printing. Because Shapeways makes high-quality 3D printing technology available to people and companies, 3D printing services like Shapeways have grown in popularity. There are a lot of cutting-edge technologies and 75+ materials available via Shapeways' services for designers and companies to use. Because of the internet, 3D printing has become more widely available to anyone who wants to make use of it. More and more firms are turning to 3D printing and Shapeways to create high-quality items in-house rather than paying for expensive outside vendors [2].

4.2 THE NEEDS OF ADDITIVE MANUFACTURING

- Create parts with greater complexity: In contrast to conventional manufacturing processes, additive manufacturing may produce more complex components with increased functionality. Typical straight cooling channels in injection molds, for example, result in a slower and less uniform cooling of the finished product. A more uniform heat transmission may be achieved by re-designing the cooling channels using 3D printing to produce more complicated or curved designs. As a consequence, the cooling qualities of a mold are enhanced, resulting in better quality components and longer mold life [3].
- Minimal material waste: Tools such as lattice structures and topology optimization may help accomplish this goal to some extent. Topology optimization may be used to determine the ideal form for a component and reduce unneeded material without affecting its structural integrity. Rather than adding more material, subtractive approaches would merely remove the existing one.
- Simplified assembly: A third game-changing advantage of additive manufacturing is the ability to consolidate parts. Traditional manufacturing necessitates the production of several components before the final product is assembled. It is possible to create a full unique item in one go because of 3D printing's ability to combine several tiny parts into one larger one. As a result, the assembly process is greatly reduced or eliminated at times. A consolidated part removes the need to procure and store extra subcomponents or replacement parts, therefore decreasing inventory and maintenance expenses.
- Material innovation: In recent years, new materials have been developed thanks to advances in material science. TPU filaments and metal superalloy powders are two examples of 3D printing materials that are difficult to process or mold. 3D printing using thermoplastics developed for engineering purposes is shown here. To save weight and money, some of these high-performance materials may even be used in place of metal components.
- Cost-effective customization: By making design iterations swift and inexpensive, 3D printing opens up hitherto unimaginable vistas of personalization. It takes much less time to produce components using

additive manufacturing than with traditional subtractive methods. This implies that corporations will be able to develop customized items more quickly and cost-effectively in the future [4].

- Minimum support structures: Part orientation is one of the most advantageous features to consider while creating a design for 3D printing. A shorter printing and post-processing time and fewer supports are also possible when component orientation is carefully considered during design. To save time and materials, it's best to use as few support structures as possible when designing complicated 3D printed items. By doing so, post-processing may be faster and less wasteful.

4.3 CLASSIFICATION OF ADDITIVE MANUFACTURING TECHNOLOGIES

1. Binder jetting: Alternating layers of finely ground powdered material and liquid binder are applied by way of 3D printing-style heads that move in all three directions.
2. Directed energy deposition: Ceramics, metals, and polymers are just a few of the materials that may be utilized in additive manufacturing using direct energy deposition. It's possible to build up material by using a horizontally moving beam of an electric arc, a laser, or an electron beam cannon.
3. Material extrusion: These polymers are spooled and used in an AM process that either extrudes or draws them via a heated nozzle on a moving arm. As the nozzle and bed move vertically and horizontally, the melted material is built up layer by layer. Thermal or chemical bonding agents are used to keep the layers together.
4. Powder bed fusion: AM methods such as direct metal laser melting, selective laser sintering, selective heat sintering, and electron beam melting are all part of powder bed fusion. Using electron beams, lasers, or thermal print heads, thin layers of material are melted or partly melted, and superfluous powder is blasted away.
5. Sheet lamination: Laminated object manufacturing (LOM) and ultrasonic additive manufacturing (UAM) are two methods of sheet lamination. In order to create visually appealing products, laminated object manufacturing employs alternating layers of paper and glue. It is possible to weld aluminum, stainless steel, and titanium using UAM, a low-energy, low-temperature procedure that employs ultrasonic welding to unite metal sheets.
6. Vat polymerization: The item is built layer by layer in a vat of liquid resin photopolymer. In order to cure consecutive layers of resin using photopolymerization, mirrors are utilized to focus ultraviolet light.
7. Wire arc additive manufacturing: Welding arcs are used in conjunction with manipulators to form 3D shapes using arc deposition. Using

a predefined route to construct the desired form, it is often done using wire. Robotic welding equipment is often used to execute this kind of additive manufacturing.

4.4 AM TECHNOLOGY IN PRODUCT DEVELOPMENT

One of the first steps in product development is prototyping. Prototyping a product might take a long time, depending on the complexity of the final product. A significant amount of time may be saved by using 3D printing. A printed object speeds up the product development process significantly, regardless of whether it is used to create a functioning or just aesthetic prototype. Picture showing an investor or prospective consumer an early prototype of your new product before it is ready to be released in the market. With 3D printing, this is possible [5]. Having something tangible in your hands, rather than just a picture on a computer screen or printed on paper, may do wonders when it comes to creating an impression on a potential employer. In addition to saving time and money, 3D printing can manufacture items in a manner that was previously unimaginable. From safety gear to prosthesis, custom-made items are available. Multiple iterations in fast succession allow for all problems to be ironed out before the product goes into mass production. Innovators would be well to embrace 3D printing as the new wave of manufacturing. Inventors who don't possess a device of their own may still benefit from the technology, and there is no excuse for not using it to its utmost potential.

4.5 MATERIALS FOR AM TECHNOLOGY

1. Biochemicals: Silicon, calcium phosphate, and zinc are examples of biochemicals that are used in AM. Additionally, bio-inks that are produced from stem cells are now being researched. These materials are often utilized for medical-related purposes.
2. Ceramics: AM makes use of a wide variety of ceramics, such as alumina, tricalcium phosphate, and zirconia, in addition to powdered glass, which, when combined with adhesives and baked, may produce new varieties of glass products.
3. Metals: The additive manufacturing technique uses a variety of metals and metal alloys, including gold, silver, stainless steel, and titanium. These are used to make jewelry and aeronautical metal parts.
4. Thermoplastics: There are a wide range of thermoplastic polymers, each with its own benefits and uses that are routinely utilized in additive manufacturing. Plastics such as acrylonitrile-butadiene-styrene (ABS), polylactic acid (PLA), and polycarbonate (PC) may be used to give brief support before being dissolved, as can water-soluble polyvinyl alcohol (PVA) [6].

4.6 APPLICATIONS OF AM

Wearable technology (WT) demands are comparable to those of AT, but product strength and fatigue resistance aren't quite as high. SLA or DLP may be used for tiny products that need high resolution and strong surface finish, whereas FFF can be used for bigger objects where low resolution and surface quality are not a problem. The more costly MJ, on the other hand, has excellent resolution and a smooth finish. MJ, SLA, and DLP products are vulnerable to UV and heat damage. Printing several materials simultaneously is possible with FFF and MJ, which is a benefit. Some SLA printers can also print multi-materials; however, this isn't normal for this kind of printing.

Hand tools: When suitable for public usage, tool handles and pHMIs need to be sturdy, fatigue resistant, and perhaps disinfection fluid resistant. High-resolution, complex-geometry printing may be necessary for pHMI, whereas tool handles need a strong strength-to-weight ratio [7]. Multimaterial printing and a high-quality surface finish may be beneficial to both industries.

Assistive technologies: When designing prosthetics, orthoses, and exoskeletons, it is important to consider their capacity to tolerate strong stresses and frequent usage, as well as their biocompatibility of portions that come into contact with skin. Additionally, it is possible that disinfection fluid resistance and huge construction volumes, complicated geometry, a good surface polish, and an affordable end-product pricing will be critical considerations. Where extremely high forces are present, PA and TPU printed with SLS or titanium produced with SLM/DMLS/EBM is most suited to meet these requirements. PA and TPU FFF is only suitable for non-load-bearing components that are not subject to recurrent mechanical loading because of the anisotropic nature of printed products.

Medical devices: The sterilizability and physical and mechanical features of medical equipment used in invasive operations must be maintained following sterilization, as well as their biocompatibility. It is critical that they be impact- and mechanical stress resistant. Surgeons' tools, on the other hand, must be fatigue resistant, strong for their weight, and have a high surface polish in order to be effective during surgery. Biological inertness is often required for medical equipment, yet intricate geometric forms are also a plus [8].

4.7 REVERSE ENGINEERING

Through the combination of 3D printing and reverse engineering, it is possible to create an exact copy of an existing item or component. Scan a 3D item and convert it into a file that can be 3D printed to produce an identical copy or a piece of spare equipment. Many old machines are sitting idle in the manufacturing industry because certain components have failed and the firm has been unable to get them from the original manufacturer. It is possible to scan and 3D-print the components in such circumstances, which would allow the equipment to work again. Some of the numerous fields that use reverse engineering include aerospace and defense as well as general engineering. As a result of reverse engineering, outdated components that are no longer manufactured, complicated existing parts that don't need to be designed in CAD software, reduced machine downtime, improved efficiency of existing parts, and cost savings may all be realized [9]. The steps for reverse engineering are

(i) Determine units and take part dimensions: The first step in reverse engineering is to figure out what units are being used. In order to determine whether or not a piece is in the Metric or Imperial system of measurement, gather information about where it was manufactured and created. Take data from the existing portion after you've figured out what units are being utilized. Begin by obtaining measurements of the main body's length, breadth, and diameter.

(ii) Plan for your design and understand the object thoroughly: The next step is to begin designing needed 3D metal printing design. Understanding the object's complicated form is essential for certain goods. If the component is part of an assembly, you must ensure that the functioning of the part does not interfere with the interrelationship of the other parts. During disassembly, keep an eye out for the pieces' fit and feel. Having a clear idea of what you're doing from the beginning will make it much simpler to plan and create.

(iii) Laser-scan and start CAD designing: 3D scanning technology such as laser scanners, structured light digitizers, or industrial CT scanning must be used to capture the actual component or item. A polygonal model is created using the point cloud or mesh that was acquired from the scanner by CAD software packages. Clean, smooth, and accurate mesh is produced as the result of mesh cleaning and sculpting. The next step is to identify any design problems in the CAD model, as well as any necessary design revisions. In addition, you must identify the regions where printing alterations are needed, such as the addition of support structures in areas where there is a danger of breaking. Stacked 2D layers are created from the 3D CAD model by converting the model to an STL file. In order to send the STL file to the printer, a specific piece of machine software is used.

(iv) 3D-print the CAD file: In the last stage, all you have to do is send the STL file to a 3D printer and wait for the 3D item to be printed out. The printer is configured with printing settings once the file and material are loaded. The material is deposited layer by layer by the 3D printer to form the model. Second layer: The powder is added to the first layer after it has been built. Repeat this procedure until the project is finished [10]. Finally, post-processing takes place, and the final product is sent out.

4.8 DATA PROCESSING FOR AM

There are several procedures that may be utilized to make a three-dimensional item using the phrase 3D printing. Layers of material are laid down by a machine that is directed by a computer. Virtual designs and all the measurements required to make a real thing are created using CAD software. Alternatively, a 3D scanner may be utilized to build the template for printing if there is already an item in existence. The usage of 3D scanners and 3D printers is on the rise. They are used in gadgets like the Kinect for Xbox or smartphones that employ scanners to collect hand gestures as command functionalities by companies like Microsoft and Google. This technology will soon make it possible to create a 3D item by taking a photo and uploading it to an online 3D printer. Even while some types of 3D printers cost thousands of dollars, there are currently several that are inexpensive enough for home usage to be practical. The computer divides the picture into a large number of horizontal layers, which are subsequently transferred to the 3D printer. The printer builds a 3D picture by laying down layers of material. To suggest that this technology is a significant deal in product development would be an understatement [11].

4.8.1 CAD model preparation

Prior to printing a product, the designer must first construct a 3D model of it using CAD software or a 3D scanner. Since the component is a reproduction of the 3D model, it is critical that all of its exterior geometry be described precisely [12]. Designing for additive manufacturing may be more flexible than designing for traditional manufacturing techniques, but there are still constraints and standards to follow when designing for the best outcomes. AM can print complicated pieces. Each additive manufacturing process and material has its own design guide. AM technology service providers and equipment manufacturers both provide substantial design guidelines.

Part orientation and support generation: The CAD file is converted by the user into a standard AM file format known as standard tessellation language (STL). This format was established by 3D Systems in the late 1980s specifically for use in their stereolithography (SLA) manufacturing equipment.

4.8.2 Model slicing

In order to produce the item, the 3D printer uses the STL file's coarseness setting. Of course, as the triangles become smaller, the approximation gets better, resulting in higher-quality prints. However, the number of triangles required to cover the surface grows as the triangle's size decreases. There is a chance that this may produce enormous STL files, which certain slicing tools will find difficult to deal with as sharing or uploading such large files is difficult as well. The appropriate balance between file size and print quality is thus very essential. This cannot be done forever, since the human eye will be unable to distinguish between the many print quality variations. In most CAD programs, you may choose from a variety of options for exporting STL files. The size of the facets, and hence the print quality and file size, may be adjusted using these options. On the left, a perfect spherical surface is approximated by tessellations, as seen in this image [13]. The coarse model shown in the image to the right is the consequence of using large triangles as shown in Figure 4.1. The image in the middle, on the other hand, makes better use of smaller triangles and so provides a more accurate representation.

4.8.3 Tool path generation

CAD tools are required to model and integrate the linkages between construction processes and structures and their associated properties for each kind of material and procedure. Deposition route topology and fairness have a significant impact on the quality of AM-produced materials. Vacuum or gaps may form between neighboring passes or layers when a poorly designed route is used. Two common toolpath planning procedures in AM technology are seen in Figure 4.2: they are zigzag pattern and the

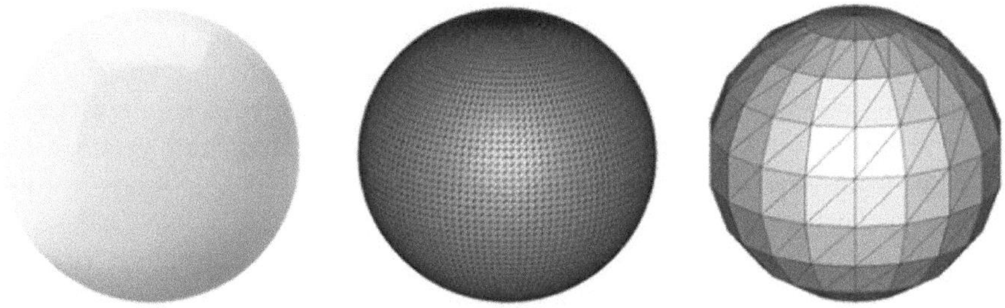

Figure 4.1 Model slicing.

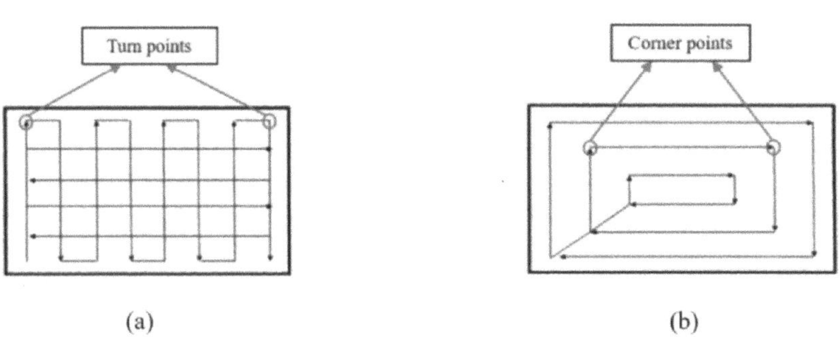

Figure 4.2 Tool path in AM technology.

contour offsetting pattern. The zigzag route pattern and the contour offsetting path pattern are both discontinuous, which may lead to overfilling at turn points and uneven surfaces during deposition. An acceleration and slowing occurs at the turning locations as a result of the irregularity of the zigzag and contour offsetting paths [14].

4.9 SOFTWARE FOR AM – MIMICS, MAGICS

The use of additive manufacturing software, or AM software, facilitates effective 3D printing tasks by preparing all work, optimizing designs, lowering design to production lead times, and reducing the total cost of operations via minimum printing time and material usage.

For 3D printing, additive manufacturing software transforms CAD data into a format that can be used by the printers. When preparing for printing, a number of steps are taken: optimization techniques, constructing adequate supports to avoid build failure, simulations of different printing strategies, determining the optimal print strategies and scan-path calculations. Then, everything is sent to the printer and printed! Adhesive manufacturing software may also assist your firm manage the order turnaround time and optimize machine utilization rates by tracking the workflow of an additive manufacturing process.

Printing three-dimensional products from a digital design is referred to as additive manufacturing, or 3D printing. At the conclusion of the design cycle, you may use additive manufacturing software to try out different ideas and see how they print. Metals, ceramics, foams, and gels are the most often utilized additive manufacturing materials, as are polymer composites and metals.

Features of CAD software for additive manufacturing include the following: Use numerous elements and specs to create a 3D model of the final product. Simulation: Use topology optimization and lattice structures to uncover a material arrangement that can be optimized within a particular design space, utilizing specified loads, boundary conditions, and restrictions. Simulation: Keep track of which 3D printing projects have been ordered, how long they've been printing, and whether or not they're ready to be dispatched. Consider the cost of each work, the time it will take to complete, and the supplies required. Offer an estimate for a certain order of 3D printing parts. Orders: Manage them. Get an overview of all of the current production orders and any shipments of materials such as alloys, elastomers, and plastics required to finish the project. Make sure that all machines are set up and ready to go at different times of the day to optimize throughput. Improve machine planning by ensuring that the personnel, equipment, and materials required for the specified construction time are available at the moment they are needed. The amount of material generated during a manufacturing run may be managed using batch management software. Ascertain that sufficient materials are on hand to fulfill all manufacturing runs. To keep manufacturing costs down, group together comparable production tasks. Post-processing: Ensure orders are processed and sent to the correct parties for distribution.

4.10 TECHNOLOGIES IN ADDITIVE MANUFACTURING

4.10.1 Liquid-based system (SLA)

Among stereolithography machine's components are the following: in order to create the component in layers, a laser source is needed. A container is filled with liquid resin, and when the laser lands on the resin, it solidifies. In order to construct parts, you need a platform. The platform is able to move in both directions. Thus, the platform travels down throughout component creation so that fresh layers of liquid resin solidify on top of the previous hardened layers. Figure 4.3 depicts the SLA's process flow diagram. When the laser arrives, it will fall onto the liquid resin above the platform, where it will vaporize. Liquid resin on top of the platform is hardened as a result of the laser action. A fresh layer of liquids solidifies on top of the previous one as the platform descends. It is via this process of layering that the platform is lowered and the product is raised to the surface [15]. It is possible to make precise parts with a smooth surface using the SLA method because of

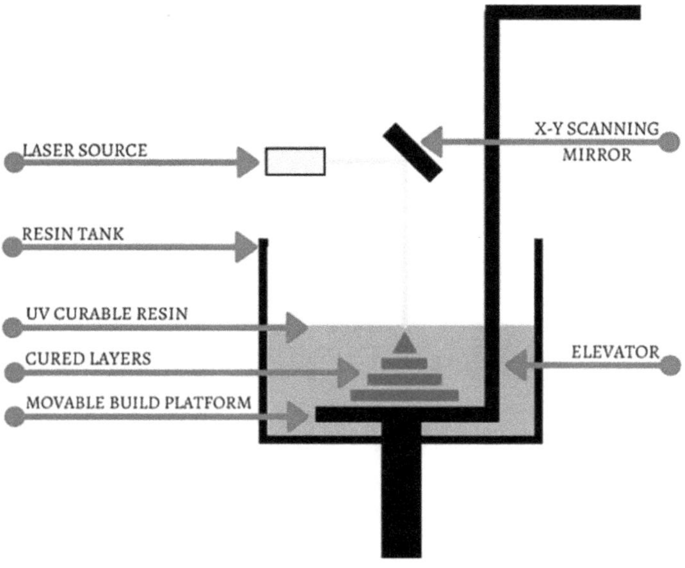

Figure 4.3 SLA process.

the reduced material waste. As opposed to more traditional methods, this one needs much more effort and money up front. Among the uses of SLA are investment casting patterns, rapid tooling, jigs, and fixtures; scale and exhibition models; optical and transparent coverings; molds and casting patterns; and rapid tooling.

4.10.2 Solid-based system (FDM)

Using a polymer filament inserted into an extruder nozzle and heated to a high temperature, FDM is an additive manufacturing process that creates a component as shown in Figure 4.4. Layers of filament are applied to a building platform and heated before being extruded into an object. This technology, which is both inexpensive and dependable, has witnessed a surge in popularity in recent years, thanks to the proliferation of desktop consumer versions. Its advantages include inexpensive cost, ease of usage, and little post-processing requirements. It is true that bad surface quality and low resolution are the main drawbacks of FDM [16]. Among the possible uses are: models for product dimensions and functional validation; parts made from certified aeronautics materials; the ability to chromate

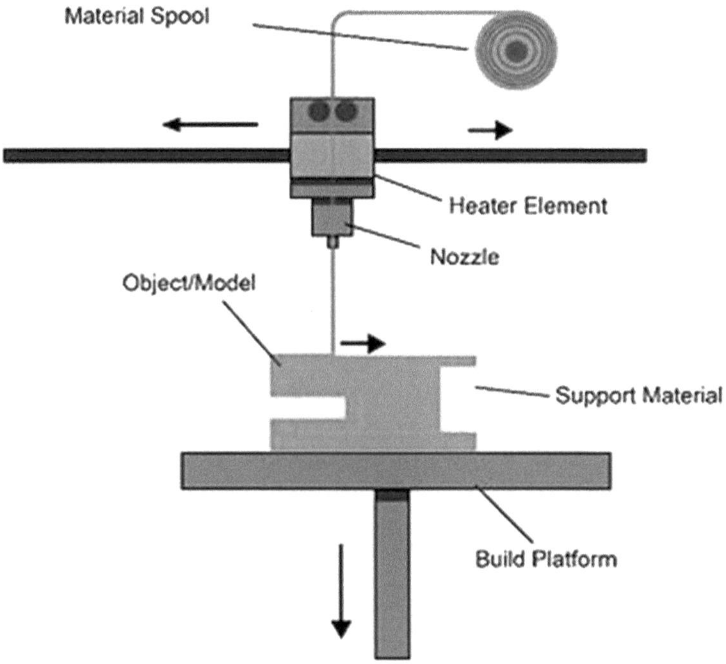

Figure 4.4 Fusion deposition modeling process.

ABS components; the production of small series (low-volume manufacturing) of pieces; and the production of lower-cost components.

4.10.3 Laminated object manufacturing

The construction platform is illustrated in Figure 4.5. The illustration is used to roll out sheets of material for laminated item manufacture. To melt the glue, a feeding roller is used to cover the materials with an adhesive coating. Using this method, each layer may then be bonded to the one before it in order to construct an item from scratch. The object's shape is drawn out using a blade or laser, and extra material is cross-hatched to make waste removal easier.

Layers are bonded in place and measurements are drawn before moving the construction platform down to insert another layer of material using the hot roller. This procedure is continued until the model or prototype is finished. Sanding may be necessary if a printed product is made of layers of paper, which have wood-like qualities. Paint or lacquer is often used to seal paper things to keep out moisture. It is possible to produce prototypes and other items quickly and cheaply with the help of the LOM. With the use of a computer-generated model of a product, companies may rapidly produce a prototype from affordable materials like paper, which makes it superior to other production techniques, such as 3D printers [17]. With the LOM technique, it is possible to build both solid and hollow items, including huge components, quicker and at a lower cost than with ordinary additive manufacturing. Large pieces may be produced with relative ease since no

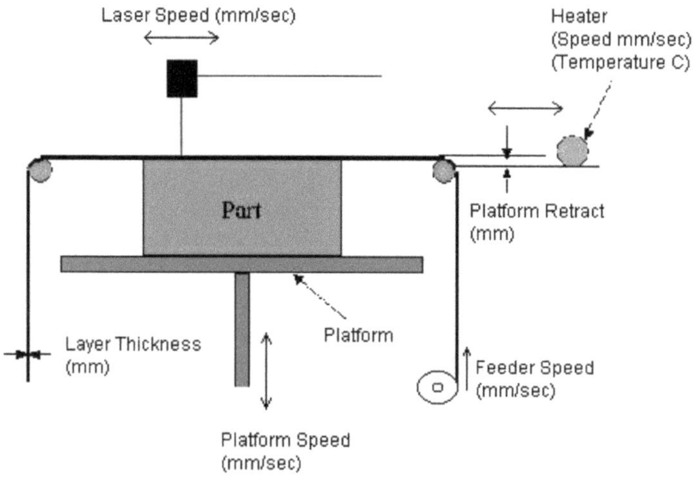

Figure 4.5 LOM process.

chemical reaction occurs during the construction process, and no support material is required because the laminated material supports itself while curing before the finished product is cut off. Paper models have wood-like properties, which means that they may be manipulated and polished in the same manner as wood models. However, there are still some drawbacks. Due to the fact that LOM is a subtractive process, it is more difficult to create complicated geometric designs than it is with other 3D printing techniques. This is due to the fact that accessing and removing extra material from inside an item is not always viable. As a result, interior structures and undercuts might be challenging to implement when working with LOM. In comparison to stereolithography and selective laser sintering, LOM delivers less dimensional precision. In addition to having a poor surface polish, LOM components and delicate paper parts may have limited strength, depending on the material used. Untreated paper LOM components may also easily absorb moisture. The speed and cheap cost of LOM make it a good choice for conceptual prototyping in spite of its drawbacks. Color inks and paper may also be used to make affordable but eye-catching 3D advertising goods, while laminated object manufacturing can be used to construct scale models.

4.10.4 Selective laser sintering (SLS)

An additive manufacturing process known as selective laser sintering (SLS) may be used to create powder bed printing devices. By layering nylon powder over a laser-fused CAD model, it builds up the part's shape from the bottom up, working its way up from the part's center. An industrial 3D printing technology known as selective laser sintering (SLS) is capable of producing accurate and quick prototypes and functioning production components within a matter of hours or even days [18]. Various nylon-based materials may be used to make very durable end products. Layer by layer, the SLS machine starts sintering each piece of geometry into a hot bed of nylon powder. Rolling over the bed of powder is done after each layer has been bonded. Layer by layer, the construction progresses is illustrated in Figure 4.6.

To break the powder bed apart and remove its assembled components, a breakout station is used. This station raises the powder bed so that it may be smashed apart. The bulk of the loose powder is removed manually with a first brushing. Bead blasting is used to remove any leftover powder residue from the parts before they are sent to the finishing department for final polishing.

Selective Laser Sintering's advantages include its low cost, high degree of size and form precision, suitability for certain functional tests, and ease with which it can copy complicated geometries. The most important ones are included below [18]. There are a vast number of products and uses for selective laser sintering, which is employed in a wide range of fields. Aerodynamic components, fans, and smaller turbines are often made using this material. Interior parts of automobiles are often made of this material as also hinged,

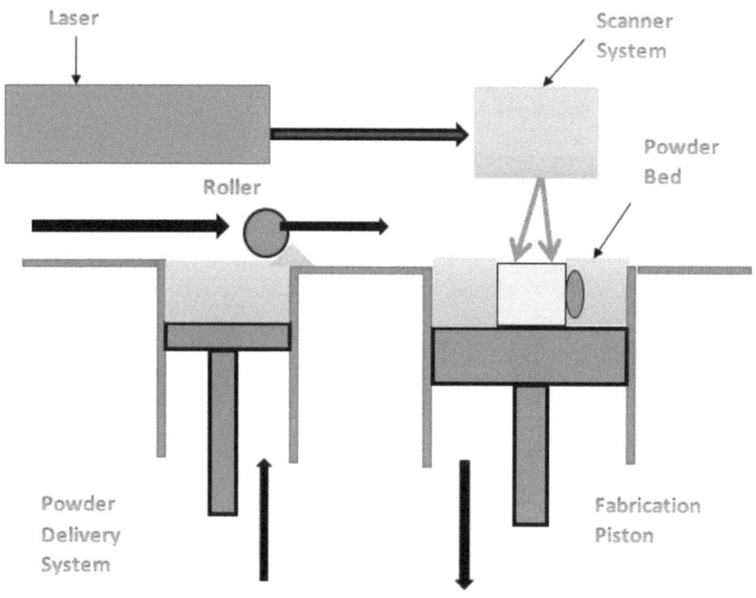

Figure 4.6 Selective laser sintering process.

electrical housings, and sporting equipment. In addition to being a common option for tubing in a wide variety of sectors, such as automotive, aerospace, medical, and oil and gas, it has a wide variety of materials and qualities.

4.10.5 Laser-engineered net shaping (LENS)

On top of the moving substrate, the deposition nozzle is shown to be stationary as shown in Figure 4.7. The deposition nozzle forms a melt pool on the substrate surface, then quickly distributes powdered material into the region through a tight dry powder stream. There is a molten pool that develops around the deposition nozzle axis, which solidifies as the substrate passes over it. Control over grain structure enables the method to be used to repair high-quality, functioning components. Although repair applications may sometimes compromise speed for high precision and a predetermined microstructure, a balance is required between surface quality and speed [19].

4.10.6 Electron beam melting

Using an electron beam to melt a powder bed is known as Electron Beam Melting (EBM). In contrast to Laser Powder Bed Fusion (LPBF), an electron beam is used to fuse metal particles and build the required portion layer by layer. Since the heat source utilized in LPBF technology is different, this is the major difference. An electron beam is generated by an electron cannon

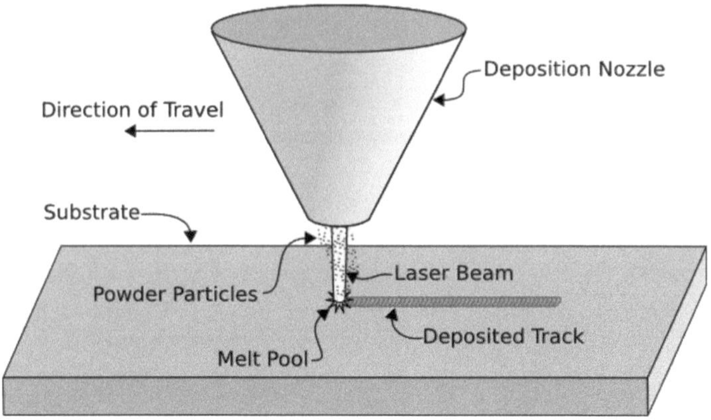

Figure 4.7 LENS deposition.

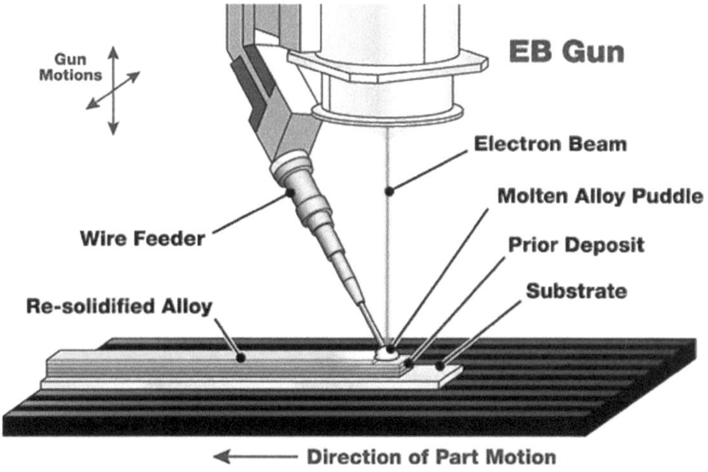

Figure 4.8 Electron beam melting process.

in this instance of EBM technology. Under vacuum, electrons are extracted from the filament and projected at high speed onto the metallic powder layer formed on the 3D printer's build plate. Those electrons will then be able to selectively fuse the powder, resulting in the creation of the component. The first step is to generate a 3D model of the desired component. 3D scanning, CAD software, or downloading a model of your choosing are all options for creating a model of it. This model is then submitted to a slicer, which will cut it according to the physical layers of material that have been deposited during the course of the 3D model. It will then transfer this information straight to the 3D printer, which may then begin the production process itself. The machine's tank may be filled with metal powder in the manner seen in Figure 4.8.

The electron beam will fuse the tiny layers of material that have been deposited. It's at this point that the 3D-printed part's cantilever sections get a boost in support. As many times as required, the machine repeats these operations in order to produce the full component. Using a blowgun or brush, the operator takes the finished product from the machine and disposes of the unmelted powder. Afterward, the printing supports may be removed and the component detached from the construction platform as shown in Figure 4.8. Surface machining, polishing, and other post-printing procedures are possibilities. Aeronautics and medical implants are two of the most common uses of EBM technology [20]. Titanium alloys are especially noteworthy because of their biocompatibility and mechanical qualities, which allow them to be both light and strong. To create turbine blades or engine elements is a common usage of CAD technology. Parts can be made more quickly using Electron Beam Melting than with Laser Powder Bed Fusion (LPBF), but the process is less precise and the finish is worse since the powder is more grainy, making things quickly. The electron beam is capable of heating the powder in many locations at once, resulting in a considerable increase in manufacturing speed. A laser, on the other hand, has to go across the whole surface in one pass. When the power is preheated before melting, it is less likely to distort, which means fewer reinforcements and supports are needed throughout production. Compared to the laser, the electron beam is a bit less precise at the powder level.

4.11 CONCLUSIONS AND FUTURE OUTLOOKS

The new technology and materials are developed constantly for additive manufacturing to keep the demand of the market to get the desired characteristics based on the application requirement. With this technological advancements, single-step fabrication of machines and components is possible. At present, 4D material printing is in its beginning stage in most of the countries. Researchers are involved in developing smart materials which are capable of adopting the requirements based on the surroundings. A dedicated 4D printing technique and printers may be developed for some critical applications in the field of medical applications. Biocompatible and biodegradable smart materials may be printed with a new and dedicated printing setup in organ and tissue engineering followed by cell culturing and implantations.

REFERENCES

[1] Lee, H., Lim, C.H.J., Low, M.J., Tham, N., Murukeshan, V.M., Kim, Y.J.: Lasers in additive manufacturing: A review. *Int. J. Precis. Eng. Manuf. Technol.* 2017, 43. 4, 307–322 (2017). https://doi.org/10.1007/S40684-017-0037-7.

[2] Abdulhameed, O., Al-Ahmari, A., Ameen, W., Mian, S.H.: Additive manufacturing: Challenges, trends, and applications. *Adv. Mech. Eng.* 2019, 11, 1–27. https://doi.org/10.1177/1687814018822880.
[3] Uriondo, A., Esperon-Miguez, M., Perinpanayagam, S.: The present and future of additive manufacturing in the aerospace sector: A review of important aspects. *Proc. Inst. Mech. Eng. G J. Aerosp. Eng.* 2015, 229, 2132–2147. https://doi.org/10.1177/0954410014568797.
[4] Guo, N., Leu, M.C.: Additive manufacturing: technology, applications and research needs. *Front. Mech. Eng.* 2013, 83. 8, 215–243. https://doi.org/10.1007/S11465-013-0248-8.
[5] Dilberoglu, U.M., Gharehpapagh, B., Yaman, U., Dolen, M.: The role of additive manufacturing in the era of industry 4.0. *Procedia Manuf.* 2017, 11, 545–554. https://doi.org/10.1016/J.PROMFG.2017.07.148.
[6] Lindgren, L.E., Lundbäck, A.: Approaches in computational welding mechanics applied to additive manufacturing: Review and outlook. *C. R.* Méc. 2018, 346, 1033–1042. https://doi.org/10.1016/J.CRME.2018.08.004.
[7] Ngo, T.D., Kashani, A., Imbalzano, G., Nguyen, K.T.Q., Hui, D.: Additive manufacturing (3D printing): A review of materials, methods, applications and challenges. *Compos. Part B Eng.* 2018, 143, 172–196. https://doi.org/10.1016/J.COMPOSITESB.2018.02.012.
[8] Huang, S.H., Liu, P., Mokasdar, A., Hou, L.: Additive manufacturing and its societal impact: A literature review. *Int. J. Adv. Manuf. Technol.* 2012, 675. 67, 1191–1203. https://doi.org/10.1007/S00170-012-4558-5.
[9] Li, N., Huang, S., Zhang, G., Qin, R., Liu, W., Xiong, H., Shi, G., Blackburn, J.: Progress in additive manufacturing on new materials: A review. *J. Mater. Sci. Technol.* 2019, 35, 242–269. https://doi.org/10.1016/J.JMST.2018.09.002.
[10] Singh, S., Ramakrishna, S., Singh, R.: Material issues in additive manufacturing: A review. *J. Manuf. Process.* 2017, 25, 185–200. https://doi.org/10.1016/J.JMAPRO.2016.11.006.
[11] Bikas, H., Stavropoulos, P., Chryssolouris, G.: Additive manufacturing methods and modelling approaches: A critical review. *Int. J. Adv. Manuf. Technol.* 2015, 831. 83, 389–405. https://doi.org/10.1007/S00170-015-7576-2.
[12] Beaman, J.J., Bourell, D.L., Seepersad, C.C., Kovar, D.: Additive manufacturing review: Early past to current practice. *J. Manuf. Sci. Eng. Trans. ASME* 2020, 142(11), 110812. https://doi.org/10.1115/1.4048193.
[13] Frazier, W.E.: Metal additive manufacturing: A review. *J. Mater. Eng. Perform.* 2014, 23, 1917–1928. https://doi.org/10.1007/S11665-014-0958-Z/FIGURES/9.
[14] Wong, K.V., Hernandez, A.: A review of additive manufacturing. *Int. Sch. Res. Netw. ISRN Mech. Eng.* 2012, 2012, 208760. https://doi.org/10.5402/2012/208760.
[15] Stereolithography (SLA) 3D Printing Guide, https://formlabs.com/asia/blog/ultimate-guide-to-stereolithography-sla-3d-printing/.
[16] What is FDM (Fused Deposition Modeling) 3D Printing? Explained by Hubs | Hubs, https://www.hubs.com/knowledge-base/what-is-fdm-3d-printing/.
[17] Laminated Object Manufacturing, LOM | Find suppliers, processes & material, https://www.manufacturingguide.com/en/laminated-object-manufacturing-lom.
[18] Selective Laser Sintering (SLS) For 3D Printing | Protolabs, https://www.protolabs.co.uk/services/3d-printing/selective-laser-sintering/.

[19] Introduction to LENS Additive Manufacturing - LENS-type Additive Manufacturing Process-Structure Linkages, http://materials-informatics-class-fall 2015.github.io/MIC-LENS/2015/09/24/Intro_LENS/.
[20] The Complete Guide to Electron Beam Melting (EBM) in 3D Printing - 3Dnatives, https://www.3dnatives.com/en/electron-beam-melting100420174/.

Chapter 5

Biocompatibility of additively manufactured materials

Mugilan Thanigachalam, Aezhisai Vallavi Muthusamy Subramanian, Madhuvanthi Sigamani, Sridhar Nagarajan, and Sridharan Vadivel
Government College of Technology

CONTENTS

DOI: 10.1201/9781003430186-5

5.1 INTRODUCTION

The long-term use of a biomaterial in a system to treat or replace tissue, organ, or physiological function is a significant concern. From a physiological standpoint, few materials are fully inert; most materials have a spectrum of components that may be harmful or irritating. Settling the substance may potentially result in adverse effects due to chemical reactions. There are a number of conditions that must be met before a biomaterial may be employed in an application. Different organizations and researchers suggest various tests to validate this functionality [1]. These tests should be done as indicated to verify this feature. Tests like this are used to determine the safety of experimental substances for human clinical use by following a series of research protocols published and controlled in various countries. A biomaterial's biocompatibility can be defined as the "ability of a biomaterial to perform its desired function concerning a medical therapy, without eliciting any undesirable local or systemic effects in the recipient or beneficiary of that therapy, but generating the most appropriate beneficial cellular or tissue response to that specific situation and optimizing the clinically relevant performance of that therapy" [2].

AM includes the computer-controlled layering of material layers to produce complex-shaped structures. CAD (Computer-Aided Design) software or biomedical imaging techniques can help. It may be used to generate 3D models, allowing the building of 3D structures unique to each person. Initial application of STL, which stands for "Stereolithography" and "Standard Tessellation Language." This is done for the AM machine to use it. After mesh data is given to the machine, the sequence of material deposition is established by digitizing and slicing the mesh data into 2D layers [3].

The ISO/ASTM 52900:2015 standard describes additive manufacturing as follows: Methods that may be employed include directed energy deposition, material extrusion, sheet lamination, binder jetting, powder bed fusion, material jetting, and vat photopolymerization [4]. This group of methods permits the manufacturing of TE structures with patient-precise designs that are likewise customizable to the surgeons who will use them. Due to the capacity to control the size and shape of the pores, these scaffolds may be constructed to possess the same structural, physical, and biological properties as native tissue [5].

5.1.1 Levels of biocompatibility assessment of AM materials

The goal of biocompatibility testing has been to make a broad set of biological tests for biomaterials so that researchers can compare the results of different studies with more confidence and accuracy. The exams at levels I, II, and III are categorized as primary (level I), secondary (level II), and preclinical (level III) [6]. These three groups correspond to tests, which examine the cytotoxicity and irritant potential of systemic toxicity in animals via intramuscular or subcutaneous implants and use tests by observing tissue reactions after insertion of the material into human teeth.

Level I examinations might involve in vitro and in vivo testing. The qualities of a chemical may be studied in a culture of living cells utilizing in vitro testing. Several formerly believed cytotoxic substances may be changed or their use limited. For in vivo testing, substances are implanted into rats and rabbits and studied over time to see how the implanted chemical affects the tissues.

Various results were discovered in clinical investigations. In vitro studies have shown no impact, less harm, or severe damage. This highlights the value of in vivo research, but it does not diminish the significance of in vitro studies. This is a notable disparity. In vitro research may still be used to study adverse reactions seen in clinical research. In vitro cell cultures lack the complex physiology necessary for multitasking by living organisms. Consequently, a biomaterial that may buffer the fluid and cellular systems of an organism may not operate in vitro but may be safe and effective in vivo. Both in vitro and in vivo studies are necessary to accurately determine the toxicity of complex compounds through laboratory assessment of the biocompatibility of materials that have been manufactured by AM processes.

Cytotoxicity testing is one of the fastest, most accurate, and most often used procedures for identifying whether a substance is hazardous or has a high concentration of chemicals that might be harmful to people. Due to the separation of cells and the lack of physiological processes that protect cells in vivo, it is possible to achieve a high degree of sensitivity. Utilizing a physiological solution that can extract a wide range of chemical structures, not just those that are water soluble, is the ideal strategy for removing chemicals from potentially released materials. Cell lines generated from human and mouse fibroblasts, lymphocytes, keratinocytes, mouse odontoblast-like cells, mouse macrophages, and rat submandibular salivary gland acinar cells may be used in cytotoxicity research.

Level II substances are evaluated on animals by breathing a compound that may cause, among other consequences, systemic toxicity and cutaneous irritation. The drug is administered subcutaneously or intramuscularly. Skin toxicity assessments are required because people are exposed to so many toxins daily. The toxicity may be reduced by exchanging or diluting a dangerous agent with a less toxic one.

Before using novel materials on people, animal studies must be conducted to understand how they affect living beings. These studies are expensive, controversial, and challenging to carry out. Numerous features of clinical and biological processes are not replicable in vitro. These results from animal studies do not represent a risk to human health. Animal research is less expensive, takes less time, and is subject to more laws since it may be arranged similarly to human clinical trials. There is a chance that animals might be exposed to substances or breakdown products that people find undesirable. Embryos and "children" may be investigated in ways that are not possible with human subjects, and reactions that are unethically challenging or impossible to test on people can be explored on animals.

Biomaterials are primarily determined by their functionality and compatibility with living organisms. The human host has been the last test for these characteristics in several respects. If a chemical is "biocompatible," it is incapable of causing mutations or influencing inflammatory mediators in a way that causes systemic reactions such as poisoning, tissue damage, or other similar consequences. These goods must not include allergens that might trigger allergic reactions in those sensitive to them. After a chemical has cleared stages I and II, it should be tested on humans (level III) to see how it performs in the real world and also to test its harmful effects.

5.2 BIOCOMPATIBILITY EVALUATION OF AM MATERIALS

5.2.1 Direct cytotoxicity assay

To assess direct cytotoxicity, mouse fibroblast 3T3 cells were grown for 24 hours on the surface of samples. The same circumstances and medium were used for cell growth in the indirect cytotoxicity test. Following a 10-minute pre-incubation in culture media, 4,104 cells were spread over each plate. According to the manufacturer's instructions, the live/dead cells were assessed through a viability/cytotoxicity test to evaluate if the cells were still alive after 24 hours using equations (5.1 and 5.2) [7,8]. The samples with attached mouse fibroblast 3T3 cells were rinsed with PBS and stained for 30 minutes in the dark at room temperature with 2 mol/ethidium homodimer-1 and 4 mol/calcein-AM dissolved in PBS. Using fluorescence microscopy, the samples were then investigated. Using a comparison of images of living (green) and dead (red) cells, the toxicity of the chemical was determined. Before SEM analysis, the cell-seeded samples were washed in PBS and then fixed at room temperature for 30 minutes in a glutaraldehyde solution containing 2.5% glutaraldehyde. After dehydrating the samples in a gradient of ethanol/PBS solutions for about 10 minutes, they were air-dried in the room. The SEM was then used to investigate the shape of the adherent cells on the surface of the sample.

$$\text{Cytotoxic reactivity } (\%) = \left\{ \frac{\text{Control} - \text{treated}}{\text{Control}} \right\} \times 100 \tag{5.1}$$

$$\text{Viability of MG} - 63 \text{ cells } (\%) = \left\{ \frac{\text{Treated}}{\text{Control}} \right\} \times 100 \tag{5.2}$$

All compounds displayed moderate cytotoxicity against MFCs, with a discernible trend toward increasing cell viability from day 1 to day 7. E-Guard clear and Dental LT had the most significant cytotoxicity among the groups, while the material exhibited the least [9]. On day 7, Invisalign demonstrated the highest degree of cell viability and no cytotoxicity (94.07% of cell viability). There was no statistically significant difference between Dental LT and E-Guard material regarding viability percentage. The material (polyurethane) was more biocompatible than directly printed aligner materials (polymethylmethacrylate) [10]. All substances exhibited a higher level of cytotoxicity on the first day and a decreasing trend after that. Despite the low degree of cytotoxicity, the results indicate an increase in material leaching throughout the initial term of usage. Direct interaction of the tested material with the ADT test resulted in the absence of harmful effects on any cell culture. The tested substance had no cytotoxic effect on both cell lines after 3 days of eluate extraction using the MTT assay. Still, it exhibited a moderate cytotoxic impact after 5, 7, and 21 days.

5.2.2 Indirect cytotoxicity assays

Using indirect cytotoxicity assays, the samples' toxicity was determined. Each sample was put through three tests. In each well of a 96-well plate, 3,000 murine fibroblast 3T3 cells were placed. Following 12 hours of cell growth, the sample extracts were applied. Using the MTT test, the cell viability of samples collected at 0, 24, 48, and 72 hours was determined (Sigma). As a negative control, an unmodified DMEM culture media was used. The MTT test was then conducted by the manufacturer's instructions. Then, a reagent, 20 µL of MTT (5 mg/mL in Phosphate-Buffered Saline, or PBS), was put to each well of the 96-well plate containing 200 µL of culture media at various intervals, and the dish was allowed to rest for 3 hours. The medium was withdrawn to dissolve the formazan crystals, and 300 µL of dimethyl sulfoxide (DMSO) was added to each well [11]. Then, 200 µL of the colored solution was transferred to a fresh 96-well plate. After measuring the absorbance of each well with an EL 800 plate reader (Biotek, USA) at 570 nm, the results were analyzed using GEN 5 software (Ver 1.08.4).

5.2.3 Antibacterial evaluation of AM scaffolds

Using the agar contact method and a growth-inhibition experiment in a liquid media, bacterial activity against *Staphylococcus aureus* and *Escherichia coli* was evaluated in vitro. The agar diffusion technique may demonstrate that scaffolds possess antibacterial properties by examining the bacterial inhibition zone. Ag@Polycaprolactone/Gel and Ag/Zn@Polycaprolactone/Gel exhibit prominent antibacterial activity against *S. aureus* and *E. coli*, while PCL/Gel and Zn@Polycaprolactone/Gel do not exhibit any evident antibacterial behavior [12]. For Ag@Polycaprolactone/Gel, the diameters of the zones of inhibition for *E. coli* were 1.82 cm, and for *S. aureus*, they were 2.43 cm. Ag/Zn@Polycaprolactone/Gel inhibited the growth of *E. coli* (2.290 cm) and *S. aureus* (4.190 cm) more effectively than Ag@Polycaprolactone/Gel. A similar conclusion was reached in the growth-inhibition test: magnetron sputtering significantly enhanced the antibacterial action. In terms of destroying bacteria, Ag/Zn@Polycaprolactone Gel was 97% efficient against *E. coli* and 97.2% effective against *S. aureus* (Figure 5.4d). Ag/Zn@Polycaprolactone/Gel is the most effective antibacterial agent against *S. aureus* and *E. coli.*

The agar disc diffusion test (Figure 5.1) showed that PLA-Col-MH and PLA-Col-MH-cHA effectively stopped bacterial growth, with inhibition zone diameters of about 26 mm, which is about the same as the diameter of the positive antibiotic control (30 g of MH), which was about 29 mm. The sample without antibiotic, PLA-Col, and the negative control both had no inhibition zone (distilled water).

5.2.4 Cell adhesion capability of 3D-printed biomaterials

MCT3T-E1 cells were submitted to MTT tests after 1, 3, 5, and 7 days of growth on 3D PCL scaffolds to assess their viability. A shorter ECP scaffold with the same pore size as CP800 was used for a more realistic biocompatibility evaluation. The cells attached to and changed the

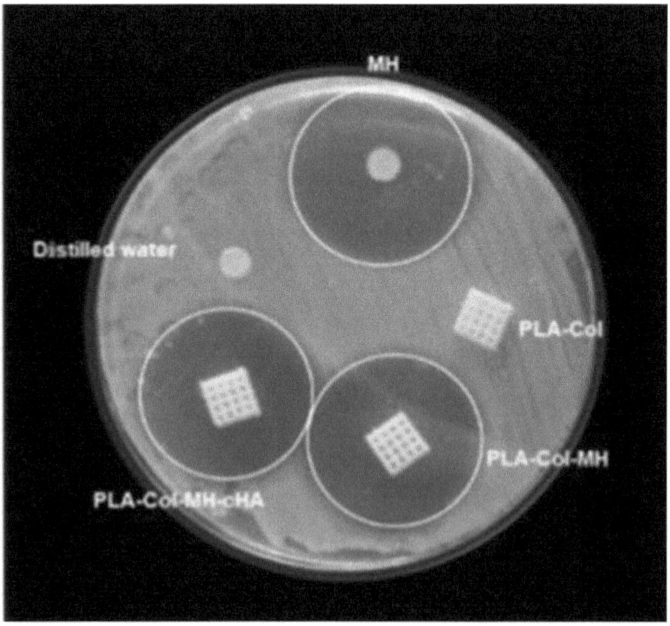

Figure 5.1 Evaluation of the activity against S. aureus (ATCC 25923) of the PLA-Col, PLA-Col-MH, and PLA-Col-MH-cHA scaffolds; (a) Agar disk diffusion test with distilled water as a negative control and MH (30 μg) as a positive control [13].

scaffolds' geometry to show the biocompatibility of the scaffolding. MTT test results indicated that CP800 scaffolds contained more cells than MP800 scaffolds over both time periods. Three days later, an FESEM examination of cells from the CP800 group that were connected to porous filaments indicated that they had a diversity of morphologies, as shown in Figure 5.2a–d. However, there were much fewer cells at the connections between the MP800 filaments. Even though the cells in the EMP group were more stretched after 7 days, the ECP scaffolds included more cells, and their shape was more stretched. The use of ECP-fabricated PCL scaffolds instead of EMP-fabricated scaffolds may improve biocompatibility, cell adhesion, and cell proliferation.

Using the following approaches, the viability of cells on 3D-printed surfaces was evaluated:

Short-term cell growth: For 6 days, the total number of cells in each dish was counted every 24 hours to measure the growth rate of 3T3 (100,000 cells per plate, 0.1×10^6 cells per cm^2). For hemocytometer counts, small volumes of Trypan blue-stained detached cell solution were utilized [15].

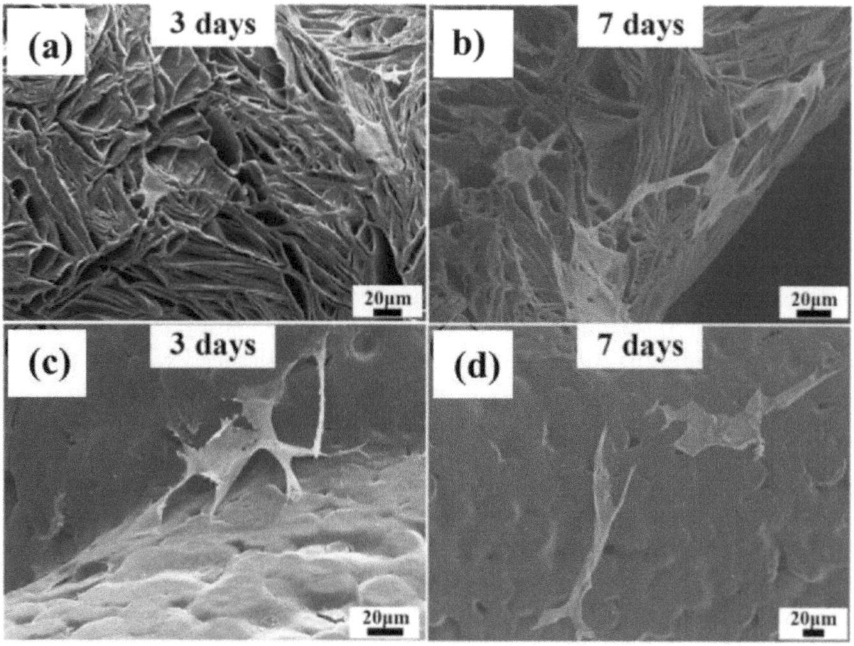

Figure 5.2 Cell adhesion results. The SEM investigates cell adhesion (a) 3 days on the MP800 scaffolds (b), the cell morphology (in red) is seen on the CP800 scaffolds (c) and after 7 days on the MP800 scaffolds (d) [14].

Cumulative population doubling level (cPDL): Three sets of 3T3 cells were seeded (0.05×10^6 cells per cm^2 or 50,000 cells per 3D-printed plate), and the medium was replaced every 2–3 days to determine the rate of cell growth. Since 3T3 cells divide every 21 hours, they attain confluence in a few days. Each week, the cells were transferred to a new culture dish of the same kind so that a shortage of room would not inhibit their development [16]. This is known as "passage." During passes, the total number of cells was determined, and an equation was used to calculate the cPDL by equation (5.3),

$$n = \frac{(\log 10F - \log 10I)}{0.031} \tag{5.3}$$

where F represents the number of cells at the end of one passage, and I represents the number of cells planted at the start of one passage. The number of times the population doubles is denoted by n.

5.2.5 Osteogenic integration of 3D polymer scaffolds

The level of alkaline phosphatase activity increased significantly throughout the testing period. The PCL/HA/BG and PCL/HA groups and their scaffolds exhibited a significant increase on day 7 compared to the control group, TCPs. Alkaline phosphatase assay activity was more effective in PCL scaffolds containing nanoparticles on day 14 compared to TCPs and PCL scaffolds alone. On day 21, PCL/HA/BG scaffolds had the greatest enzyme concentration.

The capacity of cell-filled hydrogels to differentiate in the laboratory was evaluated using ARS staining. The prints grew in a DMEM medium for 2–3 days after printing. The printed hydrogels were then washed with PBS buffer, given 200 mL of media for osteogenic differentiation, and allowed to develop for 7–14 days. Figure 5.3 shows the SEM pictures of osteoblast morphology on 3D-printed PEEK after 24 hours incubation under different conditions. Mineralization potential was greatly enhanced when 5 V–1 Hz (0.62 mT) EMFs were utilized as stimuli. However, mineralized nodule formation was enhanced in the presence of a PAI-1-loaded 3D scaffold [17]. This indicated that growth factors and electromagnetic field (EMF) stimulation might synergistically transform SCAPs into osteoblasts.

Cells were planted on Bio or C-Bio scaffolds for the assays. To establish that cAD-MSCs were transforming into osteoblast-like cells, ALP activity (reported as U/mL) is the amount of enzyme that contributes to the hydrolysis of 1 mole of p-nitrophenyl phosphate (pNPP) per unit mL, tested under all cellular settings. At 21 and 28 days following growth, the scaffolds containing graphene exhibited significantly greater extracellular ALP concentrations than the controls (p 0.05). After 28 days of development, the

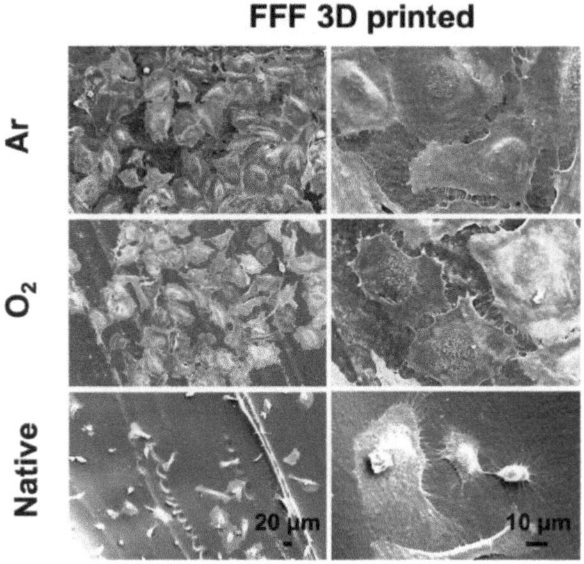

Figure 5.3 SEM pictures of osteoblast morphology on 3D-printed PEEK after 24 hours incubation under different conditions [18].

expression of the ALP, OSX, RANKL, and RUNX genes was considerably (p 0.05) greater on the graphene surface than on the control surface [19].

After 1 week of osteogenic differentiation, hAD-MSCs cultured with hPL displayed an abundance of ALP-positive cells. The HS lineage included less ALP-positive cells. A modest number of ALP-positive cells existed in the absence of FCS. While 2D hAD-MSCs were growing on the cell culture surface, hPL-induced osteogenic differentiation of hAD-MSCs was almost twice as high as that of 2D hAD-MSCs. Unlike the cells on the culture plate, the cells on the HS-coated constructs did not differentiate [20].

5.2.6 Stem cell response to 3D printed biomaterials

Bio-ink is a biomaterial having viscoelastic qualities that may be utilized for cell printing. Cell-filled three-dimensional structures are created for implantation into patients so that desired tissue may regenerate. Medical and engineering fields contribute to the rapid progression of cell bioprinting. Bioremediation (e.g., removing phenol waste) and medical applications (e.g., in situ biosynthesis) may benefit from the innovative use of a 3D bioprinting method that places bacteria in hydrogel to create "functional bio-ink." Printing cells in three dimensions may be used for several purposes in tissue engineering shown in Figure 5.4. However, it is most often utilized to create three-dimensional constructs that resemble human tissue. Extracting

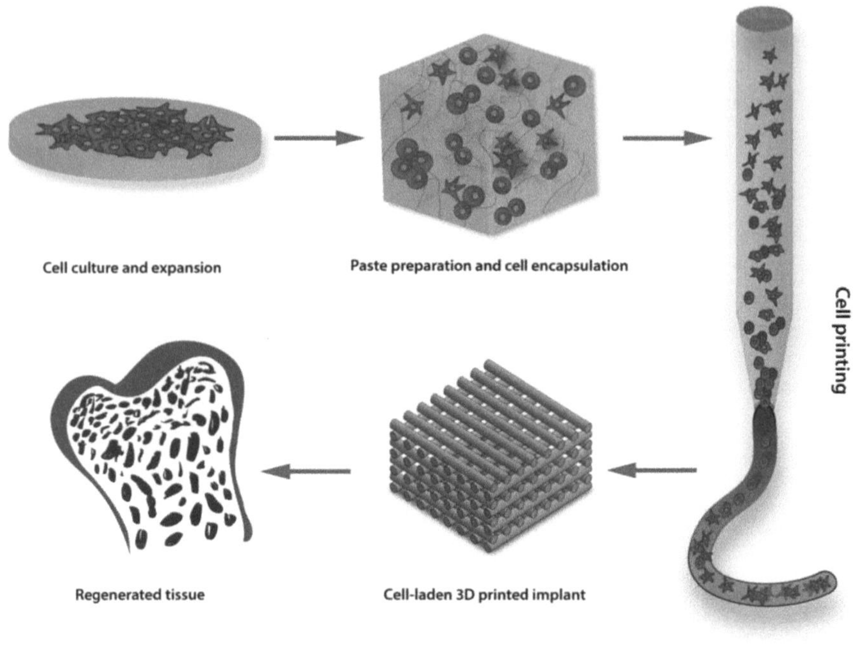

Figure 5.4 The cell printing paradigm [23].

cells from donor tissue, placing them in the proper polymeric matrix, and then printing at a 10–100 nm precision replicate how heterogeneous tissue components are assembled [21].

It took a long time to print 3D models since they could not be made with great accuracy and detail. Additionally, it was discovered that the viability of the cells increased from 33.8% to 68.1% after 24 hours. Because gel cross-linking must occur in quiet settings for 45 minutes, cell injury during printing and the necessity for stillness may contribute to a poor cell survival rate (45 minutes) [22].

5.2.7 Simulated body fluid response on 3D bio-printed materials

Numerous frequent and deadly osseous tissue abnormalities are the result of trauma. Even while conventional clinical methods such as allografts and autografts are effective, they are not without flaws. As a result, we need improved methods for treating bone tissue abnormalities. The difficulties with host tissue integration will be resolved by using a novel elastomeric filament to produce bioactive scaffolds from FDM. The present technology was used to capitalize on the many advantages of fast prototyping and

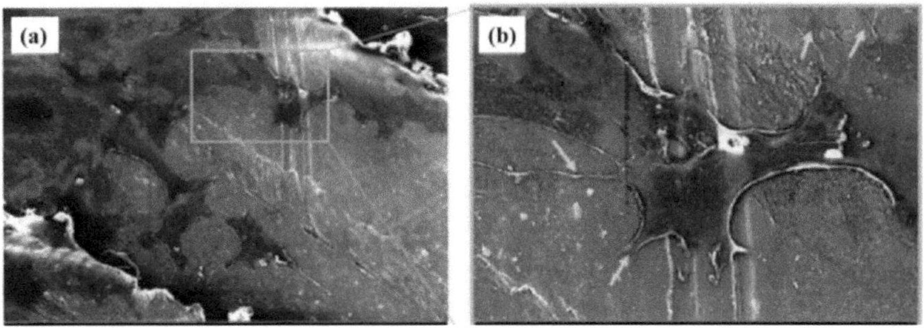

Figure 5.5 Fourteen days after implantation on porous PEEK, cells had proliferated and had extended lamellipodia and filopodia [25].

investigate the application of biomimetic nanomaterials for osteogenic regulation in human fetal osteoblasts [24]. SBF nucleation was also used to enhance the functionality of hFOB cells and the development of osseous tissue. As shown in Figure 5.5, SEM analysis revealed that 14 days after implantation on porous PEEK, cells had proliferated and had extended lamellipodia and filopodia.

5.3 CONSIDERABLE FACTORS FOR IMPROVING THE BIOCOMPATIBILITY OF 3D-PRINTED IMPLANTS

3D printing was used to test new and improved alloys for orthopedic, dental, fracture therapy, spinal, and cardiovascular implants. The automotive and aerospace industries manufactured stainless steel, titanium, and cobalt–chrome for their strength, fatigue resistance, and corrosion resistance, not for their physical performance. Frequently, metal implants are overly sensitive to metal ions and do not function well with the body's tissues to promote quicker healing. Various coatings have been tried to enhance the surface quality of implants. However, poor bonding with the underlying metal has resulted in many interface failures and repair operations. This prompted the group of specialists to propose new alloys with enhanced biocompatibility (i.e., biological performance).

Here are some ways that 3D-printed implants can be more compatible with the body. Existing tantalum-coated implants demonstrate that the metal is biocompatible and may be utilized to enhance the interaction between tissues and materials. Attempts were made to combine tantalum with titanium via 3D printing. This was done because tantalum has a very high density, a very high melting point, and is somewhat more costly than titanium. Tantalum is challenging to manipulate, but an alloy composed of tantalum and titanium is easier to manage and maintains the same density. Tests demonstrated that a combination of 90% titanium and 10% tantalum was as toxic to living organisms as pure tantalum. Therefore, we would only require a minimal quantity of tantalum [26].

5.4 RECENT APPROACHES USING CERAMIC BIOMIMETIC AM

Because its chemical components resemble bone, hydroxyapatite (HA) is well known and often used in medicine. Nano-sized HA is more biocompatible than micron-sized HA due to its structural similarity to HA crystals in bone. HA nanorods that resemble bone were created utilizing a 3D bioprinter and the hydrothermal technique in a type I collagen matrix. It was reported, however, that the nanorods were weaker than typical. $Ca_7Si_2P_2O_{16}$ was manufactured by using a coaxial 3D printing approach. Polymeric infiltration was then applied to these scaffolds to replicate the flexible collagen structure of genuine bone. This scaffold was loaded with biodegradable polymers such as gelatin, PVA, and PCL, making it more potent, more bioactive, and more effective in facilitating bone movement [27].

Collagen hydrogels have been created utilizing numerous different processes. Electrospinning is used to make aligned wall composite fibers from collagen and other protein fibers because mixing them with electrospun fibers makes them more biocompatible. Biocompatibility of PLA-based scaffolds including bovine amniotic epithelial stem cells, osteoblast-like human bone fibroblast (MG-63) cells, and nano-hydroxyapatite/PLA scaffolds with human bone marrow-derived mesenchymal stem cells has been studied in recent years (hMSCs) [28].

5.5 RECENT TRENDS OF AM IMPLANTS IN BIOMEDICAL APPLICATION

Currently, AM technologies such as Inkjet Printing (Polyjet), 3D Printing (3DP), Stereolithography (SLA), Selective Laser Melting (SLM), and Bioprinting, which is their category, have been utilized to manufacture bio-implants. They are separated depending on the sort of energy they consume and how they are manufactured, among other factors. The emerging biomedical areas where the AM techniques are needed are discussed below.

5.5.1 Drug delivery

To optimize the synthesis or functioning of precursors, composition conditions of production, and drug encapsulation technology, a customized drug delivery strategy must suit the desired release kinetics. The drug or other nutrient should be supplied at a regulated rate and dose by ensuring that size, shape, surface morphology, bioavailability, and biodegradability are suited for the target site. Biomimetic nanoparticles of different sizes were loaded with medicinal substances. Such precise nanomedicine carriers helped inflammatory image locations in molecular form and resolved inflammations and immune responses [29].

5.5.2 3D-printed bone prostheses and scaffolds

A high-performance scaffold is required for a method to be successful in scaffold-based bone tissue creation. A key objective of the discipline is to develop scaffolds for bone tissue engineering with the ideal shape, structure, physical, chemical, and biological characteristics for improved biological performance and regeneration of complex bone tissues. Because they are physically, structurally, and chemically comparable to bone apatite, which is the primary component of normal bone tissue, 3D-printed bioceramic scaffolds have been utilized extensively as bone tissue engineering scaffolds [30].

5.5.3 3D printing of vascularized tissue

The most challenging aspect of maintaining the viability of 3D-printed tissues is ensuring that they have intricate circulatory networks for transferring nutrients and eliminating waste. In the last several years, 3D-printed organs and tissue have made significant stride toward the perfection of circulatory networks. This is due to the rapid rate of science and technology advancing and making discoveries. Currently, there are two primary methods for including blood arteries in bio-printed material. The first of these depends on the regulated release of angiogenic factors, which aid in the growth of blood vessels in 3D-printed tissues. This technology has been used in current research whose primary objective is to implant printed material into the patient. The second strategy includes printing vascular scaffolds directly. The target tissue cells will then be printed directly onto or around the scaffold to produce an organoid with blood vessels [31].

5.5.4 Tissue engineering

3D bioprinting gives us a new way to industrialize the bio-fabrication of tissue, making it possible to make tissue constructs with different kinds of cells and structures. It was first utilized to construct scaffolds for bone tissue engineering. It is being used to produce cartilage, skin, and heart valves. Using 3D bioprinting, hollow channels have been created in the hydrogels to aid in developing blood arteries. After 28 days of conditioning in a bioreactor, the Organovo company has printed a blood vessel conduit.

In vivo bioprinting, in which human tissues are produced in three dimensions outside the body, has made reversible progress. Layers of cells may be 3D-printed directly onto a wound or even within the body. This is accomplished using modern surgical procedures [32].

5.6 FUTURE TRENDS FOR BIOCOMPATIBILITY OF AM IMPLANTS

Various in vitro approaches are used to discuss the advantages and disadvantages of a variety of potential novel biomaterials produced via "additive manufacturing." It may be claimed that the existing materials have limitations and that more refined multifunctional materials are required to meet the biocompatibility of various medical applications. Bioactive compounds are favored due to their compatibility with the bone around them. However, biotolerant materials can also be used to make implants. When implants come into contact with human tissues and fluids, there are several host–implant material interactions. These reactions also influence the biocompatibility of the substance. Concerns about biocompatibility include (i) thrombosis, which is when blood clots and blood platelets stick to the surface of a biomaterial, and (ii) fibrous tissue wrapping around biomaterials implanted in soft tissues. When the surface of an implant cannot osseointegrate with the surrounding bone or other tissues due to minute motions, the implant becomes loose [33].

5.6.1 Poly (methyl methacrylate), PMMA

PMMA is a hard, brittle polymer that isn't good for most therapeutic uses, but it does have a few good qualities. It can be produced at room temperature, allowing it to be utilized in an operating room or dentist's office. Therefore, it is used in dentures and bone cement. Many joint prostheses rely on the performance of PMMA cement. During surgery, this cement is created by combining powdered polymer with monomeric methyl methacrylate to form a dough that can be inserted into the bone [34].

5.6.2 Ultra-high-molecular-weight polyethylene (UHMWPE)

Ultra-high-molecular-weight polyethylene possesses more mechanical strength, less wear rate, and biocompatibility; ultra-high-molecular-weight polyethylene is one of the most often used polymers for orthopedic implants. Numerous studies are being conducted on the wear characteristics of ultra-high-molecular-weight polyethylene. In total joint arthroplasty, it is used as the bearing surface. In the past 15 years, metal on polyethylene has had a 90% success rate. However, submicron particles were detected in periprosthetic tissues in the presence of polyethylene wear. The wear resistance characteristics of ultra-high-molecular-weight polyethylene may be enhanced by increasing its crystalline structure and cross-linking. Even while cross-linking makes the material more resistant to deterioration, it also reduces its tensile strength, fracture toughness, and resistance to the propagation of fatigue cracks. Increasing the crystallinity of ultra-high-molecular-weight polyethylene enhances its elastic modulus and fracture resistance [17].

5.6.3 Zirconia (ZrO_2)

Zirconia is one of the promising biomaterials because of its excellent mechanical property and resistance to fracture. Zirconia ceramics have better benefits than other ceramics because of how their microstructure changes and makes them stronger. This is seen in the components manufactured from them. More than 20 years ago, researchers began investigating the viability of zirconia ceramics as biomaterials. Today, zirconia is used in complete hip replacements, although it is still undergoing development for usage in other medical devices [35].

5.7 CONCLUSION

Implementing high-resolution additive 3D printing in the biomedical area is a valuable technique with several advantages, including the customization of medical instruments, implants, and equipment; improved accessibility; better cost-effectiveness; and vertical design-to-manufacturing integration. Metals and ceramics made with 3D printing are already used in the medical and life science fields. However, 3D-printed polymers, which make up 82% of the 3D printing market and are the most accessible, fast, and cheap 3D-printed material, have not yet been used in biological devices because they are toxic to cells. Even though 3D printing was invented in the early 1980s, plastic polymers show that they can't interact with biologically active systems, and key properties like biodegradability, cell attachment, cytotoxicity, and biocompatibility haven't been well defined. Recent researches have revealed the good cytotoxicity and cell viability in 3D-printed and other additively manufactured scaffolds and various medical implant application materials. Improvement in structures of the implants through AM processes provides excellent biocompatible results. Thus, biocompatibility of the AM materials is an excellent candidate for cell viability and minimum toxic effect as well as improved cell adhesion properties and stem cell responses.

REFERENCES

[1] M. Kirsch, A.C. Herder, C. Boudot, A. Karau, J. Rach, W. Handke et al., Xeno-free in vitro cultivation and osteogenic differentiation of hAD-MSCs on resorbable 3D printed RESOMER®, *Materials* 13 (2020), pp. 1–17.

[2] G. Grigaleviciute, D. Baltriukiene, V. Bukelskiene and M. Malinauskas, Biocompatibility evaluation and enhancement of elastomeric coatings made using table-top optical 3D printer, *Coatings* 10, no. 3 (2020), p. 254. https://doi.org/10.3390/coatings10030254

[3] A.K. Pal, A.K. Mohanty and M. Misra, Additive manufacturing technology of polymeric materials for customized products: recent developments and future prospective, *RSC Advances* 11 (2021), pp. 36398–36438.

[4] C.B. Ayyanar, K. Marimuthu, B. Gayathri and Sankarrajan, Characterization and in vitro cytotoxicity evaluation of fish scale and seashell derived nano-hydroxyapatite high-density polyethylene composite, *Polymers and Polymer Composites* 29, no. 9 (2021), pp. 1534–1542. doi:10.1177/0967391120981551.

[5] K.B. Sagomonyants, M.L. Jarman-Smith, J.N. Devine, M.S. Aronow and G.A. Gronowicz, The in vitro response of human osteoblasts to polyetheretherketone (PEEK) substrates compared to commercially pure titanium, *Biomaterials* 29 (2008), pp. 1563–1572.

[6] I.C.C. de Moraes Porto, Polymer biocompatibility, in *Polymerization* (Edition: 1,Chapter: 3) (2012). InTech. http://dx.doi.org/10.5772/47786

[7] A.V. Muthusamy Subramanian and M. Thanigachalam, In-vitro antibacterial study and bone stress prediction of ceramic particulates filled polyether ether ketone nanocomposites for medical applications. *Journal of Polymer Research* 29 (2022), p. 318. https://doi.org/10.1007/s10965-022-03180-6.

[8] M. Thanigachalam and A.V. Muthusamy Subramanian, Evaluation of PEEK-TiO2- SiO2 nanocomposite as biomedical implants with regard to in-vitro biocompatibility and material characterization, *Journal of Biomaterials Science, Polymer Edition* 33, no. 6 (2021), pp. 727–746. doi: 10.1080/09205063.2021.2014028

[9] M. Ayyar, M.P. Mani, S.K. Jaganathan and R. Rathanasamy, Preparation, characterization and blood compatibility assessment of a novel electrospun nanocomposite comprising polyurethane and ayurvedic-indhulekha oil for tissue engineering applications, *Biomedizinische Technik* 63 (2018), pp. 245–253.

[10] A.J.T. Teo, A. Mishra, I. Park, Y.J. Kim, W.T. Park and Y.J. Yoon, Polymeric biomaterials for medical implants and devices, *ACS Biomaterials Science and Engineering* 2 (2016), pp. 454–472.

[11] A. Buzarovska, S. Dinescu, L. Chitoiu and M. Costache, Porous poly(l-lactic acid) nanocomposite scaffolds with functionalized TiO2 nanoparticles: properties, cytocompatibility and drug release capability, *Journal of Materials Science* 53 (2018), pp. 11151–11166.

[12] S. Liu, S. Zhang, L. Yang, Y. Yu, S. Wang, L. Li et al., Nanofibrous scaffold by cleaner magnetron-sputtering additive manufacturing: A novel biocompatible platform for antibacterial application, *Journal of Cleaner Production* 315 (2021), 128201.

[13] V. Martin, I.A. Ribeiro, M.M. Alves, L. Gonçalves, R.A. Claudio, L. Grenho et al., Engineering a multifunctional 3D-printed PLA-collagen-minocycline-nanoHydroxyapatite scaffold with combined antimicrobial and osteogenic effects for bone regeneration, *Materials Science and Engineering C* 101 (2019), pp. 15–26.
[14] W. Zhang, I. Ullah, L. Shi, Y. Zhang, H. Ou, J. Zhou et al., Fabrication and characterization of porous polycaprolactone scaffold via extrusion-based cryogenic 3D printing for tissue engineering, *Materials and Design* 180 (2019), p. 107946.
[15] S.T.R. Aruna, M. Shilpa, R.V. Lakshmi, N. Balaji, V. Kavitha and A. Gnanamani, Plasma sprayed hydroxyapatite bioceramic coatings from coprecipitation synthesized powder: Preparation, characterization and in vitro studies, *Transactions of the Indian Ceramic Society* 77 (2018), pp. 90–99.
[16] S. Staehlke, H. Rebl and B. Nebe, Phenotypic stability of the human MG-63 osteoblastic cell line at different passages, *Cell Biology International* 43 (2019), pp. 22–32.
[17] C.M.B. Ho, S.H. Ng and Y.J. Yoon, A review on 3D printed bioimplants, *International Journal of Precision Engineering and Manufacturing* 16 (2015), pp. 1035–1046.
[18] N. Fazeli, E. Arefian, S. Irani, A. Ardeshirylajimi and E. Seyedjafari, 3D-Printed PCL scaffolds coated with nanobioceramics enhance osteogenic differentiation of stem cells, *ACS Omega* 6 (2021), pp. 35284–35296.
[19] P. Memarian, F. Sartor, E. Bernardo, H. Elsayed, B. Ercan, L.G. Delogu et al., Osteogenic properties of 3D-printed silica-carbon-calcite composite scaffolds: Novel approach for personalized bone tissue regeneration, *International Journal of Molecular Sciences* 22 (2021), pp. 1–10.
[20] G. Kwon, H. Kim, K.C. Gupta and I.K. Kang, Enhanced tissue compatibility of polyetheretherketone disks by dopamine-mediated protein immobilization, *Macromolecular Research* 26 (2018), pp. 128–138.
[21] W. Jamróz, J. Szafraniec, M. Kurek and R. Jachowicz, 3D Printing in pharmaceutical and medical applications – Recent achievements and challenges, *Pharmaceutical Research* 35 (2018), 176. https://doi.org/10.1007/s11095-018-2454-x
[22] C. Yu, J. Schimelman, P. Wang, K.L. Miller, X. Ma, S. You et al., Photopolymerizable biomaterials and light-based 3D printing strategies for biomedical applications, *Chemical Reviews* 120 (2020), pp. 10695–10743.
[23] G. Cidonio, M. Glinka, J.I. Dawson and R.O.C. Oreffo, The cell in the ink: Improving biofabrication by printing stem cells for skeletal regenerative medicine, *Biomaterials* 209 (2019), pp. 10–24.
[24] M. Kumar, E. Gnansounou and I.S. Thakur, Synthesis of bioactive material by sol–gel process utilizing polymorphic calcium carbonate precipitate and their direct and indirect in-vitro cytotoxicity analysis, *Environmental Technology and Innovation* 18 (2020), p. 100647.
[25] N.J. Castro and L.G. Zhang, Simulated body fluid nucleation of 3D printed elastomeric scaffolds for bone regeneration the aim of the current work was to evaluate a novel 3D printable, *Tissue Engineering (Part A)* 22, no. 13–14 (2016), pp. 940–948. doi: 10.1089/ten.TEA.2016.0161

[26] L.M. Anaya-Esparza, Z.V. de la Mora, J.M. Ruvalcaba-Gómez, R. Romero-Toledo, T. Sandoval-Contreras, S. Aguilera-Aguirre et al., Use of titanium dioxide (TiO2) nanoparticles as reinforcement agent of polysaccharide-based materials, *Processes* 8 (2020), pp. 1–26.
[27] K. Hu, T. Yu, S. Tang, X. Xu, Z. Guo, J. Qian et al., Dual anisotropicity comprising 3D printed structures and magnetic nanoparticle assemblies: towards the promotion of mesenchymal stem cell osteogenic differentiation, *NPG Asia Materials* 13 (2021), p. 19. https://doi.org/10.1038/s41427-021-00288-x
[28] A. Zühlke, M. Gasik, N.E. Vrana, C.B. Muller, J. Barthes, Y. Bilotsky et al., Biomechanical and functional comparison of moulded and 3D printed medical silicones, *Journal of the Mechanical Behavior of Biomedical Materials* 122 (2021), p. 104649.
[29] A. Asthana, C.M.R. White, M. Douglass and W.S. Kisaalita, Evaluation of cellular adhesion and organization in different microporous polymeric scaffolds, *Biotechnology Progress* 34 (2018), pp. 505–514.
[30] M.M. Methani, P.F. Cesar, R.B. de Paula Miranda, S. Morimoto, M. Özcan and M. Revilla-León, Additive manufacturing in dentistry: Current technologies, clinical applications, and limitations, *Current Oral Health Reports* 7 (2020), pp. 327–334.
[31] M. Yu, J. Huang, T. Zhu, J. Lu, J. Liu, X. Li et al., Liraglutide-loaded PLGA/gelatin electrospun nanofibrous mats promote angiogenesis to accelerate diabetic wound healing: Via the modulation of miR-29b-3p, *Biomaterials Science* 8 (2020), pp. 4225–4238.
[32] A.A. Khalili and M.R. Ahmad, A review of cell adhesion studies for biomedical and biological applications, *International Journal of Molecular Sciences* 16 (2015), pp. 18149–18184.
[33] S. Gupta, P. Goyal, A. Jain and P. Chopra, Effect of peri-implantitis associated horizontal bone loss on stress distribution around dental implants – A 3D finite element analysis, *Materials Today: Proceedings* 28 (2020), pp. 1503–1509.
[34] S.G. Chen, J. Yang, Y.G. Jia, B. Lu and L. Ren, Tio2 and PEEK reinforced 3d printing PMMA composite resin for dental denture base applications, *Nanomaterials* 9, no. 7 (2019), p. 1049. doi: 10.3390/nano9071049
[35] S. Weingarten, U. Scheithauer, R. Johne, J. Abel, E. Schwarzer, T. Moritz et al., Multi-material ceramic-based components – additive manufacturing of black and- white zirconia components by thermoplastic 3D-printing (CerAM - T3DP), *Journal of Visualized Experiments* 2019 (2019), pp. 1–10.

Chapter 6

Fusion deposit molding composite material used in a beam application

Piychapillai Periyaswamy
St Peter's Institute of Higher Education and Research

Alphonse Bovas Herbert Bejaxhin
Saveetha School of Engineering, SIMATS

Devasahayam Soosai Jenaris
PSN Engineering College

Easwaran Naveen
Sri Sai Ram Engineering College

Nandagopan Ramanan
Sri Jayaram Institute of Engineering and Technology

CONTENTS

DOI: 10.1201/9781003430186-6

6.1 INTRODUCTION

In recent years, technology has been in great demand in order to support the progress of future generations. As a result, a decent product that will satisfy customers must be developed in a cost-effective, high-quality manner while remaining competitive among competitors. A material for the new product is critical in engineering industries in order to meet the present scenario. Thus, there is a need for quality, economy, and feedback to meet the current needs. In the past, polymers, metals, ceramics, and composite materials are examples of massive materials that have emerged in recent times.

6.2 MANUFACTURING METHODS

In practice, VARTM is created by laying alternate layers of metal and fiber prepared in the mold. The construction is then cured at the proper temperature and pressure, or using an autoclave. However, there is a high possibility of blow holes or bonding failures in this procedure, and these fabrication processes are costly and only suitable for limited specimens. We prepared our specimens using the VARTM technique, a specialty from Anna University's Manufacturing Technology Lab, to retain high precision in bonding at a cheap cost.

Basalt fiber drilled sheets of 300×300 mm and three layers of basalt fiber are piled such that the top and bottom layers are basalt with alternating natural fiber sheets. VARTM uses polyester resin and HY5/3 hardener in a 1:10 ratio combination as a binding agent to obtain the desired basalt through cold mold buildup. The aforementioned mixture was pushed through vacuum cannon into the stacked die at 5 bar pressure and left to cure for about a day. The appropriate test specimens for ASTM standards were cut from the resulting plate using an abrasive water jet machining setup.

6.2.1 Tensile test

Universal testing equipment may be used to determine the tensile qualities. Tensile load will be greatest when the amount of stress applied to the laminate causes it to stretch initially and subsequently break down the specimen. The specimen's reaction to axial stress applied in either direction tends to shatter the specimen at a specific stage. The tensile test is performed in accordance with ASTM D638 guidelines. The produced composite laminate is blanked in accordance with the tensile test dimension criteria, and so conforms to standards. Figure 6.1 depicts the ASTM requirements for the tensile test to be performed on the composite sample.

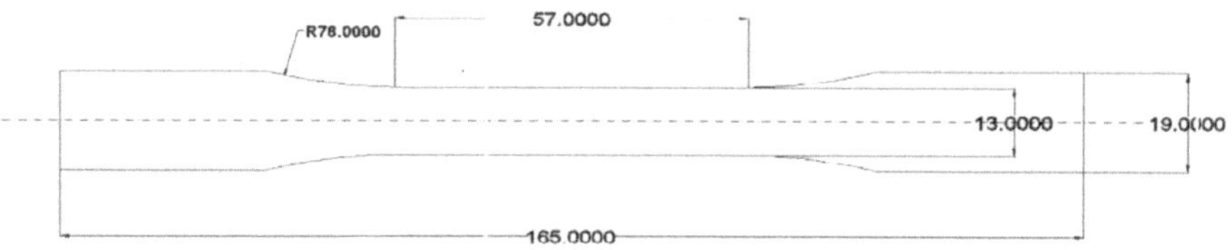

Figure 6.1 ASTM standards for tensile test [ASTM D63].

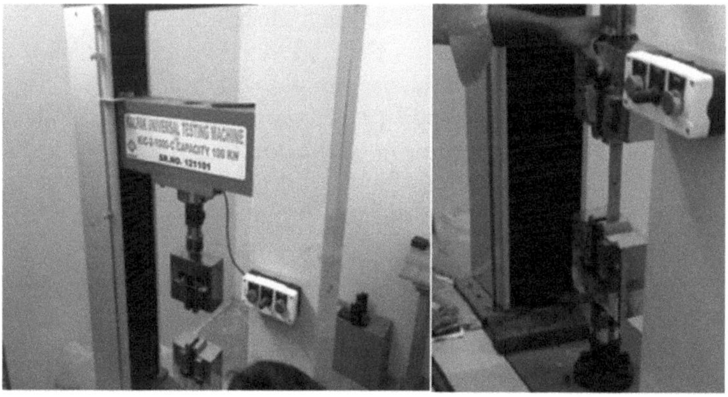

Figure 6.2 Universal testing machine.

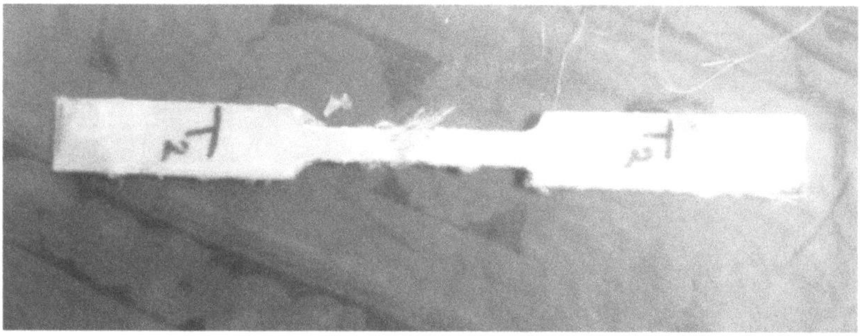

Figure 6.3 Tensile tested specimen.

Figure 6.2 depicts universal testing equipment with digitalized load and displacement data. The test is run to produce the outcome at relative humidity, and the stress versus strain is shown. The outcome of tensile breaking is recorded as a rise in gauge length, with the minimum keeping length for the test ranging from 100 to 150 mm.

During tensile test, a force is applied in either direction, and the specimen breaks at a specific point, with the related data reported and graphs drawn. The tensile tested specimen is shown in Figure 6.3.

6.2.2 Flexural test

The Universal Testing Machine is used for flexural testing, often known as three points bending testing. A tensile test can be used to calculate flexural strength. It is the capacity to bear a load when deformed. The main prerequisites for performing flexural tests are to determine its interlaminar shear strength where short beam shear tests are performed and it can be achieved by performing three point bending test in which the two ends of the specimen are supported like simply supported beam and the load is applied at the middle of the specimen, and as the load progresses continuously, it tends to bend to a maximum where it can tolerate and then the composite begins to fail due to failure. Figure 6.4 shows the ASTM D790 standards for flexural test to be carried out in the composite sample.

The behavior of specimens subjected to flexural stresses is affected by this fracture. Because of the creation of the concave shape, the highest tension will be produced in the outer layer of the fiber, while it will be minimum in the inner layer. The test is carried out in accordance with ASTM D790 guidelines. MPa is the measurement unit for flexural strength. The test is always done in a room temperature environment. The flexural specimen is shown in Figure 6.5.

6.2.3 Double shear test

To assess the composite's shear strength, a double shear test is performed. The ASTM D5379 standard is used to conduct the double shear test. When

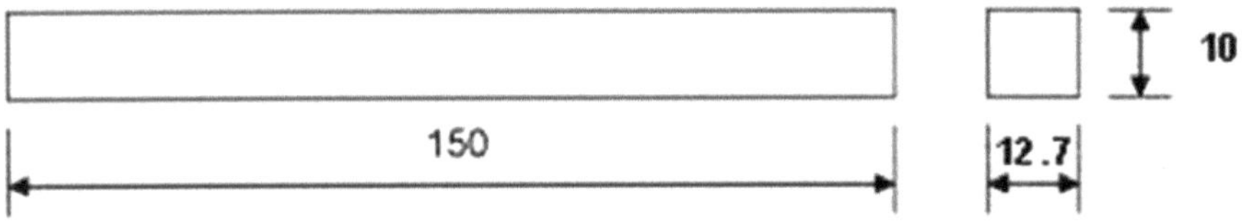

Figure 6.4 ASTM standards for flexural test [ASTM D790].

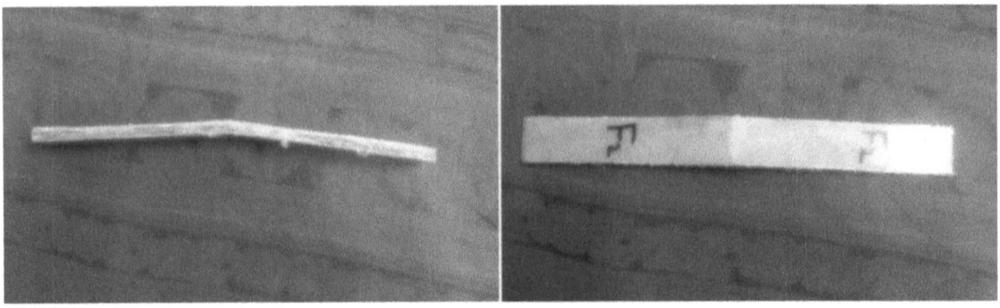

Figure 6.5 Flexural tested specimen.

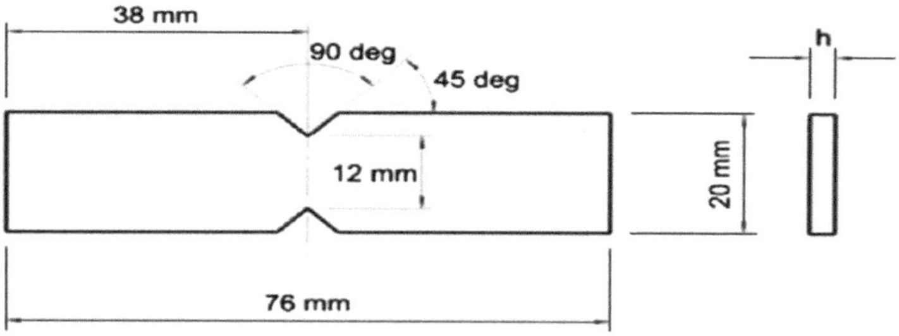

Figure 6.6 Double shear test [ASTM D5379].

Figure 6.7 (a) Double shear test. (b) Tested specimen.

bending stresses are calculated on a loaded beam, the shear stress is ignored. The ASTM requirements for the composite sample's double shear test are shown in Figure 6.6.

The test is done using a universal testing machine in which the special fixture is developed and designed to hold the work piece, thereby it can perform shear test, i.e., by applying the load axially in the opposite direction to each other until shear develops. The failure of the specimens is the ultimate breaking load and the corresponding graphs are plotted against load and displacements. Figure 6.7 shows the shear test setup and specimen.

6.2.4 Impact test

For impact testing with Charpy arrangement, an impact testing machine is employed. It is done to determine the composite's impact strength. The impact test is performed in accordance with ASTM D256 specifications as per Figure 6.8.

This test also determines the overall amount of energy absorbed in joules during the material's fracture on impact. It is the decisive factor in which the specimen composite's life is dependent. The specimen is struck suddenly

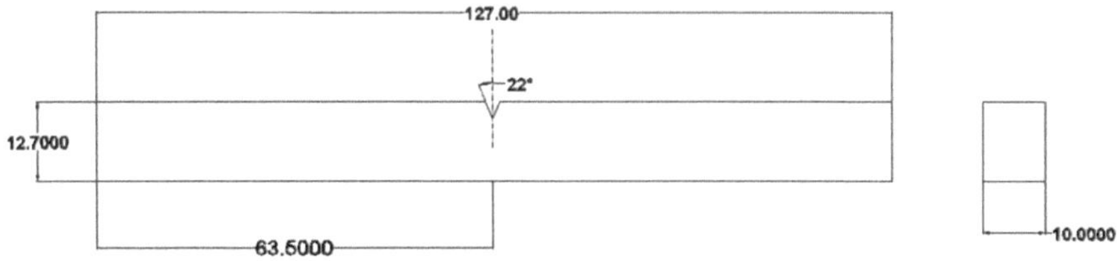

Figure 6.8 ASTM standards for impact test [ASTM D256].

(a) (b)

Figure 6.9 (a) Charpy test machine. (b) Impact tested specimen.

with the oscillating pendulum swinging back and forth, recording the greatest amount of energy absorbed during impact. The material's toughness determines the fracture. The specimen is hit with a blow that causes it to fail at a certain point. The maximum energy absorption in joules has been recorded, and the related graphs have been drawn. The ASTM requirements for impact testing in the composite sample are shown in Figure 6.9.

6.2.5 Hardness test

It is primarily determined by the material property in which the desired area tends to absorb unexpected shock, resulting in material hardness testing. ASTM D2583 is the ASTM standard for hardness testing.

6.2.6 Inter delamination test

The purpose of this test is to assess the composite's internal strength. Layer separation is the mode of failure that occurs when a structure is subjected to heavy loads. This separation leads to loss of strength internally, which delaminates physically. This test is carried out as per ASTM: D5528 standards. The load is applied on the composite laminate until it breaks and delaminates, the corresponding breaking load is determined and the corresponding graphs have been plotted. Figure 6.10 shows the delaminated test specimen.

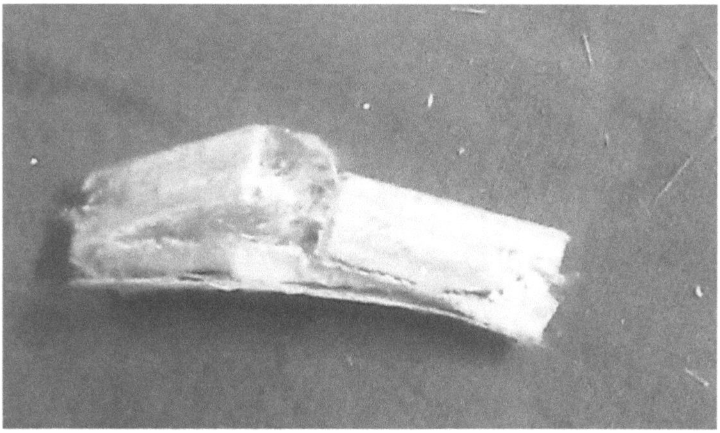

Figure 6.10 Delaminated specimens.

6.3 TENSILE ANALYSIS WITH MORPHOLOGICAL HANDOUTS

Morphological analysis is utilized to examine the specimen's interior fiber behavior. It aids in the detection of fiber fracture, voids, fractures, fiber bonding, matrix homogeneity, fiber separation, and other characteristics in composite laminates. It is possible to investigate the surface features. Each test sample is collected, dried for a while, and then coated with gold of roughly 15 nm thickness. The interior structure is next investigated using the appropriate equipment. The structure shows strong bonding, which has been explored in depth. Finally, it is expected to go to the next level of development and application (Figure 6.11).

6.3.1 Tensile fracture

SEM images show a tensile valve and a tensile SEM fracture; 107.47 N/mm^2 is the ultimate tensile strength based on Table 6.1. The material's strength is displayed as ductile characteristics. As a result, the material has a high tensile strength. The maximum displacement is also 5.52 mm. As a result of the material's high ballistic load, the displacement exceeds 4 mm. The ballistic characteristics are more than covered by the tensile displacement. As a result, the SEM was obtained at various magnifications.

The fiber appears as a ductile fracture in the SEM picture as shown in Figure 6.12. Porosity and voids are also more difficult to regulate. The hexothermal cavities were also visible in the SEM picture, which had not been visible before. As a result, the material's outstanding ballistic qualities were demonstrated.

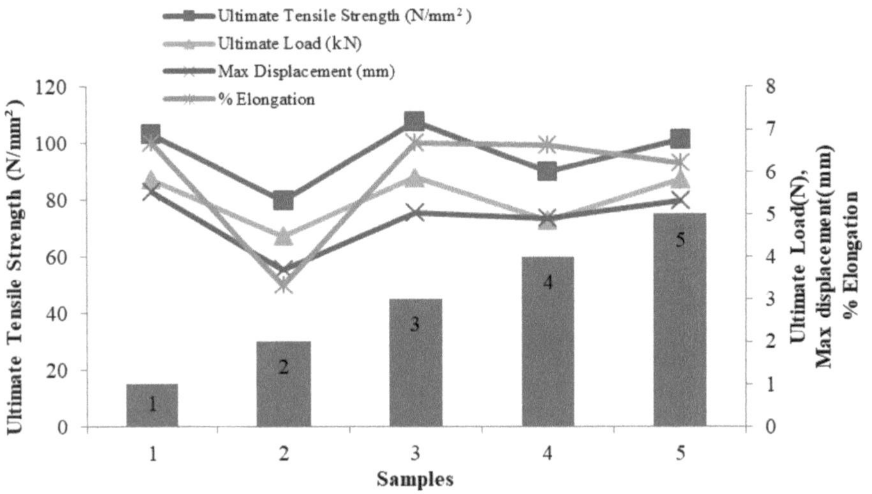

Figure 6.11 Graphical correlation of UTS, maximum displacement and percentage of elongation.

Table 6.1 Tensile test observations

Sl. no.	*Ultimate tensile strength (N/mm²)*	*Ultimate load (kN)*	*Max displacement (mm)*	*% Elongation*
1	103.215	5.81	5.52	6.68
2	79.588	4.48	3.68	3.33
3	107.479	5.85	5.03	6.68
4	89.925	4.85	4.89	6.61
5	101.315	5.82	5.32	6.21

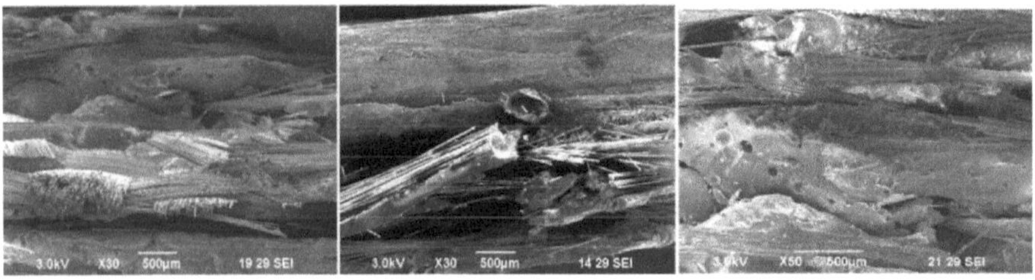

Figure 6.12 SEM images of fractured tensile specimen.

6.3.2 Flexural fracture

In a tabular column, the flexural valve is presented; 0.012 kN/m^2 is the greatest flexural stress. As a result, the material has excellent bending qualities. The maximum load is 0.58 kg. It is clear that the material has a high bending strength. The estimated formula for bending valves is 2×flexural value. As a result, the material possesses excellent bending qualities and the SEM was obtained at various magnifications. The fiber appears as a ductile fracture in the SEM picture. Porosity and voids are also more difficult to regulate. In addition, the SEM picture revealed hexothermal cavities that were not visible in the image. As a result, the material's outstanding ballistic qualities were demonstrated (Table 6.2).

6.3.3 Microform of basalt and reckoning of wear loss

Figure 6.13 depicts the macro picture. The macro photograph depicting the microscopic view of fiber inter delamination was obtained. Macro photos of tensile and flexural samples were captured. The macro photos are displayed in the order of observation. There is no magnification or resolution on the macro picture. The graphic depicts the material's great bonding strength. There is no longer any porosity or voids as depicted in Figure 6.13. There are no hexathermal characteristics in the picture. As a result, the photos were employed as high-resolution ballistic materials (Table 6.3).

The wear properties were done by pin on disc apparatus. The apparatus was made by Ducom instrument. The pin on disc test used high hardness

Table 6.2 Observations of the flexural test of the specimen

Sl. no.	*Ultimate stress (kN/m²)*	*Ultimate load (kN)*	*Max displacement (mm)*
1	0.009	0.49	0.49
2	0.01	0.48	0.48
3	0.012	0.56	0.56
4	0.011	0.52	0.52
5	0.012	0.58	0.58

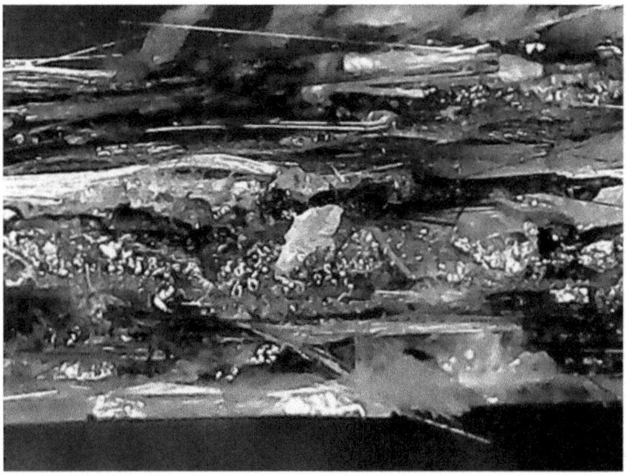

Figure 6.13 Microstructural analysis of basalt fiber.

Table 6.3 Observations of wear test using pin on disc apparatus (Frp-L27, velocity 1 m/s, track diameter 60 mm, pin diameter 10 mm)

Sl. no.	*Load (kg)*	*Disc speed (RPM)*	*Time for sliding (mm)*	*Before weight (kg)*	*After weight (kg)*	*Weight loss (kg)*
1	0.5	159	13	6.5488	6.5480	0.0008
2	0.5	159	17	6.5480	6.5475	0.0005
3	0.5	159	20	6.4014	6.4009	0.0005
4	0.5	318	7	6.4009	6.4003	0.0006
5	0.5	318	8	6.3142	6.3138	0.0004
6	0.5	318	10	6.3138	6.3134	0.0004
7	0.5	478	4	6.5394	6.5391	0.0003
8	0.5	478	6	6.5391	6.5387	0.0006
9	0.5	478	7	6.0923	6.0918	0.0005
10	1.0	159	13	6.0918	6.0911	0.0007
11	1.0	159	17	6.6605	6.6593	0.0012
12	1.0	159	20	6.6593	6.6581	0.0012
13	1.0	318	7	6.0841	6.0818	0.0023
14	1.0	318	8	6.0818	6.0801	0.0017
15	1.0	318	10	6.4334	6.4307	0.0027
16	1.0	478	4	6.4307	6.4295	0.0012
17	1.0	478	6	6.4898	6.4874	0.0024
18	1.0	478	7	6.4874	6.4854	0.0020
19	1.5	159	13	6.0223	6.0202	0.0021
20	1.5	159	17	6.0202	6.0178	0.0024
21	1.5	159	20	6.4752	6.4725	0.0027
22	1.5	318	7	6.4725	6.4702	0.0023
23	1.5	318	8	6.3410	6.3374	0.0036
24	1.5	318	10	6.3374	6.3333	0.0041
25	1.5	478	4	6.2876	6.2845	0.0031
26	1.5	478	6	6.2845	6.2807	0.0038
27	1.5	478	7	6.4816	6.4774	0.0042

disc. The wear sample was prepared by ASTM D9 or D12 standard. The disc maximum diameter is 165 mm and thickness 4 mm and concerted hole has 10 mm full depth. The pin on disc apparatus initially had a minor load applied (without load condition); and the friction properties were analyzed. Finally major load was given. Hence the test will be conducted using different samples.

The pin on disc maximum wear loss is 0.0038 kg. The tolerance limit was 0.0005 kg. In this limit the test was achieved. The weight loss properties are calculated by

$$\text{Wear loss} = (\text{Initial load} - \text{final load})\,\text{kg}$$

The initial load was taken before testing the material and final load was taken after test completion. Hence the material was proved to be a high wear resistance material.

6.4 CONCLUSION

The internal energy and effective plastic strain of sample 3 are comparatively higher using hybrid composite of jute and banana instead of FRP bumper. This indicates that the jute-banana hybrid composite may absorb tension and disperse it as gas pressure. As the headlamp bracket and chassis ends, the effective plastic strain is also reduced. As a result, we may deduce that the responses at these sites are fewer than those experienced by bumpers made of hybrid composite.

BIBLIOGRAPHY

1. Mark N. Bailey, "Automobile bumper behavior in low-speed impacts." SAE Transaction, Document Number: 930211.
2. Paul L. Sabol, "A design procedure for thermoplastic bumper," SAE Transaction, Document Number: 870109.
3. Horst Lanzerath, "Crash simulation of structural foam," SAE Transaction Document Number: 2003-01-0328.
4. Bruce R. Denison, "Advanced polyolefin bumper systems," SAE Transaction, Document Number: 930543.
5. David J. King, "Automotive bumper behavior in low speed impacts," SAE Transaction, Document Number: 930211.
6. Kay Ogiyama, "Beamless bumper system," SAE Transaction, Document Number: 2003-01-0212.
7. Sang-Ha Kim, "Design and structural analysis of bumper for automobiles," SAE Transaction, Document Number: 980114.
8. Javad Marzbanrada, "Design and analysis of an automotive bumper beam in low-speed frontal crashes," School of Automotive Engineering, Iran University of Science and Technology 16846-13114, Narmak, Tehran, Iran.

Chapter 7

A novel surface modification and mechanical characteristics on 3D-printed PLA-Al composite using laser surface irradiation

Navaneetha Krishnan Muthu Nadar
Amrita College of Engineering and Technology

Suresh Sundarraj
University College of Engineering

Emmy Prema Chellam
Bethlahem Institute of Engineering

CONTENTS

7.1 INTRODUCTION

Additive manufacturing is one of the significant manufacturing techniques because of its quick processing and applications [1–4]. Fused Deposition Modeling method is mostly utilized by many industries because of its minimum operating cost, better environment-friendly and cheap equipment, etc. [5,6]. The thermoplastic and thermosetting polymers are utilized as

DOI: 10.1201/9781003430186-7

raw materials for the FDM process. The thermoplastic polymers include Polycarbonate, Polyamide, Polylactic Acid, and Acrylonitrile Butadiene Styrene [7]. But pure polymers are not suitable for new applications [8]. Hence polymer composites solve such issues. Especially, metal fibers and glass fibers are utilized as reinforcement for polymer composites [9–11].

The tensile property of the carbon-fiber-reinforced plastic composites manufactured via FDM was analyzed by Ning et al. [12]. They analyzed the tensile behavior by varying its velocity and layer thickness. Even though FDM fabricated parts have better mechanical properties, the surface quality is the main drawback [13]. The hot cutter machine was utilized by Pandey et al. [14] to minimize the surface roughness. They obtained an 87% assured level of output.

To minimize the surface roughness of the fabricated part, laser polishing technique was introduced. This process involves heating the material surfaces with a laser source. It is a non-contact method, with adaptable polishing velocity by remelting the surfaces [15,16]. To minimize the oxidation of inert gases utilized for laser polishing technique, the material surfaces are rapidly polished by the laser source owing to fluid pressures within the melt region. The parts which are not polished by conventional techniques, such as complicated parts, is easily polished by laser irradiation [17]. Owing to these benefits, the conventional technique has been replaced by the laser polishing technique in several applications.

The surface roughness of the PLA produced via the FDM process was minimized by 68% by Chai et al. This was achieved at 150 mm/s scanning velocity and with the aid of a CO_2 laser [18]. Yung et al. [19] established an actual technique to decrease the roughness of CoCr alloys with multipart geometry surface up to 93% when the hardness of the laser-polished specimens was improved by 8% rather than the SLM apparatuses. He et al. [20] attained a supreme decrease in roughness to 0.156 nm with no impairment for bonded silica. Wang et al. [41] indicated that laser-polished AM CoCr alloy apparatuses could get 0.45 μm of roughness and around 30% resistance to corrosion compared with few mechanical specimens.

In this investigation, Polylactic Acid-Aluminum fiber (PLA-Al fiber) composites produced via the FDM technique were subjected to laser polishing. Literature from the preceding works exhibited that no work had been performed to find the surface roughness and mechanical properties of PLA-Al fiber composite. The characteristics of surface and mechanical effects of the polished PLA-Al fiber composite were analyzed at the various speeds of laser scanning. The characterization of the prepared samples was analyzed using energy-dispersive spectrum (EDS) analysis. The surface roughness of the PLA-Al composites was analyzed by profilometry. The tensile study, as well as dynamic mechanical analysis, was conducted to compare the behavior of both unpolished and polished specimens at various velocities.

7.2 EXPERIMENTAL DETAILS

7.2.1 Material and specimen preparation

The 1.7 mm diameter PLA/Al filament was utilized to fabricate the samples. The PLA/Al filament involves a 6.9 wt% Al fiber. The PLA/Al composites were fabricated using the FDM-type 3D printer. ASTM D5023-15 and ASTM D638-14 standards were followed to prepare the DMA and tensile samples separately. A temperature of 201°C was applied to the nozzle and 61°C was maintained in the bed. The nozzle feeds the required quantity of melted filament along the route with a velocity of 999 mm/min.

7.2.2 Laser polishing technique

The prepared specimen was polished with the aid of a fiber laser (λ = 1069 nm) and 199 W powder. The polishing process was carried out by pulse laser, which is controlled by employing an on/off type diode. The positions of deposition and scanning path are vertical to each other. Once the laser was applied, the PLA/Al composite surfaces were melted. The vale regions were filled by the liquid material coming from the top surfaces. Then the PLA/Al composites were solidified by stopping the laser source. This process supports minimizing the roughness of the surface. This process is carried out at three varying scan speeds such as 50, 100, and 150 mm/s.

7.2.3 EDS characterization

Figure 7.1a and b indicates the EDS characterization of unpolished PLA/Al composite and polished PLA/Al composite at 100 mm/s scan velocity, respectively. Al and Au are the two main elements identified in Figure 7.1a

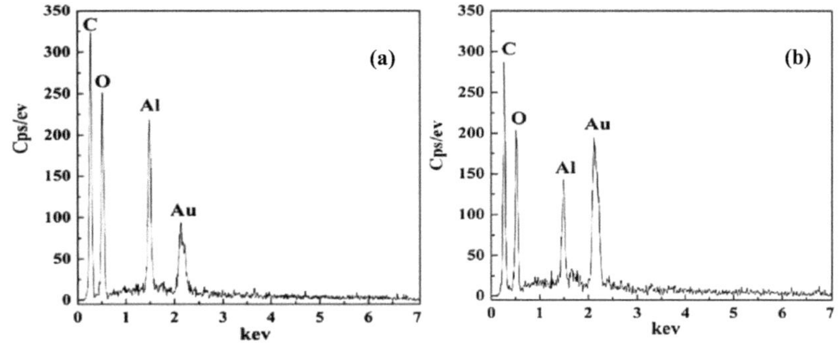

Figure 7.1 EDS of PLA/Al composites: (a) unpolished specimen and (b) polished specimen 2 (scan velocity 100 mm/s).

and b. Au appeared in the EDS because of the gold sputter. The C and O elements are also present along with Al and Au. Hence EDS results confirm the presence of the element in the PLA/Al composites.

7.2.4 Experimentation

7.2.4.1 *Surface roughness test*

The roughness of the PLA/Al composites was determined utilizing the profilometry tester. The arithmetical mean height (Sa) and arithmetic mean deviation (Ra) parameters were measured via stylus type contact profilometer. The experiments were conducted at a velocity of 0.6 mm/s with the help of a 5 μm radius containing diamond stylus probe. The conditions adopted for testing are listed in Table 7.1.

7.2.4.2 *Tensile and dynamic mechanical test*

The ASTM D638-14 standard was followed to conduct the tensile test utilizing UTM (Universal Testing Machine). The velocity of the crosshead was maintained at 1 mm/min. The DMA device is used to analyze the dynamic mechanical behavior of the PLA/Al composites. DMA analysis was carried out concerning the temperature for a distance of 30 mm. The conditions adopted for DMA analysis are listed in Table 7.2.

Table 7.1 The conditions adopted for the surface roughness test

Measuring range	:	160 μm
Radius of stylus	:	5 μm
Material of stylus	:	Diamond
Measuring speed	:	0.6 mm/s
Evaluation length	:	8 mm

Table 7.2 The conditions adopted for DMA analysis

Temperature range	:	29°C–174°C
Rate of heat	:	2°C/min
Frequency	:	1 Hz
Dynamic strain	:	1%

7.3 RESULTS AND DISCUSSION

7.3.1 Surface roughness analysis

The 2D roughness results of unpolished PLA/Al composites are shown in Figure 7.2a and b. The consequences of the laser scanning velocity in the roughness of the PLA/Al composite surfaces are expressed in Figure 7.2c–h. Figure 7.2c and d indicates the roughness profile of laser-polished PLA/Al composite surfaces at 50 mm/s velocities. Figure 7.2e and f indicates the roughness profile of laser-polished PLA/Al composite surfaces at 100 mm/s velocities. Figure 7.2g and h indicates the roughness profile of laser-polished PLA/Al composite surfaces at 150 mm/s velocities. Because of the FDM 3D printing the vale and peak were identified from Figure 7.2a and b and it

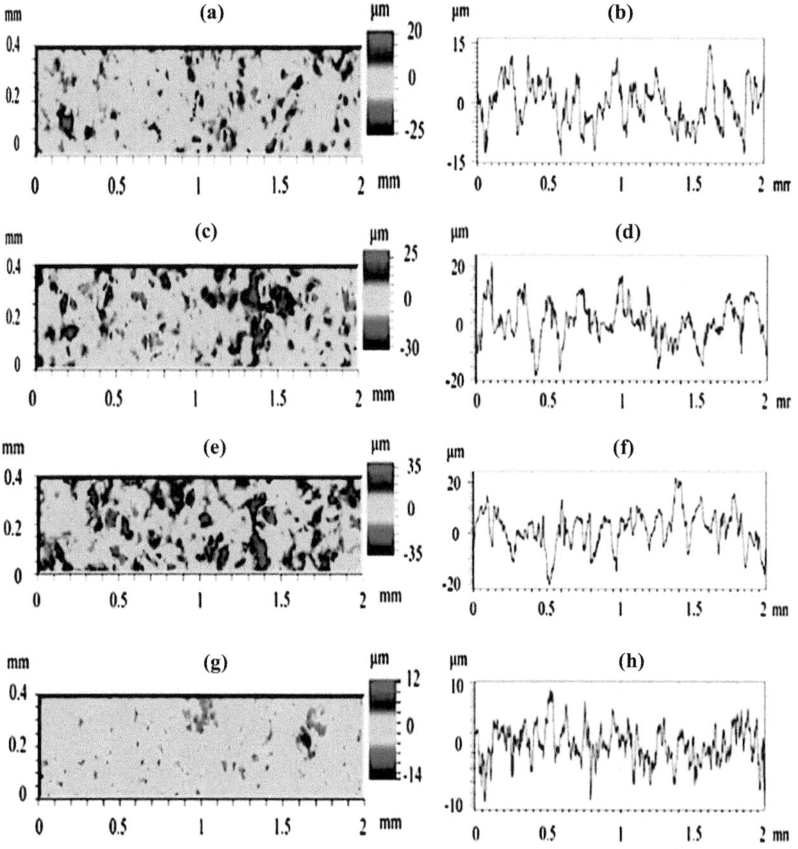

Figure 7.2 Roughness profile of the PLA/Al composite surfaces with distinct laser scanning velocity: (a and b) unpolished PLA/Al composite, (c and d) 50 mm/s, (e and f) 100 mm/s, and (g and h) 150 mm/s.

has an Ra value of 23.31 μm. Similarly, the 50, 100, and 150 mm/s laser-polished PLA/Al composites have Ra values of 4.53, 3.02, and 3.31 μm, correspondingly. It confirms that the polished PLA/Al composite surfaces have the very minimum Ra values rather than the unpolished composite. Especially, 100 mm/s laser-polished composite consists of lower Ra values.

The unpolished composite has a Sa value of 22.93 μm. But, the 50, 100, and 150 mm/s laser-polished PLA/Al composites have Sa values of 6.43, 5.44, and 5.92 μm correspondingly. It confirmed that the polished PLA/Al composite surfaces have very minimum Sa values rather than the unpolished composite. Especially, 100 mm/s laser-polished composite consists of lower Sa values. Even though the laser-polished composites had minimum roughness value, the tiny vale and peak were present. During the laser polishing process, the allowed solidified time is minimum. Hence, the tiny vale and peak were formed.

Initially, the quality of the PLA/Al composite enhances the scanning velocity. Thereafter, the quality of the PLA/Al composite diminishes the scanning velocity. This is because of the additional time obtained in the minimum scanning velocity to be solidified. These results exactly match the references [21,22].

7.3.2 Tensile behavior

The tensile behavior of PLA/Al composites at distinct laser polishing velocities is represented in Figure 7.3. The stress-strain graph confirmed that

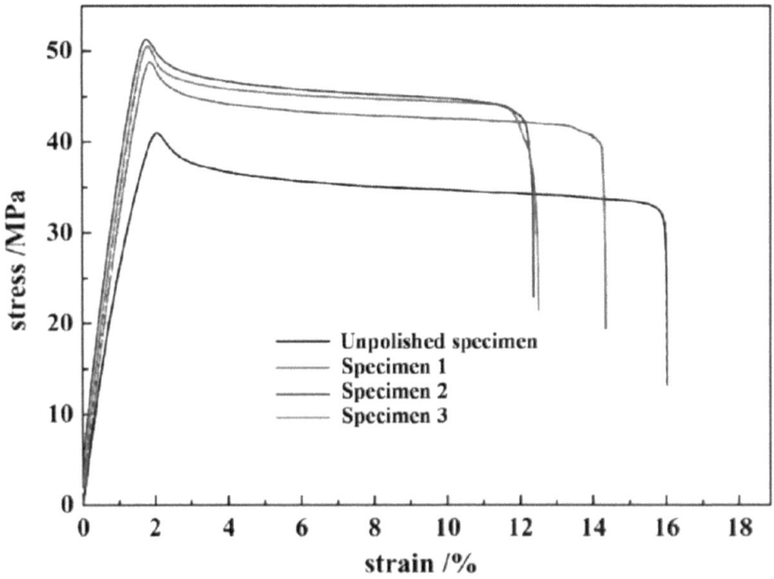

Figure 7.3 Tensile stress-strain curvature of PLA/Al composites. Laser scanning speed: 50 mm/s for specimen 1; 100 mm/s for specimen 2; and 150 mm/s for specimen 3.

the laser-polished composites have higher tensile strength rather than the unpolished composite. The tensile strength of the unpolished composite is 40.93 MPa. But the 50, 100, and 150 mm/s laser-polished PLA/Al composites have the tensile strength of 47.97, 52.03, and 51.15 MPa correspondingly. Figure 7.3 confirms that the scanning velocity has an adverse effect on the tensile property. The polished PLA/Al composites have higher tensile strength rather than the unpolished composites. This is because laser polishing provides good bonding strength among the layers. The tensile strength is minimum in the low and higher scanning velocities because at minimum scanning velocity, the liquid zone is affected by heat and at maximum laser velocity, porosity occurs in the zones.

7.3.3 Dynamic mechanical behavior

Figure 7.4 indicates the dynamic mechanical behavior of the PLA/Al composites. It expresses the disparity of E′ (modulus of storage) concerning the change in temperatures. At the minimum temperature, the E′ value is higher for the polished PLA/Al composites rather than the unpolished composites. The 100 mm/s scanning velocity PLA/Al composites contain a higher E′ value compared with the others. This is because of the strong bonding between the PLA and Al layers. The E′ of unpolished composite is 3.37 GPa. But the 50, 100, and 150 mm/s laser-polished PLA/Al composites have E′ of 4.73, 4.98, and 4.47 GPa, correspondingly. It confirms that the same consequences were found for the tensile strength.

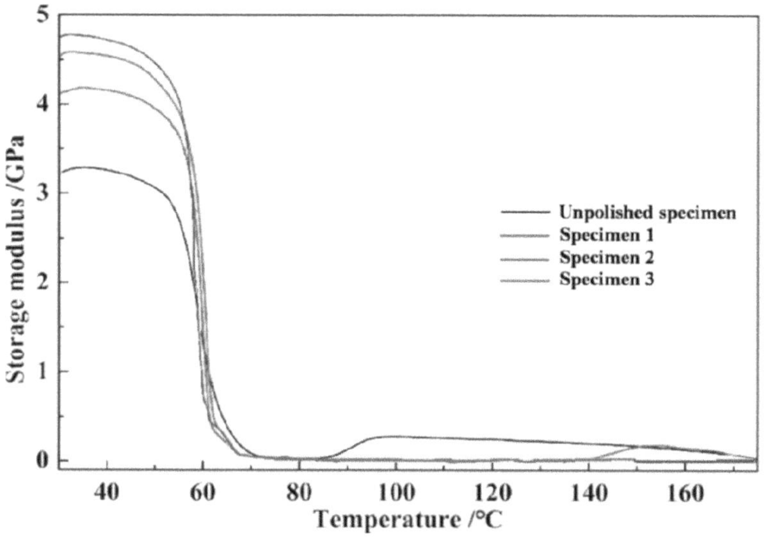

Figure 7.4 DMA analysis of the PLA/Al composites. Laser scanning speed: 50 mm/s for specimen 1; 100 mm/s for specimen 2; 150 mm/s for specimen 3.

7.4 CONCLUSION

In this investigation, the surface roughness, tensile behavior, and dynamic mechanical behavior of the unpolished and laser-polished PLA/Al composites were examined. The following conclusions were made:

- The PLA/Al composites were fabricated via the FDM 3D printing technique. The EDS examination confirms the presence of PLA and Al elements.
- The surface roughness of the PLA/Al composites was reduced by using the laser polishing process. The laser-polished composite with 100 mm/s scanning velocity exhibits better surface quality. The roughness parameter of PLA/Al composites was minimized by 87.04% using laser polishing.
- The laser-polished PLA/Al composites exhibit good tensile strength rather than unpolished composites. In particular, 100 mm/s scanning velocity composite possesses higher tensile strength of 52.03 MPa. This is due to the better bonding strength between the two layers.
- The laser-polished PLA/Al composites exhibit a good dynamic mechanical behavior rather than unpolished composites due to their higher modulus value concerning temperature.

REFERENCES

[1] W. Gao, Y. Zhang, D. Ramanujan, K. Ramani, Y. Chen, C.B. Williams, C.C.L. Wang, Y.C. Shin, S. Zhang, P.D. Zavattieri, The status, challenges, and future of additive manufacturing in engineering, *Comput. Aided Des.* 69 (2015) 65–89.

[2] D. Dimitrov, Advances in three-dimensional printing – state of the art and future perspectives, *Rapid Prototyp. J.* 12 (3) (2006) 136–147.

[3] L.E. Murr, Frontiers of 3D printing/additive manufacturing: from human organs to aircraft fabrication, *J. Mater. Sci. Technol.* 32 (10) (2016) 987–995.

[4] P. Jiang, Z. Ji, C. Yan, X. Wang, F. Zhou, High compressive strength metallic architectures prepared via polyelectrolyte-brush assisted metal deposition on 3D printed lattices, *Nano Struct. Nano Objects* 16 (2018) 420–427.

[5] I. Zein, D.W. Hutmacher, K.C. Tan, S.H. Teoh, Fused deposition modelling of novel scaffold architectures for tissue engineering applications, *Biomaterials* 23 (4) (2002) 1169–1185.

[6] Y. Liu, *A Preliminary Research on Development of a Fibre-Composite, Curved FDM System*, National University of Singapore, 2008.

[7] S. He, Z. Yu, H. Zhou, Z. Huang, Y. Zhang, Y. Li, J. Li, Y. Wang, D. Li, Polymer tubes as carrier boats of thermosetting and powder materials based on 3D printing for triboelectric nanogenerator with microstructure, *Nano Energy* 52 (2018) 134–141.

[8] G. Cicala, G. Ognibene, S. Portuesi, I. Blanco, M. Rapisarda, E. Pergolizzi, G. Recca, Comparison of Ultem 9085 used in fused deposition modelling (FDM) with polytherimide blends, *Materials* 11 (2) (2018) 285.
[9] X. Zhang, L. Chen, C. Kowalski, T. Mulholland, T.A. Osswald, Nozzle flow behavior of aluminum/polycarbonate composites in the material extrusion printing process, *J. Appl. Polym. Sci.* 136 (12) (2019) 47252.
[10] R. Singh, P. Bedi, F. Fraternali, I.P.S. Ahuja, Effect of single particle size, double particle size and triple particle size Al_2O_3 in Nylon-6 matrix on mechanical properties of feed stock filament for FDM, *Compos. B Eng.* 106 (2016) 20–27.
[11] M. Ramesh, K. Panneerselvam, PLA-based material design and investigation of its properties by FDM, in: M.S. Shunmugam, M. Kanthababu (Eds.), *Advances in Additive Manufacturing and Joining*, Springer Singapore, Singapore, 2020, pp. 229–241.
[12] F. Ning, W. Cong, Y. Hu, H. Wang, Additive manufacturing of carbon fiber reinforced plastic composites using fused deposition modelling: effects of process parameters on tensile properties, *J. Compos. Mater.* 51 (4) (2016) 451–462.
[13] C.J.L. Perez, Analysis of the surface roughness and dimensional accuracy capability of fused deposition modelling processes, *Int. J. Prod. Res.* 40 (12) (2002) 2865–2881.
[14] P.M. Pandey, N.V. Reddy, S.G. Dhande, Improvement of surface finish by staircase machining in fused deposition modelling, *J. Mater. Process. Technol.* 132 (1–3) (2003) 323–331.
[15] L. Chen, X.Z. Zhang, S.Y. Gan, Effects of laser polishing on surface quality and mechanical properties of PLA parts built by fused deposition modelling, *J. Appl. Polym. Sci.* 137 (3) (2020) 48288.
[16] L. Chen, B. Richter, X. Zhang, X. Ren, F.E. Pfefferkorn, Modification of surface characteristics and electrochemical corrosion behavior of laser powder bed fused stainless-steel 316L after laser polishing, *Addit. Manuf.* 32 (2020) 101013.
[17] A. Lamikiz, J.A. Sanchez, L.N. Lopez de Lacalle, J.L. Arana, Laser polishing of parts built up by selective laser sintering, *Int. J. Mach. Tool Manufact.* 47 (12–13) (2007) 2040–2050.
[18] Y. Chai, R.W. Li, D.M. Perriman, S. Chen, Q.H. Qin, P.N. Smith, Laser polishing of thermoplastics fabricated using fused deposition modelling, *Int. J. Adv. Manuf. Technol.* 96 (9–12) (2018) 4295–4302.
[19] K.C. Yung, T.Y. Xiao, H.S. Choy, W.J. Wang, Z.X. Cai, Laser polishing of additive manufactured CoCr alloy components with complex surface geometry, *J. Mater. Process. Technol.* 262 (2018) 53–64.
[20] T. He, C.Y. Wei, Z.G. Jiang, Y.A. Zhao, J.D. Shao, Super-smooth surface demonstration and the physical mechanism of CO_2 laser polishing of fused silica, *Opt. Lett.* 43 (23) (2018) 5777–5780.
[21] L. Gurau, A. Petru, A. Varodi, M.C. Timar, The influence of CO_2 laser beam power output and scanning speed on surface roughness and colour changes of beech (Fagus sylvatica), *BioResources* 12 (4) (2017) 7395–7412.
[22] P.-R. Jang, C.-G. Kim, G.-P. Han, M.-C. Ko, U.-C. Kim, H.-S. Kim, Influence of laser spot scanning speed on micro-polishing of metallic surface using UV nano second pulse laser, *Int. J. Adv. Manuf. Technol.* 103 (1) (2019) 423–431.

Chapter 8

3D Printable titanium alloys and their properties in biomedical applications: state of the art

Anmol Sharma and Pushpendera S. Bharti
Guru Gobind Singh Indraprastha University

Ashish Kaushik, Upender Punia, and Ramesh Kumar Garg
DeenbandhuChhotu Ram University of Science and Technology

Mohit Yadav and Deepak Chhabra
Maharshi Dayanand University

Abhi Bansal
DPG Institute of Technology and Management

CONTENTS

8.1 INTRODUCTION

3D printing is defined as a technique of joining materials layer by layer to fabricate products using 3D CAD data. The process is initiated with the designing of a 3D computed design of a part, complete with dimensions

DOI: 10.1201/9781003430186-8

and tolerances, which is then split into numerous layers computationally. An AM process is used to convert these sliced G-code layers into physical layers [1,2]. Metal 3D printing is a revolutionary path for fabricating a functional engineering metallic object. In the medical field, it has improved outcomes in the form of creating patient-specific medical implants [3,4]. As the technology grows, some image capturing techniques arise that help in creating an almost similar digital design for an implant to be produced. But the aspect that essentially needs to be considered is its cost. Generally the implants manufactured using AM technology need some post-processing to become a quality product and compared with conventional technique-based similar product it costs higher. Some examples are metal-based hip or knee implants. The need for AM-based technique arises when there are complex or intricate parts to be manufactured. Products that are needed to achieve a unique solution help in facilitating surgery so that the healing process can be accelerated and provide reduction in strain to patient. For such designs, high cost for product is even justifiable.

Although there are a range of materials for manufacturing a product, if we move towards biocompatibility, then the range is indeed limited. This chapter discusses about some of the metallic alloys that can be utilized for biomedical application using AM technology. The selection of material depends on the application which signifies what sort of property is desired from the product such as impact strength, wear resistance, compressive strength etc. They need to be compatible with bones and/or tissues, and their effect after post-processing also needs to be considered [5].

In the case of biomaterials, selection criteria also involve their chemical composition or the base material used to prepare specific design material. In the case of metal powders, their size as well as their morphology need to be accurate and precise; presence of porous properties and impurities should be properly documented. The base material and final composition need to be tested before use; their origin needs to be studied and various other factors to make outcomes as much true as possible.

After processing, the materials need to be tested for cracks or remaining porous on the micro level. Their chemical composition again needs to be verified and compared with their base material or initial alloy material. In a later stage, the micro-structural properties of the material, such as grain size and distribution, as well as the crystallographic phases and distribution of its texture, must be ensured. It can be understood by linking the microstructure with dynamic mechanical qualities such as internal stresses that are formed during the processing of an additively made component. These internal stresses then need to be relaxed utilizing desired thermal treatment such as annealing, etc. Another value, i.e. surface roughness, needs to be reduced and should be properly taken care of to avoid release of loose particles as well as improve fatigue properties.

Finally, after considering everything related to materials, the biocompatibility of metallic devices also needs to be considered for any loose particles, surface roughness, and partially attached materials (i.e. its cleanliness). If the foreign particle or loose particles on devices interact with the material, it may cause toxicity to overall component leading to healthcare problems on the short or long term.

Metallic implants prepared using additive manufactured principle can be removable or permanent (depending on bodily requirement) [6,7]. Also, there are possibilities where material can be resolved with body parts termed as bioresorbable materials. Probabilities for material are in the form of bulk or porous or can be a consolidation of both. The materials used for implants have been constructed with titanium (Ti)-based alloys or cobalt (Co)-chromium (Cr)-based alloys, Ni (nickel)-titanium (Ti) and in some cases stainless steel or tantalum. These implant materials can be studied based on their time of use or on the basis of their densities. Among permanent materials, titanium-based alloys have least density (4.3–5.1 g/cm^3). Figure 8.1 demonstrates the overview of titanium alloys used in biomedical applications.

Based on the brief draft, metallic biomaterials that have been mentioned above and are processed by AM will be studied in detail within the chapter. This explanation reviews both materials that are already available on a commercial scale for biomedical applications or those that are going to be great built-up materials in the future. They are discussed in a way that each material is been categorized with its own properties, its strength or

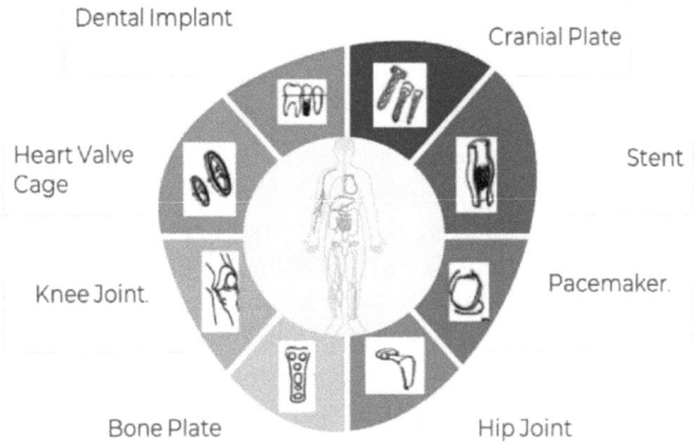

Figure 8.1 Biomedical applications of 3D-printable titanium alloys.

potential and drawbacks. They have been explained using clinical examples, where possible. This chapter aims at fulfilling below listed objectives:

i. To investigate the biocompatibility of titanium-based alloys for clinical applications.
ii. To review and analyse the functional and mechanical characteristics of 3D printable titanium and its alloys which results in selection of appropriate biomaterial after systematically reviewing the literature.
iii. To provide a convenient framework for scientific study and engineering applications of titanium-based alloys produced by additive manufacturing for researchers and professionals working in similar area.

8.2 FUNCTIONAL AND MECHANICAL CHARACTERISTICS OF 3D PRINTABLE TITANIUM AND ITS ALLOYS

Titanium is a metal with a remarkable combination of mechanical characteristics and great biocompatibility, and used as biomaterials on a large scale. Titanium which is carrying stable and protective oxides on its surface is capable of providing osseointegration and in this way it is favourable for providing a direct connection among implant surface and the part of bone attached with implant [8]. Titanium exhibits two types of crystal structure because of its allotropic nature – one is hexagonal closed packed (hcp-α-phase) and other body-centred cubic structure (bcc-β-phase) which helps it in developing a relatively huge variety of Ti alloys with explicit characteristics. Some alterations can also be performed to alter their chemical composition using alloy addition, or some heat treatment processes resulting in variation of phase combination and microstructure leading to ultimate change in mechanical properties.

Figures 8.2 and 8.3 show mechanical characteristics of first- and second-generation biocompatible titanium alloys respectively. It is seen that mechanical attributes of two-phase (α/β) titanium alloys depend upon morphology, distribution and number of phases. One such commonly known alloy of Ti is Ti-6AI-4V with two phases (α/β) yielding a special mixture of strength, low specific weight, excellent biocompatibility, toughness and corrosion resistance [9,10]. To reach certain desired characteristics of alloy, certain mechanical or thermal treatments are performed [11–14]. Figure 8.4 depicts that the nature of human bone is anisotropic and the value is higher in case of cortical bone, compared with more porous cancellous bone. Bone compatibility with the orthopaedic implant and importance of complex design is illustrated.

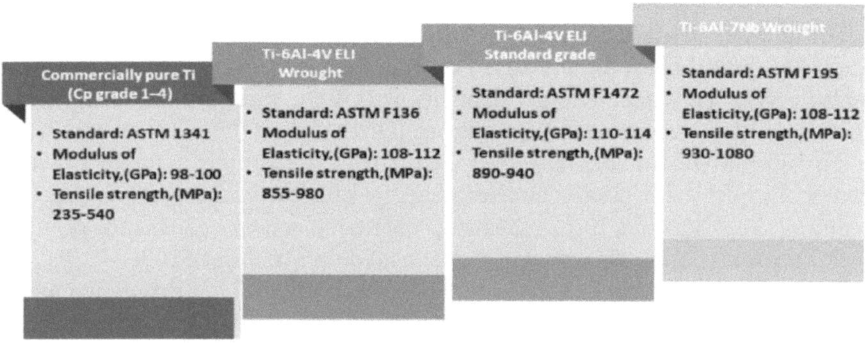

Figure 8.2 Mechanical characteristics of first-generation titanium alloys.

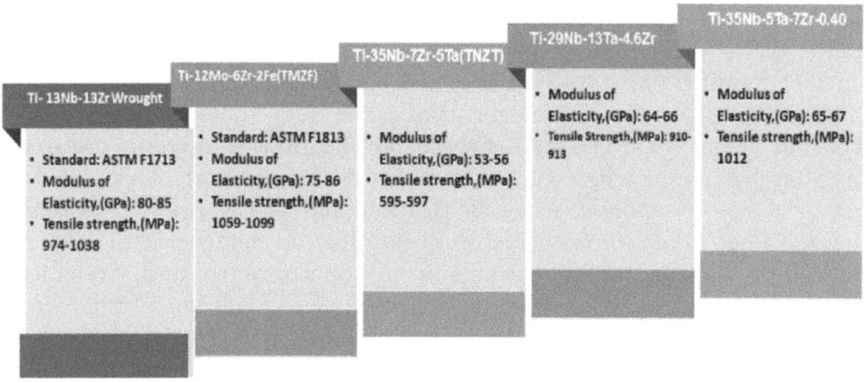

Figure 8.3 Mechanical characteristics of second-generation titanium alloys.

Property	Cortical Bone		Cancellous bone
	Longitudinal	Transverse	
Strength, Tension	79- 151 MPa	52-57 MPa	
Strength, Compression	132-225 MPa	107-134 MPa	2.1-5.1 MPa
Elastic moduli of compact bone	18-21 GPa	7-14 GPa	0.77-4.1 GPa
Apparent density	1.98 g/cm^3		0.051-1.01 g/cm^3

Figure 8.4 Mechanical characteristics of natural bone.

8.2.1 Ti-based alloy additive manufacturing

Techniques that are generally used for manufacturing of customized titanium (Ti)-based components based on rapid prototyping are electron beam melting (EBM), direct energy deposition (DED), laser power bed fusion (LPBF), and direct metal laser sintering (DMLS).

To summarize, in SLM, a laser beam travels through a metal powder composition of bed in safe environment; in EBM, a vacuum atmosphere is required, and later in a fast preheating stage, an electron beam scans powder bed. In primary step, the powder bed is heated up for about 610°C–710°C by an electron beam source. SLM and EBM methods provide different microstructures and associated properties because of various solidification conditions and connected with different cooling rates. More research work is carried on processing Ti-6AI-4V alloys compared to that of β-titanium alloys.

8.2.2 3D Printing and expansion of novel B-type titanium alloys

Current modifications on Ti-based AM materials that are being used for implant utilization involve blueprint of novel beta alloys which have less modulus of elasticity. The goal of these modifications is to reduce the stress shielding behaviour when compared with other traditional alloys that consist of a huge portion of more rigid hexagonal phase. Another criterion that has been taken into consideration for alloying elements with reducing the rigidity of alloys is that it should be non-poisonous and, preferably, biocompatible with living bone, tissues or blood. This is one of the factors that has been taken into consideration. Refractory and transition metals of group Vb, VIb and IVb, such as Ta, Hf, Mo, Nb and many others, as well as specific combinations of these elements are among the most often utilized stabilizing elements for applications in the biomedical area. Because of the high Nb content that is present (about 45 wt%), the β-phase has been completely stabilized, which has led to the formation of extremely well cellular Ti sub-grains inside laser melted tracks. Some of these β-titanium alloys were acquired right after the SLM process was completed. The mechanical characteristic is represented in Figure 8.5.

When comparing a material based on Ti-6Al-4V refined with SLM there is significantly higher tensile fracture strain and modulus of elasticity. Yield values and ultimate tensile strength for SLM-processed Ti-6Al-4V materials have been nearly 50% higher than for β-Ti alloys.

8.2.3 Biomedical aspects in 3D printing of titanium alloys

Titanium's protective and stable oxides make it capable of promoting osseointegration and minimizing discharge of potentially cytotoxic elements (like V and Al of Ti-6Al-4V), which enables anatomical and functional

Alloy composition	Young Modulus E (GPa)	Yield Strength σ_y (MPa)	Tensile Strength σ_{UTS} (MPa)	Elongation δ (%)
Ti-24Zr-4Nb-8Sn	52 ±2	561 ±40	664 ±19	13.7 ±4.2
Ti-15Ta-1.5Zr	93 ±7	891 ±50	870 ±18	16 ±1.3
Ti-15Ta-5.5Zr	70 ±6	962 ±30	930 ±30	19 ±1.9
Ti-15Ta-10.5Zr	40 ±6	809 ±10	770 ±15	15.2 ±0.8
Ti-5Al-5V-5Mo-3Cr	-	-	800 ±1	13 ±1
Ti-50Ta	74 ±2	-	924 ±10	11 ±1
Ti-6Al-4V-10Mo	72 ±2	860 ±14	924 ±5	20.4 ±1.6
Ti-6Al-4V	109 ±	1199 ±19	1270 ±3	7 ±1.5

Figure 8.5 Mechanical properties of β-titanium alloys processed using SLM.

connections to be made between bones and implants [15]. In certain ways, the most dangerous consequence that must be addressed during implantation surgery is the possibility of a bacterial disease developing at the boundary among the tissues and implants that are in close proximity to it. These diseases are not predefined, they can occur immediately or have possibility to occur years after maxillofacial or orthopaedic surgery and can cause serious implications for patient health. That is why, it is an essential requirement to functionalize implant surfaces with antibacterial properties, which is an excellent strategy to reduce chances of infections. Another way is to combine a variety of materials on a particular implant. To make a heterogeneous implant, various materials are layered onto a single implant zone that improves the body's response to the implant.

One method or possibility is to reduce bacterial infections in the use of Cu-coated implant [23], but the outcomes are not appropriate. Cu-coated pins are toxic for bacteria and also for healthy human cells. That is why, a minute amount (~1 wt%) of copper content is suggested in Ti-xCu alloys. In this manner, it efficiently restricts the release of Cu-ions to such an extent that it only destructs bacterial cells, while allowing other ions to be biocompatible with healthy human cells. One experiment used a Ti-1Cu alloy implanted into all eight of the rabbits' mature bodies during an in vivo research; the results have been positive, showing good antibacterial properties and good biocompatibility [23]. Similar to this, copper deposition onto titanium disks done using electrochemical deposition method is also performed to form an antibacterial agent [15]. Based upon these results, a study has been performed recently on Ti-6Al-4V-xCu (where x=0%–6%) with successful results. In that material 4 and 6 wt% Cu are processed using SLM method, resulting in strong antibacterial properties, and showing good cytocompatibility with bone marrow stromal cells [16].

8.2.4 Current and future trends in fabrication of Ti alloys

As per recent studies, factors that are being worked upon for new titanium alloys, more specifically for using in additive manufacturing techniques, are (i) achieving the most optimized condition of mechanical requirements as per the required design for strength, ductility with elastic stiffness values compatible with human bone, and (ii) prevention from alloying elements of potential toxicity. To achieve lower value of stiffness, beta-titanium alloys are widely studied. An alloy of titanium Ti-Nb-Ta-Zr has much lesser value for elastic modulus near 48–55 GPa, if compared with widely used Ti-6Al-4V alloys then modulus is nearly half [17]. Till now, titanium alloy that has been reported with lowest value of elastic modulus for biomedical application is Ti-35Nb-4Sn alloy [18]. For obtaining further lower value of rigidity, a specific extent of porosity has also been provided in manufactured constituent. As an example, SLM-based Ti-45Nb provided with 13% porosity will provide a completely non-poisonous alloy having compressive modulus of elastic value nearly 11 GPa (which is nearly comparative to rigidity of cortical bone) [19]. The maximum fatigue limit and the maximum fatigue life for metals that have been treated by AM is another significant attribute that needs to be examined and investigated in depth. This is especially true for lattice structures. When subjected to fatigue loading conditions, the struts have a high surface roughness because they were created in a layer-by-layer manner. This causes the struts to behave as stress concentrating points. In a compression-compression fatigue test, the fatigue strength of a porous Ti-6Al-4V scaffold structure might be improved with the use of a few mechanical techniques. The test was run for 106 cycles. The removal of material must occur concurrently with the consideration of mechanical treatment [20].

8.3 MATERIAL CONSIDERATIONS OF 3D-PRINTED TI-6AL-4V

The electron beam's energy density is high enough to melt a huge range of metals. The technique of electron beam melting possesses the capability to print a huge span of materials, including Al alloys, Co-based super-alloys and stool steel among others. On the other hand, Ti-6Al-4V is the alloy that has been investigated the most for EBM technique and several additional metal AM processes.

8.3.1 Surface characteristics of 3D-printed Ti-6Al-4V alloy

The lack of a high level of surface polish, in comparison to machined surfaces, is a major problem for various methods of metal additive manufacturing, together with EBM. This is caused by two concerns: (i) particles on the surface that have only partially melted; (ii) staircase among surface layers caused by layer-on-layer printing process. As a consequence of this, post-processing is frequently crucial in order to fix dimensional mistakes and accomplish the necessary level of surface quality [21]. EBM components have a surface arithmetic average roughness Sa value of 30–68 μm [22,23]. This is significantly higher than the 7–16 μm that SLM parts have [24,25]. Since 2003, Arcam has improved the layer thickness from 200 to 100 μm, then to 70 μm, and finally to 50 μm. As a result, using the best specifications, Sa can be decreased to 27 μm [26]. Karlsson et al. [27] provided visual confirmation that the particle size as well as the layer thickness had a substantial impact on the surface topography of EBM Ti-6Al-4V components. In addition, support structures that are installed under overhanging surfaces with the intention of reducing curling can contribute to the finish's deterioration [28].

8.3.2 Porosity of Ti-6Al-4V alloy fabrication via electron beam melting

Many researchers analysed that using Arcam-supplied GA powder, the EBM method can generate Ti-6Al-4V components having a theoretical density of 99.4%–99.8% [29–31]. Arcam's porosity results are represented in Figure 8.6 [32]. Pores are usually divided into two types: elongated and spherical. The absence of fusion between two powder layers results in the formation of pores that are elongated and aligned along the direction of the powder layers. It is possible to get rid of these pores by optimizing the parameters of the EBM process, such as the scanning techniques and the layer thickness. Spherical pores with nearly perfect spheroidicity are caused by argon gas that has become trapped within the GA nanoparticles [33,34]. When it is necessary to eliminate both types of pores from components that were produced using EBM, a hot isostatic pressing (HIP) procedure must be carried out. A common HIP approach for AM Ti-6Al-4V entails applying 100 MPa of pressure between 900°C and 920°C for 2 hours. According to Gaytan et al.'s research, even if a single typical HIP cycle is capable of removing the majority of voids, low bubble (gas) pressure and surface tension make it impossible to remove all of the remaining void completely [33].

Porosity (vol. %)	
GA Ti-6Al-4V powder	: 0.19
GA Ti-6Al-4V ELI powder	: 0.15
Ti-6Al-4V as built by SEBM	: 0.14
Ti-6Al-4V SEBM + HIP	: 0.01
Ti-6Al-4V ELI, as built by SEBM	: 0.10
Ti-6Al-4V ELI, SEBM + HIP	: 0.01

Figure 8.6 Demonstration of the porosity of Ti-6Al-4V-ELI and Ti-6Al-4V.

For AM Ti-6Al-4V, a standard HIP route with 100MPa pressure at 900°C–920°C for 120 minutes is followed. It was shown by Gaytan et al. that a single typical HIP cycle may eliminate the majority of voids; nevertheless, remnants cannot be entirely removed because of low bubble (gas) pressure and surface tension [33].

8.3.3 3D-printed Ti-6Al-4V Alloy with consideration to corrosion

As a result of production of a particularly solid and uninterrupted oxide film on the surface, conventional cast and wrought alloys of Ti-6Al-4V known for a long time to be highly corrosion-resistant materials have been used effectively in industrial domains that demand a high level of anti-corrosion ability, such as the healthcare and aquatic areas. Despite the fact that several implants made with EBM technology, such as knee and hip joints, are in widespread usage across the globe [35], their corrosion behaviour has yet to be thoroughly studied. Using an electrolytic approach and a potentiostatic polarization technique, Koike et al. analysed the corrosion characteristics of a variety of Ti-6Al-4V alloys in a refined Tani-Zucchi synthetic saliva having constituents mainly potassium chloride, baking soda, sodium dihydrogen phosphate, potassium thiocyanate and lactic acid at 37°C [36]. The results of their investigation are presented in Figure 8.7 where the values listed are the parameters' means (SD) for the alloys which have been tested. There are no significant differences if the letters are identical ($p > 0.05$). There were no discernible variations in any of the parameters between the produced and the EBM specimens, cast Ti-6Al-4V ELI specimens or any two. Within the region of oxidation state that the normal human mouth cavity possesses, the 3D-printed specimens demonstrated corrosion performance.

Further, Abdeen and Palmer used both potentiostatic and potentiodynamic tests on additively fabricated metal Ti-6Al-4V in 3.5wt% brine solution [37]. In order to determine temperature at which pitting becomes critical, several tests were carried out. For providing a point of reference, the same tests have been carried out on wrought samples. According to the

Ti-6Al-4V ELI(SEBM)	Ti-6Al-4V ELI Wrought	Ti-6Al-4V ELI Cast
• Open circuit potential(mV): -245(81)^{a} • Polarization resistance: 0.46(0.27)^{a} • Corrosion current density: 197(90)^{a} • Passive current density: 1645(583)^{a}	• Open circuit potential(mV): -157(15)^{a} • Polarization resistance: 1.35(0.48)^{ab} • Corrosion current density: 49(25)^{ab} • Passive current density: 838(249)^{ab}	• Open circuit potential(mV): -245(66)^{a} • Polarization resistance: 0.74(0.71)^{ab} • Corrosion current density: 197(158)^{a} • Passive current density: 1136(752)^{ab}

Figure 8.7 Illustration of corrosion properties of Ti-6Al-4V.

results of their investigation, both the EBM and the wrought samples show remarkable resistance to corrosion in NaCl solution and have been capable of having a huge potential at temperatures as high as 85°C. In addition to this, it was discovered that EBM samples with irregular as-built surfaces have been resistant to the effects of hostile chemicals. Immersing in NaCl solutions of various pH the condition is observed. In addition, they reached the realization that a temperature of 30°C is essential for Ti-6Al-4V in order to keep a strong pitting potential while it is immersed in NaCl solution.

8.4 CONCLUSION

The present proposed work has proficiently examined the functional and mechanical characteristics of additively manufactured titanium-based alloys on the basis of extensive literature review. The biocompatibility, corrosion resistance of titanium alloys is also studied. Additively manufactured titanium alloys utilized as a biomaterial in various biomedical applications along with current and future trends are also described. Various mechanical properties of 3D-printed Ti-6Al-4V (as-built surface characteristics, porosity, corrosion behaviour) have been successfully discussed which results in adoption of suitable material according to the requirements.

REFERENCES

[1] Sharma, A., Chhabra, D., Sahdev, R., Kaushik, A., & Punia, U. (2022). Investigation of wear rate of FDM printed TPU, ASA and multi-material parts using heuristic GANN tool. *Materials Today: Proceedings*, *63*, 559–565.

[2] Kumar, S. (2014). 10.05 Selective laser sintering/melting, in: S. Hashmi, G. Ferreira Batalha, & B. Yilbas (Ed.), *Comprehensive Materials Processing*, Volume 10. Elsevier: Amsterdam.

[3] Punia, U., Kaushik, A., Garg, R. K., Chhabra, D., & Sharma, A. (2022). 3D printable biomaterials for dental restoration: A systematic review. *Materials Today: Proceedings*, *63*, 566–572.
[4] Venuvinod, P. K., & Ma, W. (2004). Selective Laser Sintering (SLS), in: *Rapid Prototyping: Laser-Based and Other Technologies*, Springer Science & Business Media, Boston, MA. https://doi.org/10.1007/978-1-4757-6361-4_7
[5] Yadav, D., Garg, R. K., Ahlawat, A., & Chhabra, D. (2020). 3D printable biomaterials for orthopedic implants: Solution for sustainable and circular economy. *Resources Policy*, *68*, 101767.
[6] Javaid, M., & Haleem, A. (2018). Additive manufacturing applications in medical cases: A literature based review. *Alexandria Journal of Medicine*, *54*(4), 411–422.
[7] Javaid, M., & Haleem, A. (2018). Additive manufacturing applications in orthopaedics: A review. *Journal of Clinical Orthopaedics and Trauma*, *9*(3), 202–206.
[8] de Viter, V.S., & Fuentes, E. (2013). Titanium and titanium alloys as biomaterials, in: J. Gegner (Ed.), *Tribology – Fundamentals and Advancements*, InTech, Rijeka (Chapter 05). DOI: 10.5772/55860
[9] Boyer, R., Welsch, G., & Collings, E. (1994). *Materials Properties Handbook: Titanium Alloys*, ASM International, Materials Park, OH.
[10] Donachie, M. J. (2000). *Titanium: A Technical Guide*, 2nd ed., ASM International, Materials Park, OH.
[11] Ding, R., Guo, Z., & Wilson, A. (2002). Microstructural evolution of a Ti-6Al-4V alloy during thermomechanical processing. *Materials Science & Engineering A: Structural Materials: Properties, Microstructure and Processing*, *327*, 233–245.
[12] Imam, M. A., & Fraker, A. C. (1996). *Medical Applications of Titanium and Its Alloys: The Material and Biological Issues*, S. A. Brown, J. E. Lemons (eds.) ASTM Special Technical Publication. Doi: 10.1520/STP1272-EB
[13] Filip, R., Kubiak, K., Ziaja, W., & Sieniawski, J. (2003). The effect of microstructure on the mechanical properties of two-phase titanium alloys. *Journal of Materials Processing Technology*, *133*, 84–89.
[14] Punia, U., Kaushik, A., Garg, R. K., Chhabra, D., & Sharma, A. (2022). 3D printable biomaterials for dental restoration: A systematic review. *Materials Today: Proceedings*, *63*, 566–572.
[15] Jung, C., Straumann, L., Kessler, A., Pieles, U., & de Wild, M. (2014). Antibacterial copper deposited onto and into the oxide layer of titanium implants, in: *Jahrestagung der Deutschen Gesellschaftfür Biomaterialien*, 15 (S1). Dresden, Germany. DOI 10.1515/bnm-2014-9017
[16] Guo, S., Lu, Y. J., Wu, S. Q., Liu, L. L., He, M. J., Zhao, C. Q., Gan, Y. L., Lin, J. J., Luo, J. S., Xu, X. C., & Lin, J. X. (2017). Preliminary study on the corrosion resistance, antibacterial activity and cytotoxicity of selective-laser-melted Ti6Al4V-xCu alloys. *Materials Science and Engineering C*, *72*, 631–640.
[17] Li, Y. H., Yang, C., Zhao, H. D., Qu, S. G., Li, X. Q., & Li, Y. Y. (2014). New developments of Ti-based alloys for biomedical applications. *Materials*, *7*, 1709–1800.
[18] Brailovski, V., Prokoshkin, S., Gauthier, M., Inaekyan, K., Dubinskiy, S., Petrzhik, M., & Filonov, M. (2011). Bulk and porous metastable beta Ti-Nb-Zr(Ta) alloys for biomedical applications. *Materials Science and Engineering C*, *31*, 643–657.

[19] Dadbakhsh, S., Speirs, M., Yablokova, G., Krath, J.-P., Schrooten, J., Luyten, J., & Van Humbeeck, J. M. (2015). Materials, microstructural analysis and mechanical evaluation of Ti-45Nb produced by selective laser melting towards biomedical applications, in: *TMS 2015 144th Annual Meeting & Exhibition*, Springer, Cham, pp. 421–428. https://doi.org/10.1007/978-3-319-48127-2_53

[20] Van Hooreweder, B., Apers, Y., Lietaert, K., & Kruth, J. P. (2017). Improving the fatigue performance of porous metallic biomaterials produced by selective laser melting. *Acta Biomaterilia*, *47*, 193–202.

[21] Yadav, M., Yadav, D., Garg, R. K., Gupta, R. K., Kumar, S., & Chhabra, D. (2021). Modeling and optimization of piezoelectric energy harvesting system under dynamic loading, in: B.S. Sikarwar, B. Sundén, & Q. Wang (Eds.), *Advances in Fluid and Thermal Engineering*, Springer, Singapore, pp. 339–353. https://doi.org/10.1007/978-981-16-0159-0_30

[22] Jamshidinia, M., & Kovacevic, R. (2015). The influence of heat accumulation on the surface roughness in powder-bed additive manufacturing. *Surface Topography: Metrology and Properties*, *3*, 014003.

[23] Haslauer, C. M., Springer, J. C., Harrysson, O. L., Loboa, E. G., Monteiro-Riviere, N. A., & Marcellin-Little, D. J. (2010). In vitro biocompatibility of titanium alloy discs made using direct metal fabrication. *Medical Engineering & Physics*, *32*, 645–652.

[24] Strano, G., Hao, L., Everson, R. M., & Evans, K. E. (2013). Surface roughness analysis, modeling and prediction in selective laser melting. *Journal of Materials Processing Technology, 213, 589–597.*

[25] Pyka, G., Kerckhofs, G., Papantoniou, I., Speirs, M., Schrooten, J., & Wevers, M. (2013). Surface roughness and morphology customization of additive manufacture dopen porous Ti6Al4V structures. *Materials*, *6*, 4737–4757.

[26] Arcam, A. B. (2012). Review of present Ti-6Al-4V material properties, *ARCAM User Group Meeting*, Skafto, Sweden.

[27] Karlsson, J., Snis, A., Engqvist, H., & Lausmaa, J. (2013). Characterization and comparison of materials produced by Electron Beam Melting (EBM) of two different Ti–6Al–4V powder fractions. *Journal of Materials Processing Technology*, *213*, 2109–2118.

[28] Béraud, N., Vignat, F., Villeneuve, F., & Dendievel, R. (2014). New trajectories in Electron Beam Melting manufacturing to reduce curling effect. *Procedia CIRP*, *17*, 738–743.

[29] Facchini, L., Magalini, E., Robotti, P., & Molinari, A. (2009). Microstructure and mechanical properties of Ti-6Al-4V produced by electron beam melting of pre-alloyed powders. *Rapid Prototyping Journal*, *15*, 171–178.

[30] Frazier, W.E. (2014). Metal additive manufacturing: A review. *Journal of Materials Engineering and Performance*, *23*, 1917–1928.

[31] Lu, S. L., Tang, H. P., Ning, Y. P., Liu, N., St John, D. H., & Qian, M. (2015). Microstructure and mechanical properties of long Ti-6Al-4V rods additively manufactured by selective electron beam melting out of a deep powder bed and the effect of subsequent hot isostatic pressing. *Metallurgical and Materials Transactions A*, *46*, 3824–3834.

[32] Svensson, M., Ackelid, U., & Ab, A. (2010). Titanium alloys manufactured with electron beam melting mechanical and chemical properties. *Proceedings of Materials & Processes for Medical Devices Conference, 2009*, 189–194.

[33] Gaytan, S. M., Murr, L. E., Medina, F., Martinez, E., Lopez, M. I., & Wicker, R. B. (2009). Advanced metal powder based manufacturing of complex components by electron beam melting. *Materials Technology*, *24*, 180–190.
[34] Baudana, G., Biamino, S., Ugues, D., Lombardi, M., Fino, P., Pavese, M., & Badini, C. (2016). Titanium aluminides for aerospace and automotive applications processed by Electron Beam Melting: Contribution of Politecnico di Torino. *Metal Powder Report*, *71*, 193–199.
[35] Arcam, A. B. (2016). http://www.arcam.com/company/about-arcam/history/.
[36] Koike, M., Martinez, K., Guo, L., Chahine, G., Kovacevic, R., & Okabe, T. (2011). Evaluation of titanium alloy fabricated using electron beam melting system for dental applications. *Journal of Materials Processing Technology*, *211*, 1400–1408.
[37] Abdeen, D. H., Palmer, B. R., Campbell, R. I., & Bourell, D. (2016). Corrosion evaluation of Ti-6Al-4V parts produced with electron beam melting machine. *Rapid Prototyping Journal*, *22*, 322–329.

Chapter 9

Corrosion behaviour of additive manufactured materials and composites

Raghav Gurumoorthy Raaja
SCMS School of Engineering and Technology

Nagarajan K. Jawaharlal
Thiagarajar College of Engineering

Ashok Kumar Rajendran
SRM Madurai College for Engineering and Technology

Gibin George and Jenson Joseph Earnest
SCMS School of Engineering and Technology

CONTENTS

9.1 INTRODUCTION

Additive manufacturing (AM) or 3D printing has become one of the emerging fields for the manufacture of 3D and complex components [1]. The additive manufacturing (AM) process involves the deposition of powder metals or liquid polymers in layer-by-layer method to obtain the finished object [2,3]. This method is most widely used for the production of complex shapes which is very difficult to manufacture using conventional manufacturing process.

DOI: 10.1201/9781003430186-9

In additive manufacturing the powder materials or polymer wires are melted using laser or electron beam sources and deposited layer by layer as per the 3D design which is fed into the system [4,5]. Hence the main advantage of AM when compared with other traditional manufacturing processes are obvious; the foremost advantage is the ability to produce most complex components along with very minimal material wastage. The other major advantage is less production time for complex shapes compared to conventional process. Because of the abovementioned advantages the AM process is widely used for the production of complex aerospace components [6]. In the recent past, additive manufacturing (AM) process has improved a lot and a variety of alloys can be developed using AM process. However, it is necessary to explore the different properties of AM manufactured alloys such as mechanical, tribological and corrosion-resistant properties [7,8]. Even though there are more studies which help us understand the mechanical properties of AM manufactured metals and alloys, there are very few studies which let us know about the corrosion behaviour of alloys developed using additive manufacturing process.

The additive manufacturing of metals and its alloys is classified into two main categories, namely powder bed fusion systems and powder-fed systems. The powder-fed systems are also known as direct laser deposition (DLD) technique. In the DLD method metal powders and heating will be supplied to the substrate simultaneously [9–13]. The powder bed fusion systems are further classified into selective laser melting (SLM), selective laser sintering (SLS) and electron beam melting (EBM) [14,15].

Selective laser melting works in a bed in which metal powders are fed through the powder dispenser. The high-energy laser is rastered on to the powder bed as per the computer-aided design (CAD) so as to produce the components layer by layer. Figure 9.1 shows the schematic representation of SLM machine setup. Here the powders are fed into the building platform using recoater arm and then high-energy laser is used to raster the layer for consolidation. After successful consolidation of one layer, this process is repeated to form another layer. The approximate thickness of layers is 80 μm [16]. The entire process was carried out inside the vacuum chamber under argon or nitrogen atmosphere so as to avoid oxidation [17]. The alloys obtained through SLM techniques exhibit fine and smooth microstructure as the result of high cooling rates, which are greater than 6×10^{6}°C/s. The cooling rates also play a vital role in achieving good surface roughness (Ra) which is in the range of 9–16 μm in SLM produced alloys [18,19].

The electron beam machining (EBM) process is similar to LM process but utilizes electron beam as the source of heating as shown in Figure 9.2. The production of electron beams requires very high vacuum up to 10^{-6} torr. The vacuum chamber also helps in reduction of oxidation of metals and alloys [21]. It is to be noted that in all types of additive manufacturing process, the microstructures of the alloys obtained depend upon the different production parameters. Thereby the microstructure of the alloys controls the important properties such as mechanical, tribological and corrosion-resistant properties [22,23].

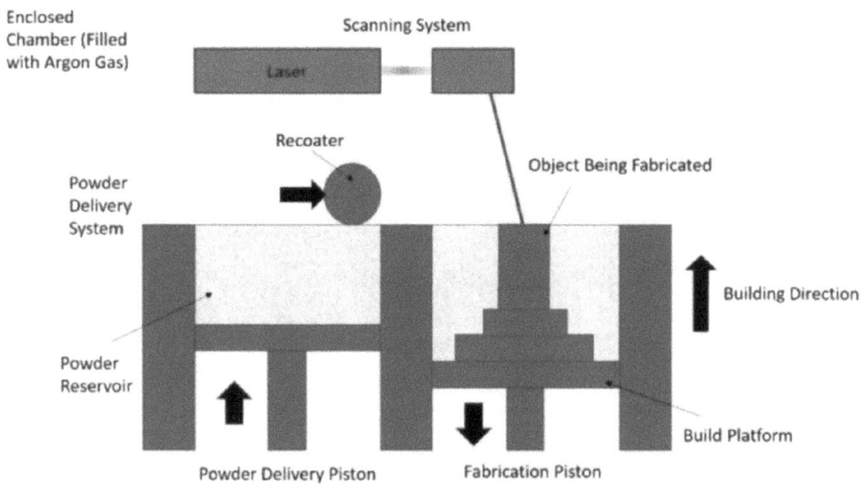

Figure 9.1 Pictorial representation of selective laser melting (SLM) process [20].

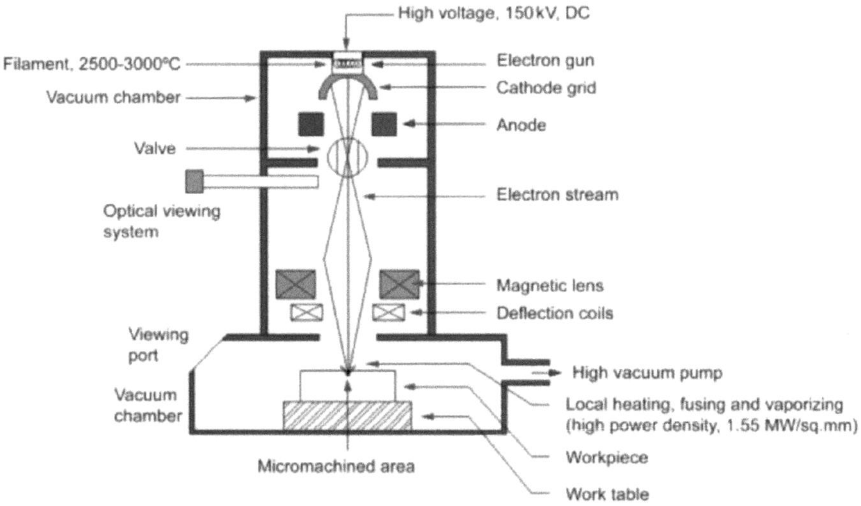

Figure 9.2 Schematic representation of electron beam machining (EBM) process [24].

In powder bed fusion such as SLM, EBM, and SLS the major parameters are the intensity of laser or electron beans, laser/electron spot size and speed of transverse motion. Other important factors which affect the alloy thermal properties are pattern of scanning, thickness of different layers and the temperature of the powder bed. On the other hand, in direct laser deposition powders of size 50–150 µm are fed into the built substrate along with heating through laser source. Also argon gas is passed into the vacuum

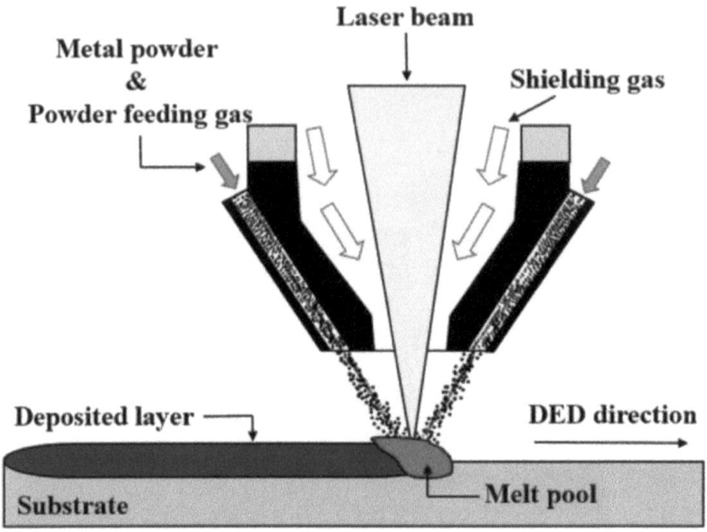

Figure 9.3 Pictorial representation of direct laser deposition (DLD) process [33].

chamber so as to maintain inert atmosphere [25,26]. The important parameters which control the microstructure of the DLD alloys are laser energy, powder size and the amount of powder injected through the nozzle. The raster speed of laser scan and laser spot size also play a vital role in refining the microstructure of the DLD alloys [27,28]. Figure 9.3 shows the pictorial representation of DLD process. It is also to be noted that SLM technique is used to machine stainless steels, titanium-based alloys, aluminium-based alloys, whereas the DLD and EBM methods are widely used to machine titanium-based alloys and stainless steels [29–32].

9.2 EFFECT OF DIFFERENT PROCESS PARAMETERS ON THE REFINEMENT OF MICROSTRUCTURES AND CORROSION-RESISTANT PROPERTIES

As discussed earlier the AM processed alloys possess a different and refined microstructure compared to other conventional manufacturing process because of the influence of different parameters involved in the AM process. In the case of SLM method, the metal powders fed into the powder bed undergo intensive heating normally greater than 2,000°C which is then followed by very fast-paced solidification process [34]. Due to this heating many thermal cycles are involved as the result of heat transfer between powder particles as well as with the surrounding. Due to the high-temperature exposure and very fast cooling rate along with the heat transfer, thermal cycles result in improved and refined microstructures. But as the result of this rapid heating and cooling there may be formation of few defects such as cracks, surface roughness and porosity [35,36]. Hence these defects might play a vital role in determining mechanical, tribological and corrosion resistance properties of AM manufactured alloys.

9.2.1 Porosity of AM Manufactured alloys

There are few studies which explain the effect of laser energy and scanning rate on the porosity of alloys developed through AM process. Selective laser melting (SLM) method was employed by Shang et al. [37] to study the effect of laser scanning rate on the porosity of 316L stainless steel specimens. In this study the laser energy was kept constant at 195 W, whereas the scanning rate of laser beam is varied in the order between 700 and 1,082 mm/s. The results show that porosity is directly proportional to the scanning rate. The increase in scanning rate results in improper melting which in turn affects the porosity of the alloys. The porosity which occurs in AM manufactured alloys may influence the corrosion-resistant properties of the metals and alloys [38]. The porosity of the additive manufactured alloys, especially Selective Laser Melting method, is classified into two types. One type of porosity can be found at the surrounding of improperly melted powders and another type of porosity is due to the presence of gases in between powder particles during atomization (gas) process, therefore by controlling the machining parameters we can reduce the porosity of additive manufactured alloys which in turn can reduce the corrosion behaviours of the additive manufactured alloys [39–41]. Another more accurate way to study the influence of machining parameters on porosity of additive manufactured of powders, also known as volumetric energy density (E_V) which is calculated by equation (9.1).

$$E_V = \frac{e}{rdt} \tag{9.1}$$

where e is the energy of laser,
r is scanning rate,
d is hatch diameter,
t is thickness of powder layer.

The energy density of laser plays a vital role in controlling the porosity of additive manufactured alloys [42,43]. The Ti6Al4V alloys were produced with porosity less than 0.1% by Hang et al. at an energy density of 120 J/mm^3. The laser density of 105 J/mm^3 was enough to produce 316L stainless steels with porosity of approximately 0.3% using Selective Laser Melting (SLM) process [36]. It is also to be noted that laser energy density is not the only factor reducing the porosity of additive manufactured alloys. Other parameters such as diameter of laser, scanning rate and hatch style are also very important in controlling the porosity of the additive manufactured (AM)

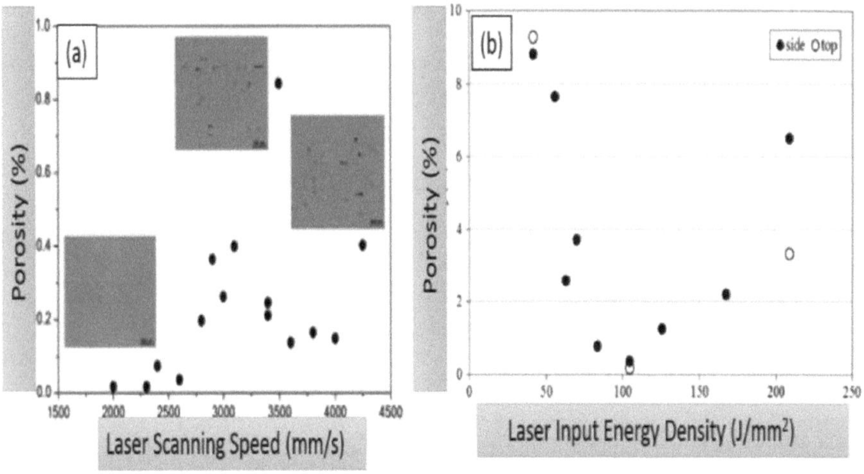

Figure 9.4 Graphical representation (a) porosity vs laser scanning speed [44] (b) porosity vs laser energy density of various SLM manufactured 316L stainless steel [45].

alloys as shown in Figure 9.4. Hence more studies should be carried out to find out the exact relationship between process parameters and porosity.

Porosity is one of the major causes for pitting corrosion in additive manufactured components, especially in selective laser melting (SLM) of 316L stainless steel in acidic media. The corrosion normally initiates at the location of pores [38]. Schaller et al. [46] employed electrochemical corrosion analysis to study the corrosion behaviour of 17-4 PH stainless steel manufactured using SLM process. The results showed that the porosity larger than 50 μm results in pitting corrosion, whereas when the porosity was around 10 μm there is no pitting corrosion. But when the normality of acids was increased and also when highly acidic acids such as sulphuric acid was utilized there is evidence of pitting corrosion in 316L stainless steels [45]. The size and shape of the pores also have a significant effect on the pitting corrosion performance; moreover, the pores which are irregular in shape will corrode easily due to the accumulation of ions at the edges and the corners. However, there are very few studies based on size distribution of pores, and hence more studies are required to be carried out to understand the effect of porosity, which includes size of the pores and *l*/*d* (aspect ratio) of the pores, on the corrosion behaviour of additive manufactured components.

9.2.2 Surface roughness of AM processed alloys

The selective laser melting (SLM) manufactured components have very high surface roughness (Ra). Wang et al. [47] reported that surface roughness (Ra) of the metals and its alloys are in the range of 10–30 μm, which is very much higher than that of surface roughness produced by conventional process such as milling. He also found that the energy density of laser (ω) plays a very important role in deciding the surface roughness of SLM manufactured components. He reported that when the ω value is around 100–160 J/mm^3, the surface roughness is low around 10 μm, but when the ω values is reduced to 70 J/mm^3, the surface roughness increases up to 15 μm.

The main reasons for high surface roughness in SLM process are the evaporation and Marangoni force that exists because of melting of powders. The expansion of entrapped gases stops the flow of melt and thereby increases the melt pool which is highly unstable. This melt pool increases the surface roughness (Ra). When the layers of powders are thick, more gas expansion takes place. Hence the surface roughness can be reduced by decreasing the powder layer thickness [45,48]. However, by decreasing the layer thickness, the overall time for completing the machining increases rapidly. The improper melting of powders and formation of metal droplets also known as balling effect are the major reason for increase in surface roughness [49]. When the power of laser is very low the metallic powders are not completely melted and few solid particles stick on to the surface of the solidified components. Hence the increase in laser power can increase the melt rate and thereby increase wettability which in turn reduces the balling phenomenon [50]. Thus surface roughness can be reduced when the energy density of laser is high enough to melt the powder particles as shown in Figure 9.5. It is also to be noted that if the laser intensity is very high it can also reduce the surface finish of the components.

Normally the morphology of the surface is very important for the corrosion resistance properties. The corrosion rate increases with increase in surface roughness of the additive manufactured alloys such as copper, Mg and Al-based alloys [51,52]. Therefore, improving the surface finish by overcoming all defects was the major challenge for 3D printed or additive manufactured components. Thus many research works should be carried out to study the effects of post-processing surface treatments on additive manufactured components.

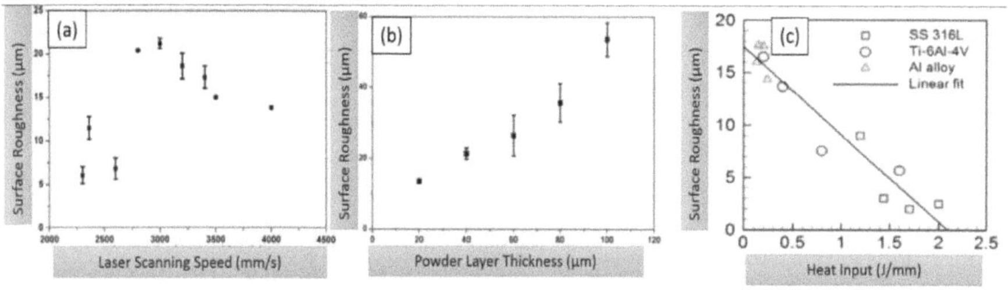

Figure 9.5 The relationship between surface roughness and (a) laser scanning speed, (b) powder layer thickness [44] and (c) heat input [45].

9.3 CORROSION BEHAVIOUR OF ADDITIVE MANUFACTURED ALLOYS

In this section various types of alloys manufactured through additive manufacturing technique are summarized and their corrosion behaviour is explored so as to provide an insight for young researchers who are trying to study the corrosion behaviour of additive manufactured alloys. Recent developments in additive manufacturing (AM) processes have made it versatile and a wide variety of metal alloys can be now prepared using additive manufacturing methods. The common types of alloys are titanium-based alloys, iron-based alloys and aluminium-based alloys.

9.3.1 Titanium-based additive manufactured alloys

Titanium-based alloys have very large industrial applications because of their properties. But a major disadvantage is their machining cost and a very large machining time when machined using conventional manufacturing processes. Hence titanium and its alloys are widely considered for manufacturing through additive manufacturing (AM) process [53–56]. Dehoff et al. [57] reported nearly 50% reduction in production cost for the titanium alloy-based engine bracket manufactured using AM process. Ti6Al4V alloy was one of the titanium alloys which was widely utilized for the production of biomedical, dental and automobile applications. It is also reported that Ti6Al4V alloys fabricated using Selective Laser Melting (SLM) have very minimal pitting corrosion, approximately around 150 mV, in sodium chloride solution. It also exhibits passivation curves which is the measure of corrosion resistance [58]. This improvement in corrosion resistance is due to the presence of a′-martensite as shown in Figure 9.6. The rapid cooling process was the reason for the formation of a′-martensite and it also possesses β-grains. Normally in SLM manufactured Ti6Al4V alloys

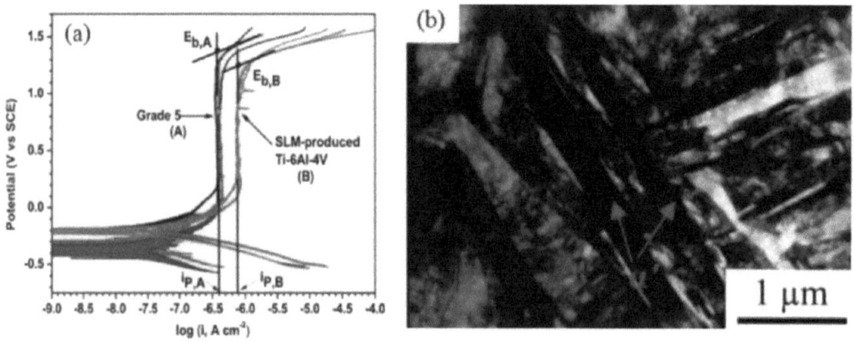

Figure 9.6 (a) Potentiodynamic polarization curves (tafel curves) of SLM manufactured Ti6Al4V alloy and Grade 5 alloy in NaCl solution (3.5 wt.%) [58]. (b) TEM image of SLM manufactured Ti6Al4V alloy [61].

the β-phase contains more vanadium presence along with oxides; hence, the β-phase is more stable compared to the α-phase. The stable β-phase plays a vital role in increasing the corrosion resistance of Ti6Al4V alloys [59]. But the percentage of β-phase present in additive manufactured Ti6Al4V alloys is very less compared to conventional manufacturing process. So it can be concluded that the SLM manufactured titanium alloys show very poor corrosion resistance [60].

9.3.2 Aluminium-based AM alloys

The selective laser melting (SLM) technique was broadly utilized for the manufacture of various aluminium-based alloys such as Al-Zn, Al-12Si, Al-50Si, Al-Cu and Al-10Si-M alloys [62–64]. Among these alloys Al-10Si-Mg alloys were widely studied by the researchers [65,66]. It is also noted as shown in Figure 9.7a and b. The corrosion potential of Al and Si particles differs, that is, Si has higher corrosion potential compared to that of Al which has low corrosion potential. This difference in corrosion potential leads to galvanic corrosion as shown in Figure 9.7c. Hence to overcome

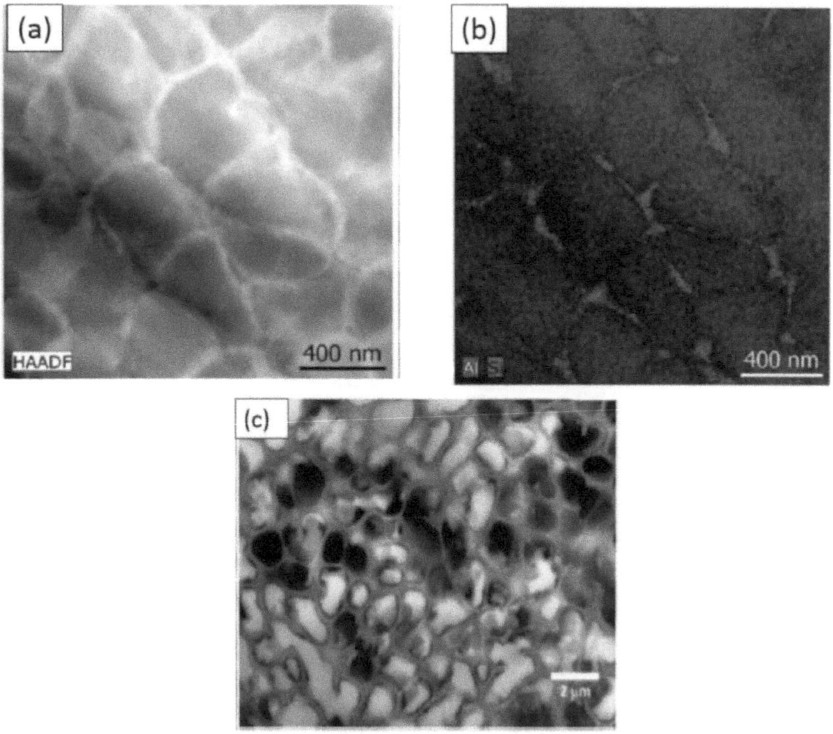

Figure 9.7 (a)STEM image of SLM manufactured Al-10Si-Mg alloy, (b) EDS mapping of Al and Si [68], (c) SEM image of corroded surface [45].

this galvanic corrosion the SLM manufactured Al-10Si-Mg alloys are subjected to heat treatment so as to improve the bonding of alloys and also to enable the formation of intermetallics, which thereby improve corrosion resistance [67].

9.3.3 Iron-based AM alloys

Stainless steels (austenitic) such as 316L and 304L will have austenitic phase when machined using SLM process, whereas only α-phase is formed if it is machined using Direct Laser Deposition (DLD) process [69–71]. The dislocation in grains also plays an important role in improvement of hardness of the alloy steel. It is also to be noted that nanoscale oxide formation has influence in deciding the mechanical and corrosion resistance of iron-based alloys [72]. Some studies also show that there is not much impact of porosity in corrosion behaviour of 316L stainless steel manufactured using SLM process as shown in Figure 9.8. Sander et al. [39] reported the corrosion behaviour of SLM-fabricated 316L stainless steel. They fabricated the 316L specimens at different scan rates and laser energies. The results exhibit that the scan rate and laser energy do not have any effect on corrosion resistance of 316L stainless steel, whereas the increase in porosity due to the faster scan rate and improper melting resulted in reduction in passivation potential, and hence increase in corrosion rate of 316L stainless steel samples as shown in Figure 9.9. The corrosion analysis of normal 316L SS samples and

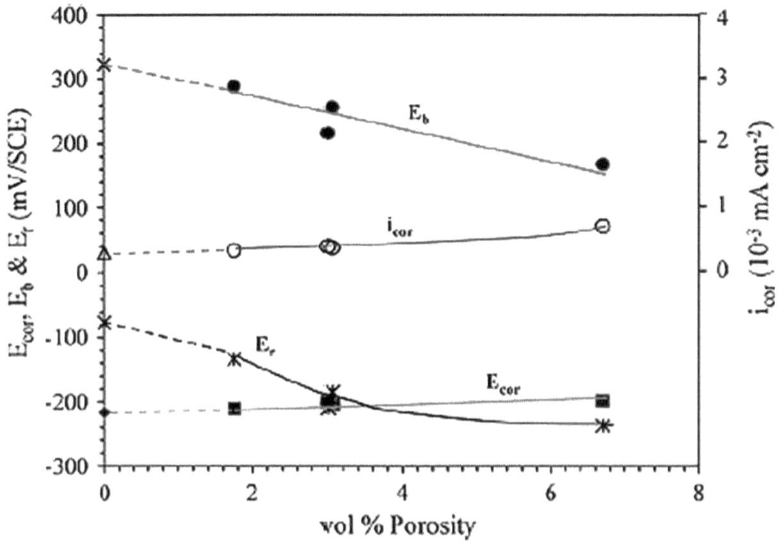

Figure 9.8 Relationship between Vol % of porosity and corrosion potential (E_{corr}), repassivation potential (E_r), breakdown potential (E_b) and corrosion current density (i_{Corr}) of SLM manufactured alloys [74].

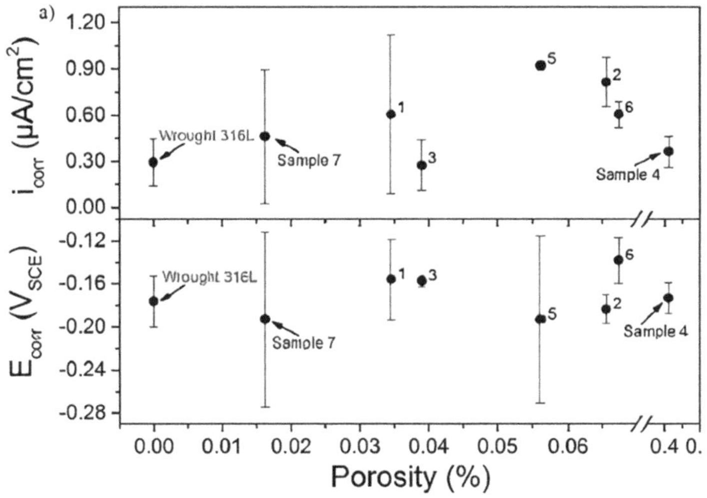

Figure 9.9 Porosity vs corrosion potential (E_{corr}) and corrosion current density (i_{Corr}) [39].

heat-treated SLM manufactured 316L SS was carried out by Hemmasian Ettefagh et al. [73] who reported that the heat treatment process has eliminated the residual stress, thereby increasing the corrosion potential. The corrosion behaviour of Laser Powder Bed Fusion (LPBF) manufactured 316L samples in 0.1 M HCl solution was studied by Trelewicz et al. [4]. Corrosion current density of LPBF manufactured 316L SS samples was much higher compared to wrought samples, when studied using potentiodynamic polarization test. The main reason for the decrease in corrosion resistance was the microstructure of LPBF manufactured 316L SS.

9.4 SCOPE FOR FUTURE WORKS

- The review gives an insight into the corrosion behaviour of AM alloys, many research articles are explored and their findings are reported. Most of the studies are not systematic and concentrate on one specific area and lack in-depth analysis of corrosion phenomena. Hence based on the studies, the following gaps have been identified:
- Corrosion analysis of additive manufactured alloys was carried out in different acidic and brine media and also with various concentration and pH levels, and hence it is very difficult to compare studies on one alloy with other.
- In the same way, corrosion analysis of additive manufactured alloys was carried out using different corrosion studies such as weight loss method, potentiodynamic polarization method, electrochemical

impedance analysis etc.; hence, it is very difficult to compare studies on one alloy with other.

- The lack of standards for carrying out corrosion test was also one of the important factors to be addressed. Many studies compared wrought or cast irons with AM counterparts but few studies show the difference in properties between cast and AM alloys. Similarly, few studies compared only the different types of alloys produced by AM process but not discussed the alloys prepared by other methods. Hence the conclusions derived from the study may not be conclusive.
- There are very few studies on the exposure of AM alloys to nuclear radiation and also there are very minimal studies on the effect of gases especially hydrogen in AM manufactured alloys.
- Similarly, the corrosion analysis of AM manufactured alloys is limited to materials such as titanium alloys, aluminium alloys and iron-based alloys. But AM process can be utilized to even wide range of metals and their alloys.
- The work has a deficiency in parting the variables, i.e., it is very tedious to show trends from changes in porosity and also chemical factors simultaneously.

9.5 CONCLUSION

An outline of the recent status of some metal matrix alloys manufactured employing additive manufacturing was presented with a focus on correlating the relationship between the defects caused due to the microstructure and their effects on corrosion resistance properties. We can conclude that the high temperature evolved during manufacturing using SLM results in high dislocation densities and refinement of grain size, which in turn improves the tensile strength. Corrosion properties depend on the formation of alpha and beta phases and their structure compared to that of alloys fabricated using conventional methods. In the coming days, many materials can be manufactured using AM process, and hence the optimization of various parameters involved such as laser density, raster velocity and size of powder particles is very important in reducing the surface roughness, porosity and also to increase the strength of the alloys. However, the intrinsic relationship between the microstructural characteristics and the corrosion behaviour of the AM-fabricated components should be actively focused as well. The qualities of the input powder material and their effects on the fabrication process should be the first focus of research. It is critical to include the following three major components when describing a powder: particle microstructure, particle morphology and particle chemistry [75]. The focus of the current research is on the morphological characterization of powders and their impact on the characteristics of manufactured parts.

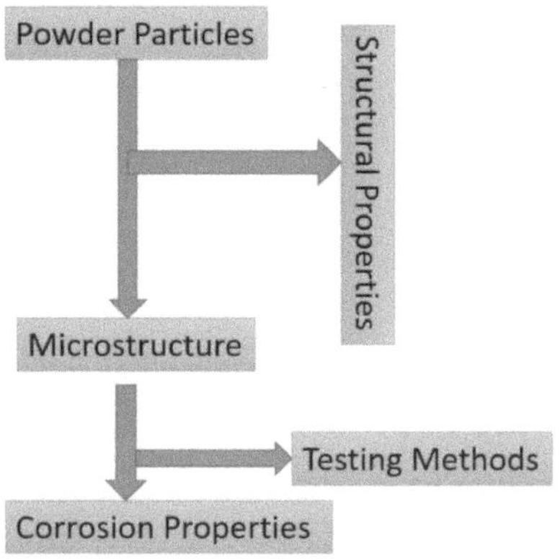

Figure 9.10 Pictorial representation of relationship between powder, microstructure and corrosion properties of AM manufactured alloys.

The mechanical and anti-corrosive qualities of the final consolidated components may be impacted by whether the feedstock powders are argon or nitrogen atomized, and whether the construction chambers are in argon or nitrogen atmosphere [76]. Figure 9.10 is a pictorial representation of AM-produced component powder, microstructure and related corrosion behaviour. The relationship between important structural properties and corrosion resistance must be established. For instance, typical MnS additions formed in wrought 316L SS were exchanged by Mn-Si oxides of nano regime in the SLM manufactured components, which reduces the vulnerability to pitting [77] and also the microstructural irregularities in the SLM manufactured components which otherwise lead to diverse growth rates (SCC) [78]. For the corrosion testing methodologies, there's an egregious lack of norms for which standardized experimentations are enforced, and presently, a wide range of distinct corrosion experimentation techniques (weight loss method, impedance analysis, potentiodynamic polarization) are indeed very difficult to compare. The standardization of testing methods and procedures as formulated by some professional bodies will be a solution for standardization problems. In general, the defects in the SLMed parts (such as pores and MPBs) usually comprise the corrosion resistance; therefore, a heat treatment process combining the hot isostatic pressing should be carried out to homogenize the composition and refine the microstructure, thereby reducing the porosity of the alloys. Thus, further exploration in this area is also warranted. Another post-processing method involves surface treatment, but surface treatment has lot of challenges which have to be addressed with the SLM manufactured metals. The various other techniques such as alkali-acid heat treatment, sandblasting, electrochemical etching and electrochemical deposition can be considered according to the properties of the raw material used. Conversely, the porous materials manufactured using AM process cannot be easily surface modified compared to solid materials. Therefore, we have very little choice for selecting the methods of manufacture. Therefore, additional exploitation in this area is also required.

REFERENCES

[1] D. R. Eyers and A. T. Potter, "Industrial additive manufacturing: A manufacturing systems perspective," *Computers in Industry*, vol. 92–93, pp. 208–218, 2017, doi: 10.1016/j.compind.2017.08.002.
[2] D. L. Bourell, "Perspectives on additive manufacturing," *Annual Review of Materials Research*, vol. 46, no. 1, pp. 1–18, 2016, doi: 10.1146/annurev-matsci-070115-031606.
[3] D. W. Rosen, "A review of synthesis methods for additive manufacturing," *Visual and Physical Prototyping*, vol. 11, no. 4, pp. 305–317, 2016, doi: 10.1080/17452759.2016.1240208.
[4] J. R. Trelewicz, G. P. Halada, O. K. Donaldson, and G. Manogharan, "Microstructure and corrosion resistance of laser additively manufactured 316L stainless steel," *JOM*, vol. 68, no. 3, pp. 850–859, 2016, doi: 10.1007/s11837-016-1822-4.
[5] M. Attaran, "The rise of 3-D printing: The advantages of additive manufacturing over traditional manufacturing," *Business Horizons*, vol. 60, no. 5, pp. 677–688, 2017, doi: 10.1016/j.bushor.2017.05.011.
[6] P. Han, "Additive design and manufacturing of jet engine parts," *Engineering*, vol. 3, no. 5, pp. 648–652, 2017, doi: 10.1016/J.ENG.2017.05.017.
[7] J. K. Telford, "A brief introduction to design of experiments," *Johns Hopkins APL Technical Digest*, vol. 27, no. 3, p. 9, 2007.
[8] J. Gong, Y. Li, Z. Hu, Z. Zhou, and Y. Deng, "Ultrasensitive NH_3 gas sensor from polyaniline nanograin enchased TiO_2 fibers," *The Journal of Physical Chemistry C*, vol. 114, no. 21, pp. 9970–9974, 2010, doi: 10.1021/jp100685r.
[9] Y. Hu and W. Cong, "A review on laser deposition-additive manufacturing of ceramics and ceramic reinforced metal matrix composites," *Ceramics International*, vol. 44, no. 17, pp. 20599–20612, 2018, doi: 10.1016/j.ceramint.2018.08.083.
[10] M. K. Mallik, C. S. Rao, and V. V. S. K. Rao, "Effect of heat treatment on hardness and wear behavior of weld deposited Co-Cr-Mo alloy," *Matéria (Rio de Janeiro)*, vol. 20, no. 2, pp. 544–549, 2015, doi: 10.1590/S1517-707620150002.0054.
[11] Y. Shi, Z. Lu, Y. Ren, and G. Yang, "Microstructure and tensile properties of laser engineered net shaped reduced activation ferritic/martensitic steel," *Materials Characterization*, vol. 144, pp. 554–562, 2018, doi: 10.1016/j.matchar.2018.08.010.
[12] Y. Li, Y. Hu, W. Cong, L. Zhi, and Z. Guo, "Additive manufacturing of alumina using laser engineered net shaping: Effects of deposition variables," *Ceramics International*, vol. 43, no. 10, pp. 7768–7775, 2017, doi: 10.1016/j.ceramint.2017.03.085.

[13] M. Ziętala et al., "The microstructure, mechanical properties and corrosion resistance of 316L stainless steel fabricated using laser engineered net shaping," *Materials Science and Engineering: A*, vol. 677, pp. 1–10, 2016, doi: 10.1016/j.msea.2016.09.028.
[14] L. E. Murr et al., "Metal fabrication by additive manufacturing using laser and electron beam melting technologies," *Journal of Materials Science & Technology*, vol. 28, no. 1, pp. 1–14, 2012, doi: 10.1016/S1005-0302(12)60016-4.
[15] S. Singh, V. S. Sharma, and A. Sachdeva, "Progress in selective laser sintering using metallic powders: A review," *Materials Science and Technology*, vol. 32, no. 8, pp. 760–772, May 2016, doi: 10.1179/1743284715Y.0000000136.
[16] R. S. Keefe, and P. D. Harvey. Cognitive impairment in schizophrenia. *Handbook of Experimental Pharmacology*, 213, pp. 11–37, 2012. doi: 10.1007/978-3-642-25758-2_2.
[17] P. K. Gokuldoss, S. Kolla, and J. Eckert, "Additive manufacturing processes: Selective laser melting, electron beam melting and binder jetting—Selection guidelines," *Materials*, vol. 10, no. 6, 2017, doi: 10.3390/ma10060672.
[18] J. Suryawanshi, K. G. Prashanth, S. Scudino, J. Eckert, O. Prakash, and U. Ramamurty, "Simultaneous enhancements of strength and toughness in an Al-12Si alloy synthesized using selective laser melting," *Acta Materialia*, vol. 115, pp. 285–294, 2016, doi: 10.1016/j.actamat.2016.06.009.
[19] T. DebRoy et al., "Additive manufacturing of metallic components – Process, structure and properties," *Progress in Materials Science*, vol. 92, pp. 112–224, 2018, doi: 10.1016/j.pmatsci.2017.10.001.
[20] L. Jiao, Z. Chua, S. Moon, J. Song, G. Bi, and H. Zheng, "Femtosecond laser produced hydrophobic hierarchical structures on additive manufacturing parts," *Nanomaterials*, vol. 8, no. 8, p. 601, 2018, doi: 10.3390/nano8080601.
[21] B. Vayre, F. Vignat, and F. Villeneuve, "Metallic additive manufacturing: state-of-the-art review and prospects," *Mechanics & Industry*, vol. 13, no. 2, pp. 89–96, 2012, doi: 10.1051/meca/2012003.
[22] H. Gu, H. Gong, D. Pal, K. Rafi, T. L. Starr, and B. E. Stucker, "Influences of energy density on porosity and microstructure of selective laser melted 17-4PH stainless steel," 2013. http://dx.doi.org/10.26153/tsw/15572
[23] G. Sander et al., "Corrosion of additively manufactured alloys: A review," *Corrosion*, vol. 74, no. 12, pp. 1318–1350, 2018, doi: 10.5006/2926.
[24] B. Bhattacharyya, "Chapter 1 - Introduction," in *Electrochemical Micromachining for Nanofabrication, MEMS and Nanotechnology*, B. Bhattacharyya, Ed. William Andrew Publishing, 2015, pp. 1–23. doi: 10.1016/B978-0-323-32737-4.00001-3.
[25] R. Koike et al., "Evaluation for mechanical characteristics of Inconel625–SUS316L joint produced with direct energy deposition," *Procedia Manufacturing*, vol. 14, pp. 105–110, 2017, doi: 10.1016/j.promfg.2017.11.012.
[26] Q. Chao, T. Guo, T. Jarvis, X. Wu, P. Hodgson, and D. Fabijanic, "Direct laser deposition cladding of AlxCoCrFeNi high entropy alloys on a high-temperature stainless steel," *Surface and Coatings Technology*, vol. 332, pp. 440–451, 2017, doi: 10.1016/j.surfcoat.2017.09.072.
[27] L. Song, V. Bagavath-Singh, B. Dutta, and J. Mazumder, "Control of melt pool temperature and deposition height during direct metal deposition process,"

The International Journal of Advanced Manufacturing Technology, vol. 58, no. 1, pp. 247–256, 2012, doi: 10.1007/s00170-011-3395-2.
[28] N. Shamsaei, A. Yadollahi, L. Bian, and S. M. Thompson, "An overview of Direct Laser Deposition for additive manufacturing; Part II: Mechanical behavior, process parameter optimization and control," *Additive Manufacturing*, vol. 8, pp. 12–35, 2015, doi: 10.1016/j.addma.2015.07.002.
[29] K. Guan, Z. Wang, M. Gao, X. Li, and X. Zeng, "Effects of processing parameters on tensile properties of selective laser melted 304 stainless steel," *Materials & Design*, vol. 50, pp. 581–586, 2013, doi: 10.1016/j.matdes.2013.03.056.
[30] W. Xu, E. W. Lui, A. Pateras, M. Qian, and M. Brandt, "In situ tailoring microstructure in additively manufactured Ti-6Al-4V for superior mechanical performance," *Acta Materialia*, vol. 125, pp. 390–400, 2017, doi: 10.1016/j.actamat.2016.12.027.
[31] W. H. Kan et al., "A critical review on the effects of process-induced porosity on the mechanical properties of alloys fabricated by laser powder bed fusion," *Journal of Materials Science*, vol. 57, pp. 9819–9865, 2022, doi: 10.1007/s10853-022-06990-7.
[32] P. Wang, H. C. Li, K. G. Prashanth, J. Eckert, and S. Scudino, "Selective laser melting of Al-Zn-Mg-Cu: Heat treatment, microstructure and mechanical properties," *Journal of Alloys and Compounds*, vol. 707, pp. 287–290, 2017, doi: 10.1016/j.jallcom.2016.11.210.
[33] J.-S. Lim, W.-J. Oh, C.-M. Lee, and D.-H. Kim, "Selection of effective manufacturing conditions for directed energy deposition process using machine learning methods," *Scientific Reports*, vol. 11, no. 1, p. 24169, 2021, doi: 10.1038/s41598-021-03622-z.
[34] E. Liverani, S. Toschi, L. Ceschini, and A. Fortunato, "Effect of selective laser melting (SLM) process parameters on microstructure and mechanical properties of 316L austenitic stainless steel," *Journal of Materials Processing Technology*, vol. 249, pp. 255–263, 2017, doi: 10.1016/j.jmatprotec.2017.05.042.
[35] J. H. Martin, B. D. Yahata, J. M. Hundley, J. A. Mayer, T. A. Schaedler, and T. M. Pollock, "3D printing of high-strength aluminium alloys," *Nature*, vol. 549, no. 7672, pp. 365–369, 2017, doi: 10.1038/nature23894.
[36] J. A. Cherry, H. M. Davies, S. Mehmood, N. P. Lavery, S. G. R. Brown, and J. Sienz, "Investigation into the effect of process parameters on microstructural and physical properties of 316L stainless steel parts by selective laser melting," *The International Journal of Advanced Manufacturing Technology*, vol. 76, no. 5–8, pp. 869–879, 2015, doi: 10.1007/s00170-014-6297-2.
[37] Y. Shang, Y. Yuan, D. Li, Y. Li, and J. Chen, "Effects of scanning speed on in vitro biocompatibility of 316L stainless steel parts elaborated by selective laser melting," *The International Journal of Advanced Manufacturing Technology*, vol. 92, no. 9, pp. 4379–4385, 2017, doi: 10.1007/s00170-017-0525-5.
[38] E. Otero, A. Pardo, M. V. Utrilla, E. Sáenz, and F. J. Perez, "Influence of microstructure on the corrosion resistance of AISI type 304L and type 316L sintered stainless steels exposed to ferric chloride solution," *Materials Characterization*, vol. 35, no. 3, pp. 145–151, 1995, doi: 10.1016/1044-5803(95)00099-2.

[39] G. Sander et al., "On the corrosion and metastable pitting characteristics of 316L stainless steel produced by selective laser melting," *Journal of the Electrochemical Society*, vol. 164, no. 6, pp. C250–C257, 2017, doi: 10.1149/2.0551706jes.
[40] A. L. Maximenko and E. A. Olevsky, "Pore filling during selective laser melting - assisted additive manufacturing of composites," *Scripta Materialia*, vol. 149, pp. 75–78, 2018, doi: 10.1016/j.scriptamat.2018.02.015.
[41] R. Laquai, B. R. Müller, G. Kasperovich, J. Haubrich, G. Requena, and G. Bruno, "X-ray refraction distinguishes unprocessed powder from empty pores in selective laser melting Ti-6Al-4V," *Materials Research Letters*, vol. 6, no. 2, pp. 130–135, 2018, doi: 10.1080/21663831.2017.1409288.
[42] S. M. Yusuf and N. Gao, "Influence of energy density on metallurgy and properties in metal additive manufacturing," *Materials Science and Technology*, vol. 33, no. 11, pp. 1269–1289, 2017, doi: 10.1080/02670836.2017.1289444.
[43] J. Kluczyński, L. Śnieżek, K. Grzelak, and J. Mierzyński, "The influence of exposure energy density on porosity and microhardness of the SLM additive manufactured elements," *Materials*, vol. 11, no. 11, p. 2304, 2018, doi: 10.3390/ma11112304.
[44] C. Qiu, C. Panwisawas, M. Ward, H. C. Basoalto, J. W. Brooks, and M. M. Attallah, "On the role of melt flow into the surface structure and porosity development during selective laser melting," *Acta Materialia*, vol. 96, pp. 72–79, 2015, doi: 10.1016/j.actamat.2015.06.004.
[45] D. Kong, C. Dong, X. Ni, and X. Li, "Corrosion of metallic materials fabricated by selective laser melting," *npj Materials Degradation*, vol. 3, no. 1, p. 24, 2019, doi: 10.1038/s41529-019-0086-1.
[46] R. F. Schaller, J. M. Taylor, J. Rodelas, and E. J. Schindelholz, "Corrosion properties of powder bed fusion additively manufactured 17-4 PH stainless steel," *Corrosion*, vol. 73, no. 7, pp. 796–807, 2017, doi: 10.5006/2365.
[47] D. Wang, Y. Liu, Y. Yang, and D. Xiao, "Theoretical and experimental study on surface roughness of 316L stainless steel metal parts obtained through selective laser melting," *Rapid Prototyping Journal*, vol. 22, no. 4, pp. 706–716, 2016, doi: 10.1108/RPJ-06-2015-0078.
[48] Y. Tian, D. Tomus, P. Rometsch, and X. Wu, "Influences of processing parameters on surface roughness of Hastelloy X produced by selective laser melting," *Additive Manufacturing*, vol. 13, pp. 103–112, 2017, doi: 10.1016/j.addma.2016.10.010.
[49] N. T. Aboulkhair, I. Maskery, C. Tuck, I. Ashcroft, and N. M. Everitt, "On the formation of AlSi10Mg single tracks and layers in selective laser melting: Microstructure and nano-mechanical properties," *Journal of Materials Processing Technology*, vol. 230, pp. 88–98, 2016, doi: 10.1016/j.jmatprotec.2015.11.016.
[50] J. P. Kruth, L. Froyen, J. Van Vaerenbergh, P. Mercelis, M. Rombouts, and B. Lauwers, "Selective laser melting of iron-based powder," *Journal of Materials Processing Technology*, vol. 149, no. 1, pp. 616–622, 2004, doi: 10.1016/j.jmatprotec.2003.11.051.
[51] R. Walter and M. B. Kannan, "Influence of surface roughness on the corrosion behaviour of magnesium alloy," *Materials & Design*, vol. 32, no. 4, pp. 2350–2354, 2011, doi: 10.1016/j.matdes.2010.12.016.

[52] D. Kong et al., "Surface monitoring for pitting evolution into uniform corrosion on Cu-Ni-Zn ternary alloy in alkaline chloride solution: Ex-situ LCM and in-situ SECM," *Applied Surface Science*, vol. 440, pp. 245–257, 2018, doi: 10.1016/j.apsusc.2018.01.116.
[53] S. Pal, M. Finšgar, T. Bončina, G. Lojen, T. Brajlih, and I. Drstvenšek, "Effect of surface powder particles and morphologies on corrosion of Ti-6Al-4 V fabricated with different energy densities in selective laser melting," *Materials & Design*, vol. 211, p. 110184, 2021, doi: 10.1016/j.matdes.2021.110184.
[54] T. Majumdar, N. Eisenstein, J. E. Frith, S. C. Cox, and N. Birbilis, "Additive manufacturing of titanium alloys for orthopedic applications: A materials science viewpoint," *Advanced Engineering Materials*, vol. 20, no. 9, p. 1800172, 2018, doi: 10.1002/adem.201800172.
[55] C. Phutela, N. T. Aboulkhair, C. J. Tuck, and I. Ashcroft, "The effects of feature sizes in selectively laser melted Ti-6Al-4V parts on the validity of optimised process parameters," *Materials*, vol. 13, no. 1, p. 117, 2020, https://doi.org/10.3390/ma13010117.
[56] L.-C. Zhang, H. Attar, M. Calin, and J. Eckert, "Review on manufacture by selective laser melting and properties of titanium based materials for biomedical applications," *Materials Technology*, vol. 31, no. 2, pp. 66–76, 2016, doi: 10.1179/1753555715Y.0000000076.
[57] R. Dehoff et al., "Case study: Additive manufacturing of aerospace brackets," *Advanced Materials and Processes*, vol. 171, no. 3, pp. 19–22.
[58] N. Dai, L.-C. Zhang, J. Zhang, Q. Chen, and M. Wu, "Corrosion behavior of selective laser melted Ti-6Al-4V alloy in NaCl solution," *Corrosion Science*, vol. 102, pp. 484–489, 2016, doi: 10.1016/j.corsci.2015.10.041.
[59] J.-R. Chen and W.-T. Tsai, "In situ corrosion monitoring of Ti–6Al–4V alloy in H_2SO_4/HCl mixed solution using electrochemical AFM," *Electrochimica Acta*, vol. 56, no. 4, pp. 1746–1751, 2011, doi: 10.1016/j.electacta.2010.10.024.
[60] F. Toptan et al., "Corrosion and tribocorrosion behaviour of Ti6Al4V produced by selective laser melting and hot pressing in comparison with the commercial alloy," *Journal of Materials Processing Technology*, vol. 266, pp. 239–245, 2019, doi: 10.1016/j.jmatprotec.2018.11.008.
[61] X. Zhao et al., "Comparison of the microstructures and mechanical properties of Ti–6Al–4V fabricated by selective laser melting and electron beam melting," *Materials & Design*, vol. 95, pp. 21–31, 2016, doi: 10.1016/j.matdes.2015.12.135.
[62] T. Gu, B. Chen, C. Tan, and J. Feng, "Microstructure evolution and mechanical properties of laser additive manufacturing of high strength Al-Cu-Mg alloy," *Optics & Laser Technology*, vol. 112, pp. 140–150, 2019, doi: 10.1016/j.optlastec.2018.11.008.
[63] M. L. Montero-Sistiaga et al., "Changing the alloy composition of Al7075 for better processability by selective laser melting," *Journal of Materials Processing Technology*, vol. 238, pp. 437–445, 2016, doi: 10.1016/j.jmatprotec.2016.08.003.
[64] X. Ai et al., "A high Fe-containing AlSi12 alloy fabricated by laser powder bed fusion," *Journal of Materials Research and Technology*, vol. 18, pp. 4513–4521, 2022, doi: 10.1016/j.jmrt.2022.04.008.

[65] A. Zakay and E. Aghion, "Effect of post-heat treatment on the corrosion behavior of AlSi10Mg alloy produced by additive manufacturing," *JOM*, vol. 71, no. 3, pp. 1150–1157, 2019, doi: 10.1007/s11837-018-3298-x.
[66] J. de Damborenea, A. Conde, M. Gardon, G. A. Ravi, and M. A. Arenas, "Effect of growth orientation and heat treatment on the corrosion properties of AlSi10Mg alloy produced by additive manufacturing," *Journal of Materials Research and Technology*, vol. 18, pp. 5325–5336, 2022, doi: 10.1016/j.jmrt.2022.05.021.
[67] R. I. Revilla and I. De Graeve, "Influence of Si content on the microstructure and corrosion behavior of additive manufactured Al-Si Alloys," *Journal of the Electrochemical Society*, vol. 165, no. 13, pp. C926–C932, 2018, doi: 10.1149/2.0101814jes.
[68] J. Wu, X. Q. Wang, W. Wang, M. M. Attallah, and M. H. Loretto, "Microstructure and strength of selectively laser melted AlSi10Mg," *Acta Materialia*, vol. 117, pp. 311–320, 2016, doi: 10.1016/j.actamat.2016.07.012.
[69] J. Lei, J. Xie, S. Zhou, H. Song, X. Song, and X. Zhou, "Comparative study on microstructure and corrosion performance of 316 stainless steel prepared by laser melting deposition with ring-shaped beam and Gaussian beam," *Optics & Laser Technology*, vol. 111, pp. 271–283, 2019, doi: 10.1016/j.optlastec.2018.09.057.
[70] O. O. Salman, C. Gammer, A. K. Chaubey, J. Eckert, and S. Scudino, "Effect of heat treatment on microstructure and mechanical properties of 316L steel synthesized by selective laser melting," *Materials Science and Engineering: A*, vol. 748, pp. 205–212, 2019, doi: 10.1016/j.msea.2019.01.110.
[71] K. Abd-Elghany and D. L. Bourell, "Property evaluation of 304L stainless steel fabricated by selective laser melting," *Rapid Prototyping Journal*, vol. 18, no. 5, pp. 420–428, 2012, doi: 10.1108/13552541211250418.
[72] R. F. Schaller, A. Mishra, J. M. Rodelas, J. M. Taylor, and E. J. Schindelholz, "The role of microstructure and surface finish on the corrosion of selective laser melted 304L," *Journal of the Electrochemical Society*, vol. 165, no. 5, pp. C234–C242, 2018, doi: 10.1149/2.0431805jes.
[73] A. Hemmasian Ettefagh and S. Guo, "Electrochemical behavior of AISI316L stainless steel parts produced by laser-based powder bed fusion process and the effect of post annealing process," *Additive Manufacturing*, vol. 22, pp. 153–156, 2018, doi: 10.1016/j.addma.2018.05.014.
[74] Y. Sun, A. Moroz, and K. Alrbaey, "Sliding wear characteristics and corrosion behaviour of selective laser melted 316L stainless steel," *Journal of Materials Engineering and Performance*, vol. 23, no. 2, pp. 518–526, 2014, doi: 10.1007/s11665-013-0784-8.
[75] G. Chen, S. Y. Zhao, P. Tan, J. Wang, C. S. Xiang, and H. P. Tang, "A comparative study of Ti-6Al-4V powders for additive manufacturing by gas atomization, plasma rotating electrode process and plasma atomization," *Powder Technology*, vol. 333, pp. 38–46, 2018, doi: 10.1016/j.powtec.2018.04.013.
[76] A. Aksoy and R. Ünal, "Effects of gas pressure and protrusion length of melt delivery tube on powder size and powder morphology of nitrogen gas atomised tin powders," *Powder Metallurgy*, vol. 49, no. 4, pp. 349–354, 2006, doi: 10.1179/174329006X89425.

[77] T. Kurzynowski, K. Gruber, W. Stopyra, B. Kuźnicka, and E. Chlebus, "Correlation between process parameters, microstructure and properties of 316 L stainless steel processed by selective laser melting," *Materials Science and Engineering: A*, vol. 718, pp. 64–73, 2018, doi: 10.1016/j.msea.2018.01.103.
[78] A. Strondl, O. Lyckfeldt, H. Brodin, and U. Ackelid, "Characterization and control of powder properties for additive manufacturing," *JOM*, vol. 67, no. 3, pp. 549–554, 2015, doi: 10.1007/s11837-015-1304-0.

Chapter 10

Microstructure and mechanical properties of Al7075 and Al7075-based hybrid metal matrix composites by additive manufacturing

Ashokraj Rajendran
J.J. College of Engineering and Technology
Trichy, Tamilnadu, India.

Pavendhan Rajangam
Mohamed Sathak Engineering College
Kilakarai, Tamilnadu, India.

Kumaraguruba ran Balasubramanian
University College of Engineering, Anna University
Trichy, Tamilnadu, India.

CONTENTS

10.1 INTRODUCTION

Over the last several years, matrix aluminium hybrid composite materials have steadily displaced traditional engineering materials. Researchers are developing aluminium metal matrix with a high strength-to-weight ratio, stiffness and resistance to abrasion and corrosion for structural applications, particularly in engineering aircraft and automobiles [1]. For this application, Al7075 alloy is employed; however, they have some substantial drawbacks in terms of strength and stiffness. By incorporating hard intermetallic compounds further into aluminium alloy, the requisite characteristics are

DOI: 10.1201/9781003430186-10

enhanced [2]. Silicon carbide (SiC) is a ceramic particle that is chemically compatible with aluminium. It does so by establishing a non-intermetallic link between the matrix and the reinforcement. High thermal conductivity, superior workability, simplicity of machining and lower cost are all benefits [3]. The lattice matching of titanium carbide (TiC) particles is sufficient for direct aluminium nucleation, and they have a low friction coefficient, high energy density and exceptional wettability [4]. TiC may combine different aluminium alloy matrix because of its high wettability.

Reinforcement is placed in matrix alloy using a variety of production methods. Stir casting was among the most promising methods for mass-producing large-scale components [5]. The two-step stir casting procedure was utilized in this study to create a hybrid composite with homogeneous particle dispersion [6]. The wear of Al7075 composites reinforced with B_4C was examined using the stir casting process, and this was observed by improving the weight percentage of reinforcement improved wear [7]. The tensile strength of the 5 and 13 m SiC particles was greater than that of the Al7075/SiC (5, 13 and 60 m) composite-based alloy [8]. After fixing the Al7075/SiC composite via squeeze casting, the 10% SiC-reinforced composite was discovered to have the maximum hardness, tensile strength and bending strength through both casting and heat-treated conditions [9]. Stirring casting was utilized to manufacture the Al7075 mixed metal matrix composite. Hardness, tensile strength, bending strength and wear resistance are all improved by greater than the weight percent of Al_2O_3 strengthening [10]. When the mechanical and physical characteristics of the two-phase material Al/Al7075 were examined, it was discovered that lowering the base metal boosted tensile and compressive strengths while decreasing ductility [11]. We investigated the relationship among true stress and true strain in an Al7075 composite reinforced with SiC [12].

The wear rates of Al7075 composite reinforced with TiC were investigated under various load settings [13]. Using additive manufacturing, Al-Mg alloy is combined with SiC and Zr to improve performance [14,15]. The direct energy deposition method was used to create the Al-Mg-Sc-Zr alloy (DEDM). This fabrication produced an equiaxed crystal structure. Oxides served as solidification nuclei during the SLM fabrication of Al-Mg-Sc-Zr alloy [16]. This research examined the mechanical characteristics of Al7075/TiB_2/MoS_2 composites made using Direct Energy Deposition Method.

10.2 MATERIALS AND METHODS

Aluminium 7075 alloy with 2 μm is utilized in structural applications owing to its high wear resistance, improved corrosion resistance, specific strength and fatigue resistance. Because of its poor mechanical properties, it is unsuitable for use in the aerospace, marine and automobile industries. As a consequence of this approach, Al7075 was identified as a matrix material for the present inspection job. The Al7075 matrix powder was supplied by M/s Metals Company in Madurai, Tamil Nadu, India. Table 10.1 represents the chemical composition of Al7075 alloy.

Titanium diboride and molybdenum disulphide are two reinforcing particles mixed in with the base metal. The particle size of TiB_2 and MoS_2 reinforcements are 10 and 2 μm, and the reinforcements are used to improve the material qualities of the matrix. With the help of Subra Scientific Company in Bangalore, India, the reinforced powders were bought from Sigma Aldrich. Titanium diboride (TiB_2) is a ceramic substance defined by its melting point, hardness, corrosion resistance and wear resistance. The composition of TiB_2 is given in Table 10.2.

In nature, molybdenum disulphide is an inorganic chemical with the formula MoS_2 found in the mineral molybdenite. Its hexagonal layered crystals have a graphite-like composition. Because of its purifying properties, this chemical is useful. Because of its layered structure, hexagonal MoS_2 is a good dry lubricant. MoS_2 is an excellent lubricant due to its layered nature and low friction coefficient. The composition of MoS_2 is shown in Table 10.2.

Table 10.3 displays the parameters of matrix and reinforcing materials. Direct Energy Deposition Method is used to fabricate the Al7075

Table 10.1 Material composition of aluminium alloy 7075

Elements	*Cu*	*Zn*	*Si*	*Mn*	*Ti*	*Mg*	*Fe*	*Cr*	*Al*
Wt.%	1.5	5.6	0.08	0.06	0.07	2.4	.24	0.20	Bal.

Table 10.2 Elements composition of TiB_2 and MoS_2

Elements	*Ti*	*B*	*O*	*Fe*	*C*	*N*	*Mo*	*S*	*Fe*	*Cu*
TiB_2wt.%	68.50	30.14	0.46	0.09	0.24	0.26	–	–	–	–
MoS_2wt.%	–	–	–	–	–	–	59.94	40.05	0.005	0.001

Table 10.3 Properties of matrix Al7075 and reinforcing materials

Properties	*Alloy 7075*	*MoS_2*	*TiB_2*
Particle size (μm)	2	2	10
Density (g/cm^3)	2.80	5.06	4.52
Modulus of elasticity (GPa)	72	254	410
Tensile strength (MPa)	241	226	336
Melting point (°C)	555	1,185	2,248
Purity	98.5	99	98
Poisson ratio	0.3	0.23	0.28

matrix-based composites. The Al7075/12 wt.% TiB_2/x wt.% MoS_2 ($x=0$, 1.5, 3, 4.5 and 6) composites are manufactured under the conditions mentioned in Table 10.5. The LDM8060 (YuCheng Tech. Ltd., Beijing, China) equipped with a 2,000 W fibre laser with a 1-mm spot size was used to prepare the DED samples. A semiconductor fibre laser, a powder feeder, a purification system and a control system are all included in the setup. The optimum processing parameters were established during the laser deposition process and the parameters are 2,000 W laser, 10 mm/s scanning, 0.6 mm layer thickness, 1.2 mm scanning distance and 1.2 r/min powder feeder. Under an argon atmosphere, serpentine reciprocating scanning was used to create specimens with dimensions of 60 × 60 10 mm on a hybrid composite substrate. Surface grinding and polishing were performed on the printed DED samples.

The DEDM manufactured specimens were cut into required ASTM with the help of wire cut EDM. When compared to traditional machining methods, wire cut EDM reduces dislocation of composite particles and avoids changes in the bonding structure of composite particles when machining. Microstructural analysis identifies matrix reinforcement interfacial bonding and reinforcement distribution. Each specimen end is polished with abrasive grade sheets of 600, 800 and 1,000. The contrast is enhanced by Keller's reagent. At various magnifications observed by scanning electron microscope, the microstructure of reinforced Al7075 composites and unreinforced Al7075 was identified. SEM (HITACHI Model S-3000) is used to guarantee that the reinforcements in the matrix material are spread uniformly and that defect-free composites are generated. EDAX tests are also used to assess the presence of Al7075 matrix material and reinforcements in manufactured composites. The density of Al7075/TiB_2/MoS_2 composites was determined using Archimedes' principle. The specimens are weighed in open and submerged in filtered distilled water that used a Contech Precision Scale CA Series type electronic weighing equipment with a 0.0001 g accuracy. All produced specimens' theoretical density is determined using the mixing rule. The microhardness, tensile and compressive tests are conducted on all manufactured samples. Each sample composite was tested for hardness using Vickers hardness tester. The hardness is tested by applying a weight of 0.5 kg for 15 s. During hardness test, three trials were conducted on Al7075-based composites. The compressive strength is determined using a computerized FMI-F100 series UTM with a capacity of 200 kN. The ASTM E09 standard was followed for compressive tests. The load and displacement resolutions are 5 N and 0.01 mm, respectively. Tensile tests are performed on Al7075 and its composites according to ASTM E-08 [17]. A small Universal Testing Machine by way of a load-carrying capacity of 200 kN is used for the tensile test (TUE-C model, Fine manufacture).

10.3 RESULTS AND DISCUSSION

The Al7075 matrix-based hybrid composites are manufactured with the help of Direct Energy Deposition Method. The fabricated Al7075 matrix-based hybrid composites are subjected to EDAX test and SEM test.

Density test, microhardness test, tensile and compression test are used to confirm the weight percentage of elements, study the microstructure, determine the density, find the microhardness and determine the tensile and compressive strength of the Al7075 matrix-based composites. EDAX examines the elements in both reinforced and non-reinforced specimens (Al7075, TiB_2 and MoS_2). It's a subjective and quantitative investigative approach for separating the components in samples.

In general, Al7075 alloy contains various chemical constituents such as aluminium (94.05%) and zinc (5.95%), as shown in Figure 10.1 and the values are presented in Table 10.4.

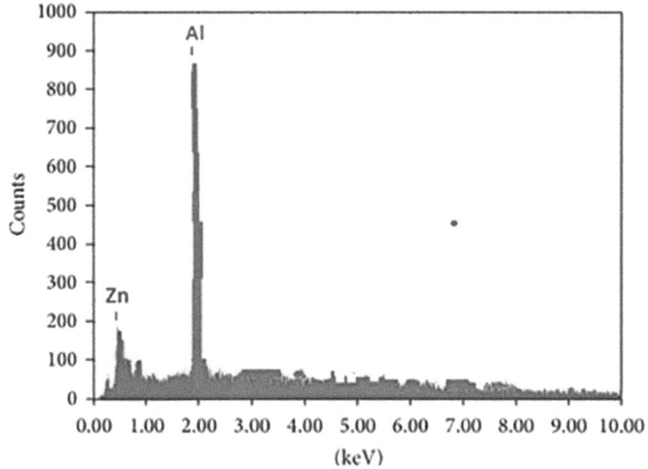

Figure 10.1 EDAX of Al7075 alloy.

Table 10.4 Al7075 EDAX results

El	*AN*	*Series*	*Norm.C (wt.%)*	*Unn.C (wt.%)*	*Atom.C (wt.%)*	*Error (%)*
Zn	30	K. Series	5.9	5.70	5.2	2.19
Al	13	K. Series	94.1	92.53	94.8	6.97
Total			100	98.23	100	–

All aluminium alloys have these peaks. Multiple constituents of the Metal Matrix Composite are shown in Table 10.5, as Al7075 alloy is strengthened with 12% TiB_2. Al (84.94%), Ti (9.23%), B (3.98%) and Zn (2.65%) are the distinct elements contained in this hybrid composite, as shown in Figure 10.2.

The different elements included in Al7075/12 wt.% TiB_2/1.5 wt.% MoS_2 hybrid composite are Al (83.77%), Ti (5.23%), Mo (2.19%), B (4.18%), S (2.83%) and Zn (1.80%), as represented in Table 10.6 and indicated in Figure 10.3.

The various components detected in Al7075/12 wt.% TiB_2/3 wt.% MoS_2 hybrid composite are Al (79.65%), Ti (6.23%), Mo (3.29%), B (3.88%), S (3.83%) and Zn (3.12%), as shown in Table 10.7 and Figure 10.4.

Table 10.5 Al7075/12 wt.% TiB_2 EDAX results

El	*AN*	*Series*	*Norm.C (wt.%)*	*Unn.C (wt.%)*	*Atom.C (wt.%)*	*Error (%)*
Al	13	K. Series	84.09	84.94	84.09	5.36
B	5	K. Series	3.98	3.31	3.63	2.65
Ti	22	K. Series	9.23	9.40	9.09	3.01
Zn	30	K. Series	2.65	2.35	2.94	0.16
Total			100	96.60	100	–

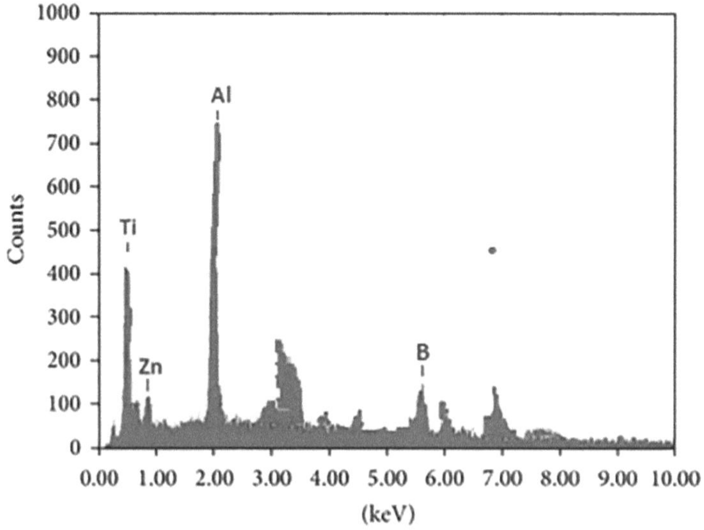

Figure 10.2 EDAX of Al7075 alloy 12 wt.% TiB_2.

Table 10.6 Al7075/12 wt.% TiB_2/1.5 wt.% MoS_2 EDAX results

El	*AN*	*Series*	*Norm.C (wt.%)*	*Unn.C (wt.%)*	*Atom.C (wt.%)*	*Error (%)*
Al	13	K. Series	83.77	82.79	81.45	4.17
Mo	42	K. Series	2.19	2.04	2.68	0.83
S	13	K. Series	2.83	1.16	1.79	0.57
B	5	K. Series	4.18	3.29	2.72	2.10
Ti	22	K. Series	5.23	8.32	7.26	2.31
Zn	30	K. Series	1.70	2.40	4.10	0.22
Total			96.5	100	100	–

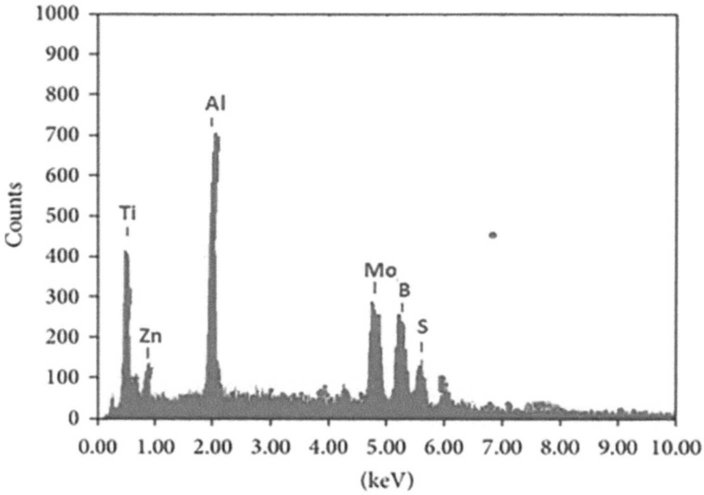

Figure 10.3 Al7075/12 wt.% TiB_2/1.5 wt.% MoS_2 EDAX.

Table 10.7 Al7075/12 wt.% TiB_2/3 wt.% MoS_2 EDAX results

El	*AN*	*Series*	*Norm.C (wt.%)*	*Unn.C (wt.%)*	*Atom.C (wt.%)*	*Error (%)*
Al	13	K. Series	79.65	78.61	78.01	3.93
S	13	K. Series	3.83	1.56	1.99	0.51
Mo	42	K. Series	3.29	1.94	3.58	0.85
B	5	K. Series	3.88	3.98	3.73	2.01
Ti	22	K. Series	6.23	7.63	8.06	2.51
Zn	30	K. Series	3.12	2.88	4.69	0.19
Total			100	96.0	100	–

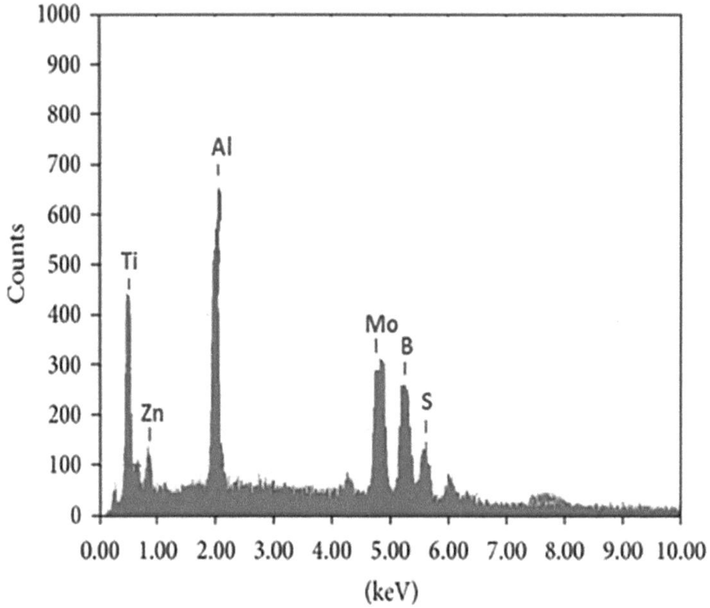

Figure 10.4 Al7075/12 wt.% TiB_2/3 wt.% MoS_2 EDAX results.

Table 10.8 Al7075/12 wt.% TiB_2/4.5 wt.% MoS_2 EDAX results

El	*AN*	*Series*	*Norm.C (wt.%)*	*Unn.C (wt.%)*	*Atom.C (wt.%)*	*Error (%)*
Al	13	K. Series	80.72	78.18	78.13	3.97
S	13	K. Series	2.08	2.14	1.95	0.54
Mo	42	K. Series	2.15	2.19	2.17	0.87
B	5	K. Series	4.08	4.02	3.98	2.05
Ti	22	K. Series	8.01	7.63	9.06	2.65
Zn	30	K. Series	2.86	2.54	4.71	0.19
Total			100	96.0	100	–

The various components detected in Al7075/12 wt.% TiB_2/4.5 wt.% MoS_2 hybrid composite are Al (79.65%), Ti (6.23%), Mo (3.29%), B (3.88%), S (3.83%) and Zn (3.12%), as shown in Table 10.8 and Figure 10.5.

The various components detected in Al7075/12 wt.% TiB_2/6 wt.% MoS_2 hybrid composite are Al (79.65%), Ti (6.23%), Mo (3.29%), B (3.88%), S (3.83%) and Zn (3.12%), as shown in Table 10.9 and Figure 10.6.

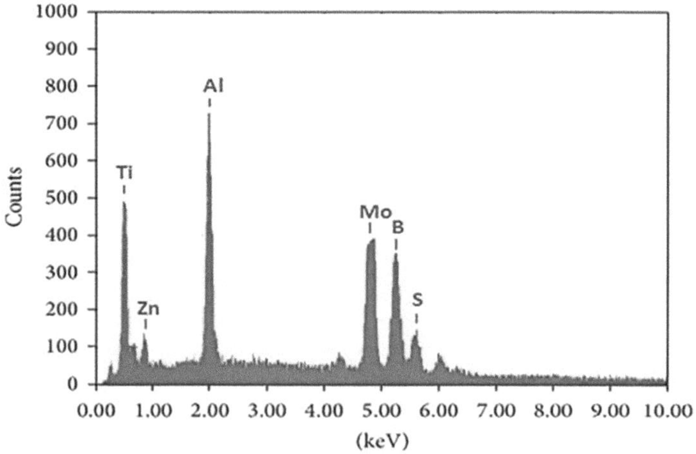

Figure 10.5 Al7075/12 wt.% TiB_2/4.5 wt.% MoS_2 EDAX results.

Table 10.9 Al7075/12 wt.% TiB_2/6 wt.% MoS_2 EDAX results

El	*AN*	*Series*	*Norm.C (wt.%)*	*Unn.C (wt.%)*	*Atom.C (wt.%)*	*Error (%)*
Al	13	K. Series	80.72	78.18	78.13	3.67
S	13	K. Series	2.98	2.94	2.95	0.64
Mo	42	K. Series	2.95	2.89	2.97	0.77
B	5	K. Series	4.08	4.02	3.98	2.15
Ti	22	K. Series	8.01	7.63	9.06	2.55
Zn	30	K. Series	2.86	2.54	4.71	0.18
Total			100	96.0	100	–

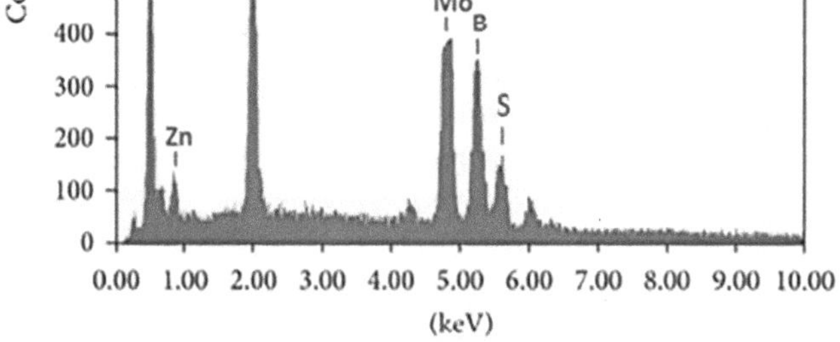

Figure 10.6 Al7075/12 wt.% TiB_2/4.5 wt.% MoS_2 EDAX results.

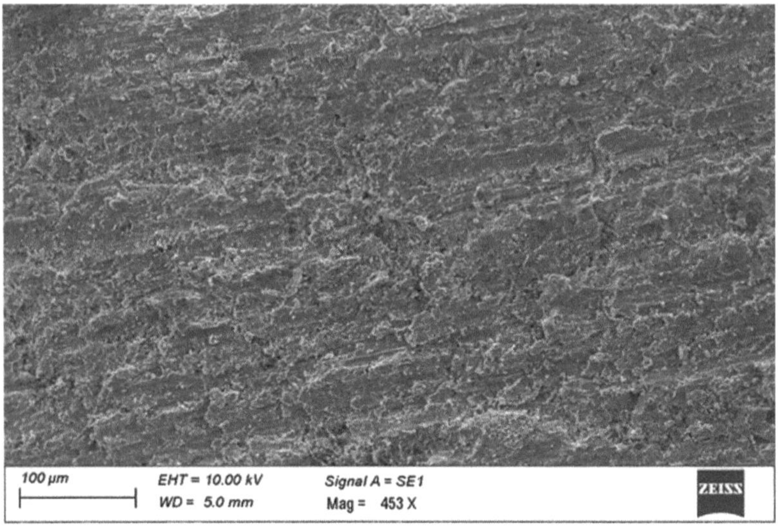

Figure 10.7 SEM picture of Al7075.

Microstructural analysis of newly fabricated is performed by scanning electron microscope, and the diversity of the samples is determined by scanning electron microscope. SEM images of Al7075 alloy and composites are depicted in Figure 10.7. The SEM images reveal that TiB_2 and MoS_2 particles are evenly dispersed throughout the matrix. Furthermore, SEM pictures appear to show that the particles TiB_2 and MoS_2 are all bonded to the aluminium matrix and that their concentrations are extremely low in each composite. Each specimen's end has been polished with a 600, 800 and 1,000 grade abrasive board. The surfaces of the reinforced and unreinforced specimens were captured in SEM at various scales and magnifications. At a magnification of 100 µm, Figure 10.8 shows a SEM photo of Al7075 aluminium alloy. According to microstructural testing, particle distribution in the matrix is stable. Due to a shrinking matrix during solidification, the substance is porous. The Al7075, Al7075/12% TiB_2/MoS_2 (X=0, 1.5, 3 and 4.5) specimens are analysed using scanning electron microscopy. The SEM photograph of Al7075-strengthened composites is shown in Figure 10.8. The elements in the aluminium matrix are bonded with titanium diboride and Molybdenum Disulphide, according to the SEM images of all composites. The SEM images show that titanium diboride and Molybdenum Disulphide are distributed uniformly in the aluminium matrix. Sulphur escapes from the hybrid composite during the solidification process, resulting in porosity. Furthermore, during cooling in a humid environment, the conversion of MoS_2 to MoO_3 generates an oxide on the

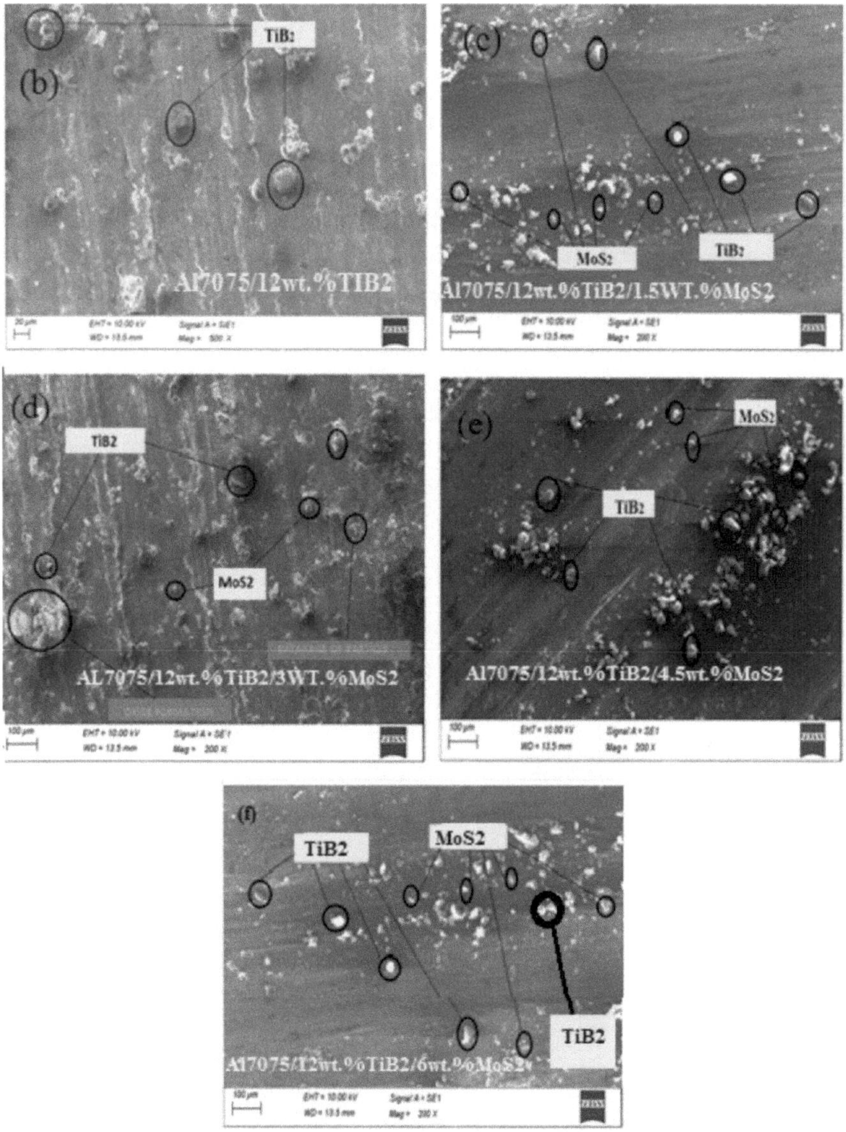

Figure 10.8 SEM picture of Al7075/12 wt.% TiB_2/x wt.% MoS_2 (x = 0, 1.5, 3, 4.5 and 6).

surface of the composites. Figure 10.8 shows how Titanium Diboride bonds strongly to the Al7075 matrix material. In the SEM photos of Figure 10.8, titanium diboride and Molybdenum Disulphide form a strong contact with the Al7075 matrix material.

The experimental density of DEDM-fabricated Al7075 hybrid composites was determined by Archimedes theory. The theoretical density of composites is calculated using the mixing law [18]. The experimental density of the generated composites is speculative at best. Table 10.10 compares the experimental and theoretical densities of aluminium alloy 7075 composites.

The addition of titanium diboride and Molybdenum Disulphide to produce composites enhances their density [19], as seen in Table 10.5. Because of the high density of molybdenum disulphide, the Al7075 composite density increases as the weight percent of molybdenum disulphide increases [20].

On Al7075, Al7075/TiB_2 and Al7075/TiB_2/MoS_2 composites, a microhardness test was performed in a Vickers hardness tester at a loading rate of 0.5 kg for 15 s. Each sort of composite formed is put through three trials before an average value is calculated. The microhardness results are shown in Table 10.6. The effects of the TiB_2 and MoS_2 reinforcement particles on the composite hardness obtained by the hardness test are shown in Table 10.11. The addition of particles like TiB_2 and MoS_2 considerably improved the hardness [21,22].

Table 10.10 Comparison of theoretical and experimental density

S. no.	*Material*	*Theoretical density (g/cm³)*	*Experimental density (g/cm³)*
1	AL7075	2.833	2.830
2	AL7075/12 wt.% TiB_2	2.850	2.846
3	AL7075/12 wt.% TiB_2/1.5 wt.% MoS_2	2.861	2.855
4	AL7075/12 wt.% TiB_2/3 wt.% MoS_2	2.877	2.874
5	AL7075/12 wt.% TiB_2/4.5 wt.% MoS_2	2.894	2.892
6	AL7075/12 wt.% TiB_2/6 wt.% MoS_2	2.884	2.882

Table 10.11 Microhardness of Al7075-based composites

S. no.	*Material*	*Microhardness*
1	AL7075	92
2	AL7075/12 wt.% TiB_2	105
3	AL7075/12 wt.% TiB_2/1.5 wt.% MoS_2	112
4	AL7075/12 wt.% TiB_2/3 wt.% MoS_2	119
5	AL7075/12 wt.% TiB_2/4.5 wt.% MoS_2	126
6	AL7075/12 wt.% TiB_2/6 wt.% MoS_2	121

The microhardness test findings revealed that adding TiB_2 and MoS_2 to reinforced Al7075 composites increases their hardness. Because of its high hardness, the inclusion of Titanium Diboride improves the hardness of the composites formed. Due to its firm attachment to the thickness of the lattice, and Titanium Diboride fortification, molybdenum disulphide improves the material hardness. The normalized conveyance of Titanium Diboride with molybdenum disulphide in the Al7075 grid material increases the durability of the composites. During a microhardness test, the excellent bond of titanium diboride and molybdenum disulphide to Al7075 prevents particle dislocation, which increases hardness. When molybdenum disulphide is added, the microhardness benefit increases to 4.5% and then drops to 6%. The addition of 6 wt.% molybdenum disulphide to Al7075/TiB_2 composites reduces the microhardness. During microhardness testing, further MoS_2 aggregation leads particles to dislocate quickly, resulting in substantial wear loss.

For determining the compressive strength of Al7075, Al7075/TiB_2, and Al7075/TiB_2/MoS_2 composites, compressive tests were performed. The compressive strength of Al7075 is improved by adding TiB_2 and MoS_2 particles [23], as illustrated in Table 10.12.

Titanium Diboride is resistant to high compression because it bonds strongly to the Al7075 matrix components. The compression strength of Al7075-based composites is improved by better distribution and incorporation of Titanium Diboride in the matrix material. A large weight percentage of Titanium Diboride incorporation decreases the compressive strength of Al7075-reinforced composites due to the discontinuity of TiB_2 bonding with Al7075 matrix content [23].

The addition of up to 4.5 wt.% MoS_2 to Al7075/TiB_2 improves the matrix's bonding power while lowering it by 6 wt.%. Because of the high weight percentage of MoS_2 (over 4.5 wt.%) in Al7075/TiB_2, the compression strength of the composites is reduced. The load-carrying capacity of molybdenum disulphide is restricted, and it cannot traverse or resist the burden as well as Titanium Diboride. The SEM picture of

Table 10.12 Compression strength of Al7075-based composites

S. no.	*Material*	*Compressive strength (MPa)*
1	AL7075	257
2	AL7075/12 wt.% TiB_2	346
3	AL7075/12 wt.% TiB_2/1.5 wt.% MoS_2	352
4	AL7075/12 wt.% TiB_2/3 wt.% MoS_2	368
5	AL7075/12 wt.% TiB_2/4.5 wt.% MoS_2	374
6	AL7075/12 wt.% TiB_2/6 wt.% MoS_2	362

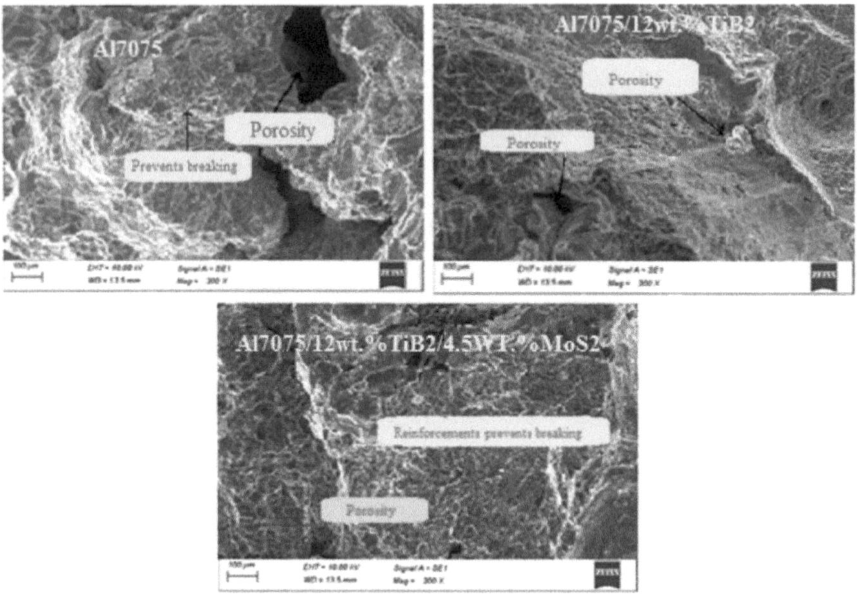

Figure 10.9 Compression-tested fractured surface.

compression-tested fractured surface of Al7075, Al7075/12 wt.% TiB_2/x wt.% MoS_2 ($x=0$ and 4.5) is shown in Figure 10.9.

Tensile strength was determined by testing Al7075, Al7075/TiB_2, and Al7075/TiB_2/MoS_2 specimens. Table 10.13 displays the tensile test results for specimens Al7075 and Al7075/TiB_2. Tensile test results for Al7075/TiB_2/MoS_2 composites are presented in Table 10.13. The Al7075/12 wt.% TiB_2/4.5 wt.% MoS_2 composites are more unique than other Al7075 composite mixtures.

The Al7075 hybrid composites' tensile strength is reduced by adding 6 wt.% MoS_2 and increased by adding 4.5 wt.% MoS_2 [24]. The large quantity of MoS_2 in the matrix or MMC decreases the composites' bonding capability. The SEM pictures of fractured tensile-tested Al7075, Al7075/12 wt.% TiB_2 and Al7075/12 wt.% TiB_2/4.5 wt.% MoS_2 are depicted in Figure 10.10.

10.4 CONCLUSION

The hybrid Al7075-based composites are fabricated successfully with the help of Direct Energy Deposition Method (DEDM). The fabricated composites are cut into required ASTM size with the help of wire cut EDM.

Table 10.13 Compression strength of Al7075-based composites

S. no.	*Material*	*Compressive strength (MPa)*
1	AL7075	191
2	AL7075/12 wt.% TiB_2	277.6
3	AL7075/12 wt.% TiB_2/1.5 wt.% MoS_2	286
4	AL7075/12 wt.% TiB_2/3 wt.% MoS_2	293
5	AL7075/12 wt.% TiB_2/4.5 wt.% MoS_2	306
6	AL7075/12 wt.% TiB_2/6 wt.% MoS_2	298

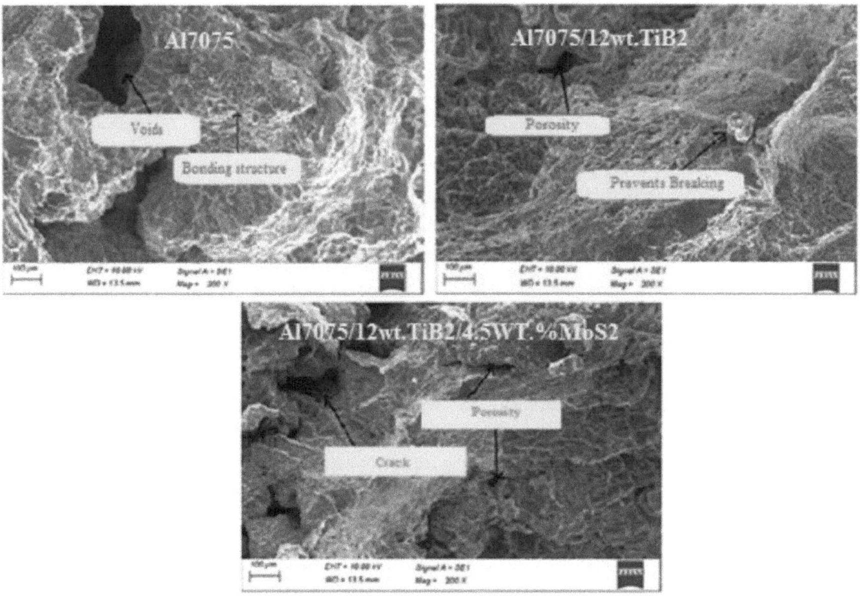

Figure 10.10 Tension-tested fractured surface.

The DEDM-fabricated composites are subjected to EDX analysis, SEM test, density test, microhardness test, compression test and tension test. The EDAX analysis confirms the presence of elements in the composite. The SEM pictures show the uniform distribution of molybdenum di sulphide in the Al7075/12 wt.% TiB_2 composites. The density, microhardness, compressive and tensile strength of composites improved by 4.5 wt.% and decreased by 6 wt.% MoS_2. The high accumulation of molybdenum disulphide decreases the density, microhardness, tensile and compression strength of the fabricated alloys.

REFERENCES

1. Moona, G., R. S. Walia, V. Rastogi, and R. Sharma. "Aluminium metal matrix composites: a retrospective investigation," *Indian Journal of Pure & Applied Physics*, 2018, 56, 164–175. https://www.researchgate.net/publication/323470523_Aluminium_metal_matrix_composites_A_retrospective_investigation.
2. Bera, S., S. G. Chowdhury, Y. Estrin, and I. Manna. "Mechanical properties of Al7075 alloy with nano-ceramic oxide dispersion synthesized by mechanical milling and consolidated by equal channel angular pressing." *Journal of Alloys and Compounds*, 2013, 548, 257–265. https://www.sciencedirect.com/science/article/abs/pii/S0925838812015678.
3. Narayanasamy, R., T. Ramesh, and M. Prabhakar. "Effect of particle size of SiC in aluminium matrix on workability and strain hardening behaviour of P/M composite," *Materials Science and Engineering A*, 2009, 504, 13–23. https://www.researchgate.net/publication/248474030_Effect_of_particle_size_of_SiC_in_aluminium_matrix_on_workability_and_strain_hardening_behaviour_of_PM_composite.
4. Kennedy, A. R., D. P. Weston, and M. I. Jones. "Reaction in Al–TiC metal matrix composites," *Materials Science and Engineering: A*, 2001, 316, 32–38. https://www.sciencedirect.com/science/article/abs/pii/S092150930101228X.
5. Hashim, J. "The production of cast metal matrix composite by a modified stir casting method," *Jurnal Teknologi*, 2001, 35, 9–20. https://www.researchgate.net/publication/43767427_The_Production_Of_Cast_Metal_Matrix_Composite_By_A_Modified_Stir_Casting_Method.
6. Madhukar, P., N. Selvaraj, C. S. P. Rao, and G. B. Veeresh Kumar. "Fabrication and characterization two step stir casting with ultrasonic assisted novel AA7150-hBN nanocomposites," *Journal of Alloys and Compounds*, 2020, 815, 152464. https://www.researchgate.net/publication/336119147_Fabrication_and_characterization_two_step_stir_casting_with_ultrasonic_assisted_novel_AA7150-hBN_nanocomposites.
7. Singh, K. M., and A. K. Chauhan. "Wear behaviour of al-7075 composites reinforced with boron carbide," *International Journal of Mechanical and Production Engineering Research and Development*, 2020, 10, 10125–10134. https://www.researchgate.net/publication/348817861_WEAR_BEHAVIOUR_OF_AL-7075_COMPOSITES_REINFORCED_WITH_BORON_CARBIDE.
8. Balaji, V., N. Sateesh, and M. Manzoor Hussain. "Manufacture of aluminium metal matrix composite (Al7075-SiC) by stir casting technique," *Materials Today: Proceedings*, 2015, 2, 3403–3408. https://www.sciencedirect.com/science/article/pii/S221478531500560X.
9. Baradeswaran, A., and A. Elaya Perumal. "Study on mechanical and wear properties of Al 7075/Al_2O_3/graphite hybrid composites," *Composites Part B: Engineering*, 2014, 56, 464–471. https://www.sciencedirect.com/science/article/abs/pii/S1359836813004216.
10. Hariprasad, T., K. Varatharajan, and S. Ravi. "Wear characteristics of B4C and Al_2O_3 reinforced with Al 5083 metal matrix based hybrid composite," *Procedia Engineering*, 2014, 97, 925–929. https://www.sciencedirect.com/science/article/pii/S1877705814034365.

11. Sherafat, Z., M. H. Paydar, and R. Ebrahimi. "Fabrication of Al7075/Al, two phase material, by recycling Al7075 alloy chips using powder metallurgy route," *Journal of Alloys and Compounds*, 2009, 487, 395–399. https://www.semanticscholar.org/paper/Fabrication-of-Al7075%2FAl%2C-two-phase-material%2C-by-Sherafat-Paydar/b54b1d220e98a81d2dfd394da1ec5c5e755e4365.
12. Rao, T. B. "An experimental investigation on mechanical and wear properties of Al7075/SiCp composites: effect of SiC content and particle size," *Journal of Tribology*, 2018, 140, 031601. https://www.researchgate.net/publication/319498551_An_Experimental_Investigation_on_Mechanical_and_Wear_Properties_of_Al7075SiCp_Composites_Effect_of_SiC_Content_and_Particle_Size.
13. Agrawal, E. S., and V. B. Tungikar. "Influence of reinforcement of TiC particles on wear behaviour of Al7075/TiC composite material," In *Advanced Engineering Forum*, 2020, vol. 37, 37–45. doi: 10.4028/www.scientific.net/AEF.37.37; https://www.semanticscholar.org/paper/Influence-of-Reinforcement-of-TiC-Particles-on-Wear-Agrawal-Tungikar/5ef50c8a2ed48a54d01d2e09bd9a87db7d690be6.
14. Herrero, P. R. "Additive manufacturing of Al and Mg alloys and composites," *Encyclopedia of Materials: Metals and Alloys*, 2022, 1, 245–255. https://www.researchgate.net/publication/350328907_Additive_Manufacturing_of_Al_and_Mg_Alloys_and_Composites.
15. Nie, X., H. Zhang, H. Zhu, Z. Hu, L. Ke, and X. Zeng. "Effect of Zr content on formability, microstructure and mechanical properties of selective laser melted Zr modified Al-4.24 Cu-1.97 Mg-0.56 Mn alloys," *Journal of Alloys and Compounds*, 2018, 764, 977–986. https://jglobal.jst.go.jp/en/detail?JGLOBAL_ID=201802248893221261.
16. Shi, Y., P. Rometsch, K. Yang, F. Palm, and X. Wu. "Characterisation of a novel Sc and Zr modified Al–Mg alloy fabricated by selective laser melting," *Materials Letters*, 2017, 196, 347–350. https://libgen.ggfwzs.net/book/64520309/1c9c04.
17. Nayak, B., R. K. Sahu, and P. Karthikeyan. "Study of tensile and compressive behaviour of the in-house synthesized al-alloy nano composite," *IOP Conference Series: Materials Science and Engineering*, 2018, 402, 012070. https://iopscience.iop.org/article/10.1088/1757-899X/402/1/012070.
18. Gudipudi, S., S. Nagamuthu, K. S. Subbian, and S. P. Rao Chilakalapalli. "Enhanced mechanical properties of AA6061-B4C composites developed by a novel ultra-sonic assisted stir casting," *Engineering Science and Technology*, 2020, 23(5), 1233–1243. https://www.sciencedirect.com/science/article/pii/S221509861932350X.
19. Krishnamurthy, K., M. Ashebre, J. Venkatesh, and B. Suresha. "Dry sliding wear behavior of aluminum 6063 composites reinforced with TiB_2 particles," *Journal of Minerals and Materials Characterization and Engineering*, 2017, 5, 74–89. https://www.scirp.org/journal/paperinformation.aspx?paperid=75038.
20. Varun, K. M., and R. Raman Goud. "Investigation of mechanical properties of Al 7075/SiC/MoS2 hybrid composite," *Materials Today: Proceedings*, 2019, 19, 787–791. https://www.researchgate.net/publication/336575749_Investigation_of_mechanical_properties_of_Al_7075SiCMoS2_hybrid_composite.

21. Rajasekaran, N. R., and V. Sampath. "Effect of in-situ TiB_2 particle addition on the mechanical properties of AA 2219 Al alloy composite," *Journal of Minerals & Materials Characterization & Engineering*, 2011, 10, 527–534. https://www.scirp.org/pdf/JMMCE20110600005_82413923.pdf.
22. Nautiyal, H., S. Kumari, O. P. Khatri, and R. Tyagi. "Copper matrix composites reinforced by rGO-MoS2 hybrid: Strengthening effect to enhancement of tribological properties," *Composites Part B: Engineering*, 2019, 173, 106931. https://www.sciencedirect.com/science/article/abs/pii/S1359836818342100.
23. Manoj, M., G. R. Jinu, J. Suresh Kumar, and V. Mugendiran. "Effect of TiB_2 particles on the morphological, mechanical and corrosion behaviour of Al7075 metal matrix composite produced using stir casting process," *International Journal of Metalcasting*, 2021, 16, 1517–1532. https://www.researchgate.net/publication/355334320_Effect_of_TiB2_Particles_on_the_Morphological_Mechanical_and_Corrosion_Behaviour_of_Al7075_Metal_Matrix_Composite_Produced_Using_Stir_Casting_Process.
24. Liu, S., Y. Wang, T. Muthuramalingam, and G. Anbuchezhiyan. "Effect of B_4C and MOS_2 reinforcement on micro structure and wear properties of aluminum hybrid composite for automotive applications," *Composites Part B: Engineering*, 2019, 176, 107329. https://www.sciencedirect.com/science/article/abs/pii/S1359836819322395.

Chapter 11

Key thermal properties of stainless steel deposits for wire arc additive manufacturing

Sowrirajan Maruthasalam
Coimbatore Institute of Engineering and Technology, Coimbatore.

Nallathambi Siva Shanmugam
National Institute of Technology Tiruchirappalli.

Rajesh Kannan Arasappan
Hanyang University, Ansan.

CONTENTS

DOI: 10.1201/9781003430186-11

11.1 INTRODUCTION

The global community is running behind the fabrication of components for meeting the needs of various sectors including aerospace industries, automobile sectors, bio-medical fields and production industries [1]. The requirement of components in all sectors has been increasing day-by-day due to the increase in population across the globe. Therefore, a standard method of fabrication of components is of primary importance and Additive Manufacturing (AM) has been identified as of primary interest by the worldwide fabrication community. AM has received ample consideration over the past few decades in the manufacturing and fabrication industries to make part models and prototypes [2]. This technique is a fabrication of desired components as per the desired shape and size including three-dimensional components. During AM fabrication, the component is built on the principle of adding one layer over another layer to produce the three-dimensional component [3]. Thus, this fabrication involves the building of materials as layers; this AM technique can also be referred to as 3D printing.

In these modern times, 3D printing or Additive Manufacturing is basically a transformative method to conventional industrial fabrication of models and prototypes that takes us to the creation of light as well as strong components simply by adding the material layers. The concept of 3D printing is famous now not only in the manufacturing sector but also in the construction field. AM is a leading process currently engaged in the widespread industries to the subtractive methods such as forming and machining [4]. AM is illustrated in Figure 11.1 for a better understanding of the principle.

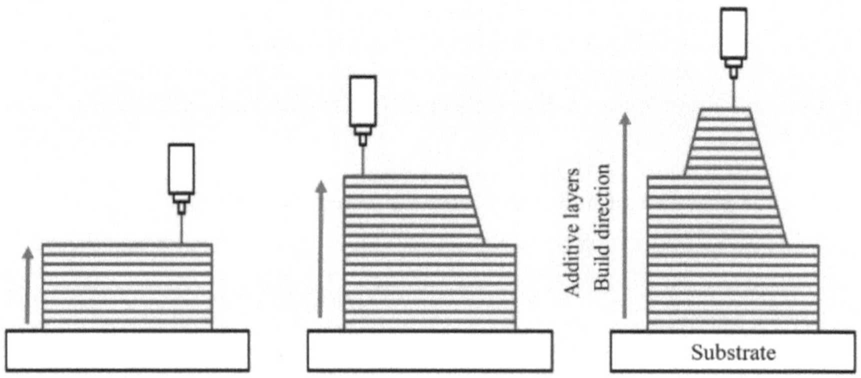

Figure 11.1 Illustration of AM.

11.1.1 Different additive manufacturing processes

Different familiar AM processes include Binder jetting, Directed Energy Deposition (DED), Material extrusion, Sheet lamination processes such as Laminated Object Manufacturing (LOM), Ultrasonic Additive Manufacturing (UAM) and Powder bed fusion methods such as Direct Metal Laser Melting (DMLM), Direct Metal Laser Sintering (DMLS), Electron Beam Melting (EBM), Selective Laser Sintering (SLS) and Selective Heat Sintering (SHS) and also the photopolymerization techniques such as sintering, melting and stereolithography techniques. DED has types such as Laser-based DED, Electron beam-based DED, Plasma or Electric arc-based DED, Powder-based DED and Wire-based DED. Further, this DED shall be used for almost entire materials including metals, ceramics and polymers [5].

11.1.2 Wire arc additive manufacturing (WAAM)

Wire Arc Additive Manufacturing (WAAM) is a subtype under DED which uses the electric arc from a welding process as an energy source for melting the metal, preferably solid wire metal, and deposits the melts in layers to fabricate the metal component. Commonly in DED, the energy sources such as electric arc, electron beam and laser are utilized for melting the feed material in which the metal component is fabricated either in the form of solid filler wire or powder [6]. Because of the reason of employing any of the welding processes for melting the metals in the form of wire, it is generally called Welding-based Additive Manufacturing (WAM). Moreover, because electric arc is used during the welding to deposit the metal, this is preferably called Wire Arc Additive Manufacturing (WAAM) [7] and Figure 11.2 illustrates the basic GMAW-based WAAM process to acquire the idea about the working of WAM.

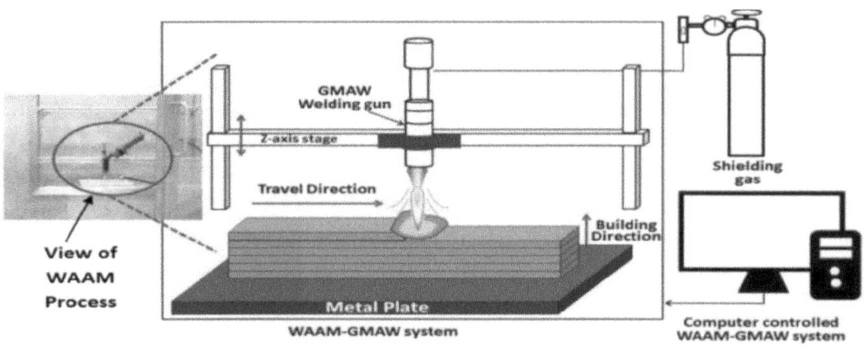

Figure 11.2 Illustration of GMAW-based WAAM process.

In WAAM, the metal component is fabricated by employing a suitable welding process. The very familiar welding processes identified for WAM include Gas Metal Arc Welding (GMAW), Gas Tungsten Arc Welding (GTAW) and Plasma Arc Welding (PAW). Thus, the metal component is fabricated by any of the suitable welding processes. One who fabricates the metal components using WAM should use a substrate for building component, suitable weld conditions and various welding process parameters [6]. Many researchers have reported that the final chemical composition of weld deposits and properties at weld layers is strongly affected by the welding process employed. This phenomenon is severe when using different substrate metals compared to weld metal. Therefore, it can also be incidental that the post-weld thermal properties of components fabricated in WAM also may be varied [6–8]. Therefore, a clear text is of fundamental importance to reach the effect of welding processes on the metal component's properties.

11.1.2.1 Welding processes for WAAM

The selection of an appropriate welding process is instrumental to the WAAM fabrication process. WAAM actually makes use of any suitable welding process to melt the metal to be fabricated in the form of metal solid wire or powder and deposits over the substrate. This procedure is repeated till the achievement of the final shape and size of the metal components. Generally, welding processes selected for WAAM would give high deposition rates and therefore, WAAM could represent the best alternative AM process for available traditional AM processes. The increasing interest of the entire world in WAAM is mostly due to the advances in robotic controls and modern electronic control circuit systems for different welding machines.

The controlled short arc processes with reduced heat input are referred to as Cold Metal Transfer (CMT) technology, which is purposefully established nowadays as a CMT-GMAW process by Fronius International, a firm that manufactures the welding machines worldwide, especially for the WAAM. Plasma Arc Welding (PAW) and Tungsten Inert Gas (TIG) welding processes could also be employed because of the separation of mass and energy input for specific materials. Laser-assisted welding is also used apart from standard arc welding processes frequently for the WAAM. Most commonly, GMAW is preferred for a high deposition rate comparing GTAW even though process stability and quality of fabrication are slightly lower. But PAW gives enough energy density arcs for welding suitable materials and provides high travel speeds and high-quality welds with minimum weld distortion. This PAW needs high capital expenses. Some of the famous welding processes employed for metal components fabrication are listed below [6,9].

1. Gas Metal Arc Welding (GMAW)
2. Plasma Arc Welding (PAW)
3. Gas Tungsten Arc Welding (GTAW)
4. Laser-Assisted Welding
5. Submerged Arc Welding (SAW) etc.

Generally this chapter provides the available AM processes and neatly illustrates the Welding-based Additive Manufacturing (WAM) called WAAM. Also, there is the need to study various thermal properties on WAAM components. The use of Laser Flash Apparatus (LFA) for the thermal property measurement is presented with clear explanation. The procedure followed for the measurement of thermal properties in LFA is given lucidly. It is believed that this text would be a better one in assisting any layperson who decides to go for the thermal property measurement of the WAAM components.

11.2 THERMAL PROPERTIES AND THEIR MEASUREMENTS

Thermal properties of a material under different thermal environments have to be revealed well before its use. These properties are majorly considered under a broad banner called thermo-physical properties. Generally most of the thermal properties would be measured during heat treatment [10]. Thermo-physical properties are defined as the properties of materials disturbing the transfer and storing of heat, which differ with the pressure, temperature, composition of mixtures etc., without changing the chemical individuality of the material [11].

11.2.1 Need for the measurement

The accurate measurement of heat energy transferred in many of the thermal equipment is needed in various thermal environments. Thermal conductivity, specific heat capacity and thermal diffusivity are believed to be the important thermal properties of a welding metal and also in the welding-based fabrication. These thermal properties have the ability to influence the residual stresses, temperature distribution and thermal distortion in the weld metal structures [10]. Final post-weld chemical composition and properties of weld metals are already proved to be affected by welding. The same phenomenon could also be extended to thermal properties of welding-based layer depositions. It is apparent from previous studies that the thermal properties of weld deposits would not be the same especially after the welding process [8]. Some reasons such as the effect of welding process parameters, metallurgical changes, chemical compositions and welding conditions have the strong ability to alter the properties of weld deposits. Therefore, the thermal properties of WAAM fabrications are essentially to be evaluated to ensure the ability for working at different thermal environments smoothly.

11.2.2 Thermal properties

Important thermal properties as addressed by thermal professionals are thermal expansion, specific heat capacity, thermal conductivity and thermal stress. The measurements of thermo-physical properties are not easy as that of some mechanical properties like tensile strength and hardness [12]. Some of the thermal properties as listed below are the crucial post-weld properties.

1. Specific heat capacity
2. Thermal conductivity
3. Thermal diffusivity etc.

11.2.2.1 Specific heat capacity

Specific heat capacity is the quantity of heat per unit mass required to raise the temperature by 1°C. The specific heat capacity is defined as the quantity of heat energy (J) absorbed per unit mass (kg) of the material when temperature increases by 1 K or 1°C. The SI unit familiarly used for specific heat is J/kg-K. The values of specific heat capacity for few commonly used metals are furnished in Table 11.1 [12].

11.2.2.2 Thermal conductivity

The thermal conductivity of a material is defined as a measure of the ability of the material to conduct heat. The thermal conductivity of a material is the rate at which heat passes through a unit area per unit time with a unit degree temperature gradient per unit distance. It is generally denoted with letter "k." SI unit of thermal conductivity is W/m^2-K.

In general, thermal conductivity, $k = \frac{\Delta T}{L}$

Table 11.1 Specific heat values of few common materials

S. no.	*Material*	*Specific heat capacity (J/kg-K)*
1	Aluminium	900
2	Stainless steel	500
3	Brass	375
4	Copper	390

11.2.2.3 Thermal diffusivity

Thermal diffusivity is the thermal conductivity divided by the density and the specific heat capacity at constant pressure. It actually measures the ability of a material to conduct thermal energy in the form of heat relative to its ability to store thermal energy. High diffusivity means that the heat is transferred very quickly. It is denoted by "α." SI unit used for thermal diffusivity is m^2/s. Especially in WAAM-fabricated metal components, the focus is essentially to be given on this property because the welding processes by virtue of process parameters and many more have high possibility to change this property and thereby other allied thermal properties also.

11.3 DEPOSITION OF WELD BEADS USING WAAM

In order to fabricate the metal components, a number of weld beads are to be deposited over another. As already discussed, WAAM is a fabrication process by which metal components of needed shape and size could be fabricated by laying required numbers of weld beads. As an example, AISI 316L grade of austenitic stainless steel weld bead is deposited to fabricate a component in GMAW-based WAAM process.

Generally there are two options readily available to fabricate metal plates. Usually in arc-welding-based AM, component is prepared by layer-by-layer concept. Another typical AM method of fabrication could be used when particularly the height of the fabricating component would be lesser than the weld bead thickness of single weld bead. For example, if one considers fabricating a thin-walled plate, the thickness of the plate could be attained with single weld bead itself but the required length of the plate shall be accomplished with multiple depositions to cover this length. Standard WAAM and typical WAAM options available for the plate fabrication could be understood easily while referring Figure 11.3, and the plate component to be fabricated in welding-based AM is indicated in red coloured box. After depositing the additive weld beads, the metal plate is to be cut using any of the suitable methods [13]. Such a fabrication and the thermal properties of the fabricated plate are included as a case in this text. Illustration of WAAM options for the thin-walled plate component fabrication is represented in Figure 11.3.

11.3.1 Thin-walled plate component fabrication by typical WAAM

A thin-walled 316L grade of austenitic stainless steel plate component is fabricated as detailed below. It is decided to go with the typical WAAM to fabricate the plate because of its lesser thickness needed to be produced. It is decided that the said plate could be fabricated with the dimensions $50 \times 30 \times 6$ mm. As the length of the plate is more than that of the possible

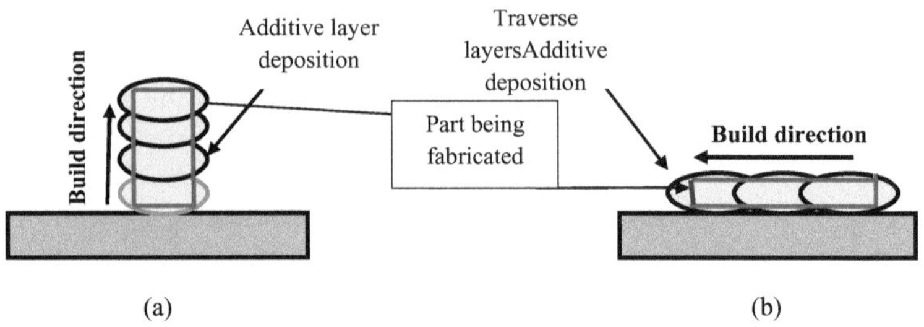

Figure 11.3 Illustration of WAAM options for thin-walled plate component fabrication.

single weld bead dimension, multiple number of weld beads are needed to be deposited. The main reason to go with the typical WAAM is that the thickness of plate seems to be less enough and this thickness could be achieved with single weld bead deposition. Therefore, three weld beads were decided to be deposited in transverse direction to cope up with typical WAAM process and breadth of 30 mm could be controlled by weld gun motion so that the required size of the plate could be well fabricated.

Gas Metal Arc Welding (GMAW) operating parameters such as open circuit voltage, wire feed rate, welding speed, nozzle-to-plate distance and electrode angle have been selected to control the process based on impacts of these process parameters on various properties of weld deposits. The experimental work for the fabrication of thin-walled plate was done using GMAW machine of Ador made at Welding Research Laboratory, Coimbatore Institute of Engineering Technology, Coimbatore, India. A three-axis welding manipulator is used for achieving the linear motions so that deposition layers could be well controlled for the fabrication.

Low-carbon high tensile structural steel (ASTM A 105/IS: 2062) was used as substrate for building component. Austenitic stainless steel 316L was unanimously used as filler wire material. The 316L stainless steel filler wire of 1.2 mm diameter was used and was fed in the wire feeder. The process parameter values used for control are given in Table 11.2. There were

Table 11.2 GMAW process parameter values used

Process parameter	*Value used*
Open circuit voltage	34 V
Wire feed rate	13 m/min
Welding speed	0.34 m/min
Nozzle-to-plate distance	21 mm
Electrode angle	15°

(a) (b)

Figure 11.4 Thin-walled plate component fabrication by typical WAAM.

three weld beads laid as typical additive layers in the traverse direction to meet the size designed for thin-walled plate component. Effective weld bead overlap was maintained with the naked eye. The inter-pass temperature between the layers is maintained carefully to avoid warping [14]. Then, the component fabricated was sent to machine the required size in milling machine. Finally, 316L stainless steel thin-walled plate component was fabricated. The final thin-walled component is shown in Figure 11.4.

11.4 THERMAL PROPERTIES' EVALUATION OF POST-WELD THIN-WALLED PLATE COMPONENT

Different methods exist for the measurement of various thermal properties of metals. Some of the famous methods are given below.

1. Guarded hot plate method
2. Hotwire method
3. Heat flow meter method
4. Laser flash method

Laser Flash Method (LFM) was identified to be more suitable for measuring the post-weld thermal properties of WAAM components because of sample size of weld metal, fabrication of the samples, temperature ranges and time duration [8]. Comparison of abovementioned methods with respect to the temperature range and metals is presented in Figure 11.5. Laser flash method is a popular method to determine the key thermal properties of a variety of metals at various temperature levels. This method is also suitable for weld metals. Thermo-physical properties such as thermal conductivity, specific heat and thermal diffusivity were evaluated by LFA in this text.

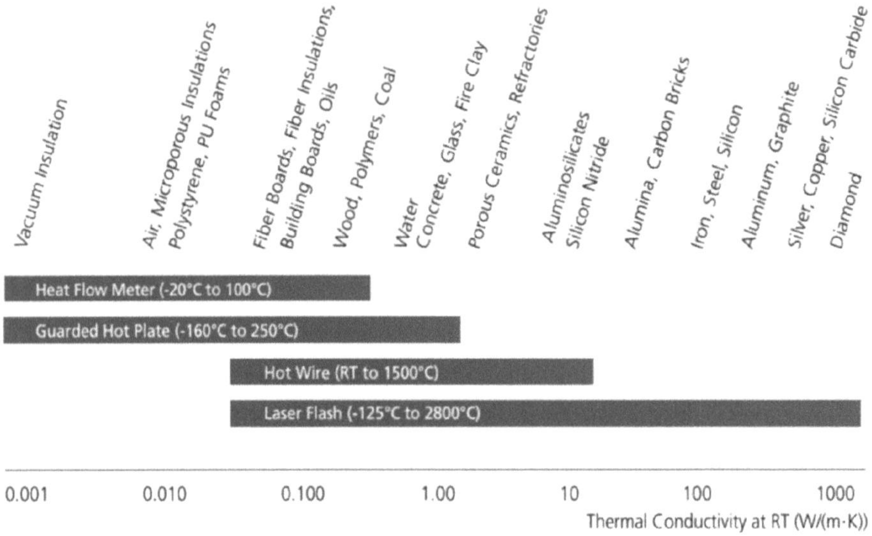

Figure 11.5 Comparison of thermal properties measurement methods.

11.4.1 Laser flash apparatus (LFA)

The LFA works on the principle of laser flash method. The schematic drawing of a typical LFA 447 apparatus is presented in Figure 11.6.

The sample holder, which could grasp 1–4 samples at a time, is situated at the centre. The heat source is generally the xenon lamp which is located at the bottom. The infrared detector is positioned on the top portion of sample holder. The temperature at the sample holder is handled by a furnace and that temperature may be set between the room temperature and high temperature. This is to be in cooled condition with the help of liquid nitrogen to avoid the warmth of IR detector. The casing of infrared detector is to be levelled with liquid nitrogen. Generally, the liquid nitrogen endures for approximately 4–6 hours of measurements, if filled once. The sample holders in the LFA are appropriate for circular samples with 12.7 mm diameter and/or $10 \times 10\,mm^2$, especially in matrix mode. The samples thickness shall be 1–3 mm [11]. A strong recommendation is that the trial samples shall be of the same thickness of reference samples as possible. The graphite spray is to be employed on the entire outer surfaces of samples to control light transmission. Special care is taken as the accuracy of specific heat measurement could be affected. The specification details of the apparatus used are given in Table 11.3.

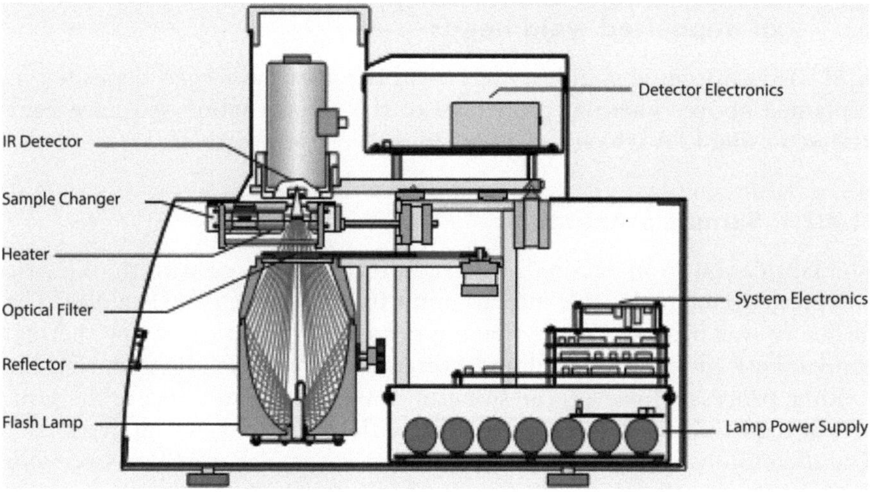

Figure 11.6 Laser flash apparatus (LFA).

Table 11.3 Specifications of laser flash apparatus

S. no.	*Factors*	*Specifications*
1	Temperature range	Ambient to 300°C
2	Standard sample size	Up to 25.4 mm diameter, or 6 mm/ 8 mm/10 mm/12.7 mm^2, up to 3 mm thick, specials on request
3	Dimensions of LFA	61 cm × 56 cm × 43 cm
4	MTX version	Scanning range: 50 mm × 50 mm, resolution: down to 0.1 mm; operation at room temperature
5	Thermal conductivity	0.1 W/(m·K) to 2,000 W/(m·K)
6	Thermal diffusivity range	0.01–1,000 mm^2/s
7	Repeatability	Thermal diffusivity: ±2%; specific heat: ±3%
8	Accuracy	Thermal diffusivity: ±3%; specific heat: ±5%
9	Flash source	Xenon flash lamp, wavelength: 150–2,000 nm Pulse energy: up to ≈10 J (selectable by voltage and pulse length)
10	Sensor type	IR detector with integrated dewar

11.4.2 Measurement of thermal properties of deposited weld beads

AISI 316L austenitic stainless steel weld deposits have been deposited as explained above. Thermal properties of the sample fabricated have been assessed using LFA which is detailed in the previous sections.

11.4.2.1 Sample preparation

The fabrication from Section 11.3.1 is further processed with the Electric Discharge Machine (EDM) into 10 mm×10 mm square plate samples. The difficulty was observed while shaping the samples because of low thermal conductivity of stainless steel and presence of some inclusions during the welding process. However, the situation was managed to prepare the samples at required dimensions to suit LFA with enough surface qualities [10]. The dimensions of the final prepared deposit samples were assessed with digital vernier caliper to calculate the bulk density of fabricated samples and the correctness was ensured.

11.4.2.2 Measurement of thermal properties in LFA

A sample as required by the apparatus is prepared for the LFA analysis and weighed by the electronic weighing balance scale and mass (m) of the metal sample is calculated. Moreover, volume (V) of the prepared square sample is calculated as per their geometric sizes. Bulk density (ρ) value of sample is derived from the equation, $\rho = m/V$ by following LFA procedures. Thus, the measurements for thermal properties such as thermal diffusivity, specific heat and thermal conductivity were made at room temperature using NETZSCH LFA 447 Nanoflash™ system shown in Figure 11.7, which is available in Indian Institute of Technology, Kanpur, India [8].

Post-weld thermal properties of weld deposit sample assessed in LFA are presented in Table 11.4. The thermal diffusivity value measured from LFA was

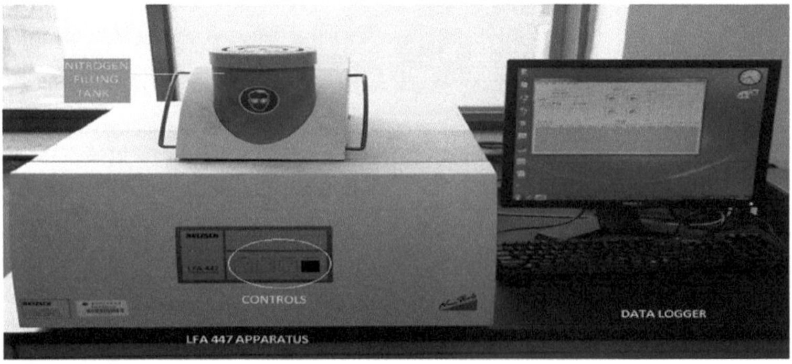

Figure 11.7 Photographic view of Netzsch LFA 447 apparatus used for the measurements.

Table 11.4 Details of measurement for assessed sample

Description	*Measured value*
Mass (*m*)	1.870 g
Thickness (*t*)	2.72 mm
Bulk density (ρ)	7.415 g/cm^3
Thermal diffusivity	5.27 mm^2/s
Specific heat	0.50 J/g-K
Thermal conductivity	17.76 W/m^2-K

5.27 mm^2/s. This is the clear indication that heat-carrying capacity of sample is quiet acceptable. The specific heat capacity value was found to be 0.5 J/g-K. The thermal conductivity value has been calculated to be 17.76 W/m^2-K. Therefore, it is quite clear from the result that the value of thermal conductivity slightly differs after the welding. Thus, it is understood that the thermal conductivity value of deposit might fall between the values of parent and filler metal as they are fusing in the weld pool [15]. Therefore, it is strongly suggested that the use of substrate material may be similar compared to the filler wire material/material of the component [16], unless this change in different thermal properties could be possible due to the weld dilution as reported in previous studies.

11.5 CONCLUSIONS

This text includes the details of different methods for AM. Also, work has been carried out by a variant WAAM process called typical WAAM and the key thermal properties have been lucidly presented by studying the few thermal properties of weld deposits in this text. Also, it is seen that the Laser Flash Apparatus (LFA) is a trendy and feasible simple technique to be used for measuring the post-weld thermal properties, especially the fabricated components by WAAM. The results mainly show that the thermal properties could be affected by the welding processes and therefore much care should be given during the fabrication of metal components using WAAM. Hope this text would help budding engineers from industries to fabricate the metal components depending upon the different environments.

REFERENCES

1. Gao, W., Zhang, Y., Ramanujan, D., Ramani, K., Chen, Y., Williams, C. B., & Zavattieri, P. D. (2015). The status, challenges, and future of additive manufacturing in engineering. *Computer-Aided Design*, *69*, 65–89.
2. Seifi, M., Salem, A., Beuth, J., Harrysson, O., & Lewandowski, J. J. (2016). Overview of materials qualification needs for metal additive manufacturing. *JOM*, *68*(3), 747–764.

3. Gibson, I., Rosen, D. W., Stucker, B., Khorasani, M., Rosen, D., Stucker, B. & Khorasani, M. (2021). *Additive Manufacturing Technologies* (Vol. 17). Cham, Switzerland: Springer.
4. Wong, K. V., & Hernandez, A. (2012). A review of additive manufacturing. *International Scholarly Research Notices*, 2012, 1–10.
5. Gibson, I., Rosen, D. W., Stucker, B., Khorasani, M., Rosen, D., Stucker, B., & Khorasani, M. (2021). *Additive Manufacturing Technologies* (Vol. 17). Cham, Switzerland: Springer.
6. Cunningham, C. R., Flynn, J. M., Shokrani, A., Dhokia, V., & Newman, S. T. (2018). Invited review article: Strategies and processes for high quality wire arc additive manufacturing. *Additive Manufacturing*, *22*, 672–686.
7. Duraisamy, R., Mohan Kumar, S., Rajesh Kannan, A., Siva Shanmugam, N., & Sankaranarayanasamy, K. (2021). Reliability and sustainability of wire arc additive manufactured plates using ER 347 wire-mechanical and metallurgical perspectives. *Proceedings of the Institution of Mechanical Engineers, Part C: Journal of Mechanical Engineering Science*, *235*(10), 1860–1871.
8. Sowrirajan, M., Mathews, P. K., Vijayan, S., & Amaladasan, Y. (2018). Effect of weld dilution on post-weld thermal conductivity of austenitic stainless steel clad layers. *Materials Research Express*, *5*(9), 096512.
9. Wu, B., Pan, Z., Ding, D., Cuiuri, D., Li, H., Xu, J., & Norrish, J. (2018). A review of the wire arc additive manufacturing of metals: Properties, defects and quality improvement. *Journal of Manufacturing Processes*, *35*, 127–139.
10. Sowrirajan, M., Vijayan, S., Arulraj, M., & Babu, N. V. (2022). Effect of austenitic stainless steel weld clad layers on conduction heat transfer across the walls of pressure vessels. *Proceedings of the Institution of Mechanical Engineers, Part C: Journal of Mechanical Engineering Science*, *236*(12), 6677–6687. doi:10.1177/09544062211072481.
11. Ley, F. H., Campbell, S. W., Galloway, A. M., & McPherson, N. A. (2015). Effect of shielding gas parameters on weld metal thermal properties in gas metal arc welding. *The International Journal of Advanced Manufacturing Technology*, *80*(5), 1213–1221.
12. Cengel, Y. A. (2008). *Introduction to Thermodynamics and Heat Transfer: Engineering*. McGraw-Hill.
13. Sowrirajan, M., et al. (2022). A new approach to the fabrication of thin-walled plate component through typical wire arc additive manufacturing. *Journal of Advanced Mechanical Sciences*, *1*, 8–13. https://doi.org/10.5281/zenodo.6330531.
14. Sowrirajan, M., Koshy Mathews, P., & Vijayan, S. (2018). Simultaneous multi-objective optimization of stainless steel clad layer on pressure vessels using genetic algorithm. *Journal of Mechanical Science and Technology*, *32*(6), 2559–2568.
15. Rodrigues, T. A., Duarte, V., Miranda, R. M., Santos, T. G., & Oliveira, J. P. (2019). Current status and perspectives on wire and arc additive manufacturing (WAAM). *Materials*, *12*(7), 1121.
16. Kandavel, N., Krishnaraj, C., Dhanapal, P., & Sowrirajan, M. (2021). Assessing the feasibility of fabrication and welding of nickel-alloyed ductile iron through the evaluation of tensile properties and mechanical characterization. *Proceedings of the Institution of Mechanical Engineers, Part E: Journal of Process Mechanical Engineering*. doi:10.1177/09544089211051639

Chapter 12

Review of properties of additive manufactured materials and composites

Francis Luther King
Swarnandhra College of Engineering and Technology

John Baruch
CK College of Engineering & Technology

CONTENTS

DOI: 10.1201/9781003430186-12

12.1 INTRODUCTION

According to ISO/ASTM 52900:2015, additive manufacturing (AM), commonly called as 3D printing, is the industrial process used to generate three-dimensional forms from 3D model data layer by layer [1]. There are many characteristics for the implementation and adoption of 3D printing in the industry, including [2]: structural flexibility, component usefulness, cost-effective low-volume production, and ecological responsibility. Computer-Aided Design (CAD) is used to generate the necessary digital data information for printing [3,4]. The method consists of printing layers of materials such that they are formed one on top of another. Charles Hull invented stereolithography (SLA) in 1986, and additional technologies such as powder bed fusion (PBF), fused deposition modelling (FDM), inkjet printing, and carving have evolved. AM technology enables the construction of lightweight and intricate structures which are difficult to replicate using conventional manufacturing methods. The extrusion-based Stratasys FDM method [5] is one of the additive manufacturing (AM) methods. This AM method is now the most popular on the market [6]. The economic and environmental advantages of additive manufacturing, such as decreased material use, saved energy, individualized production, and prolonged product longevity, are increasing the technology appeal. In the beginning, architects and designers used 3D printing to make prototypes for their projects since it was a fast and cheap way to test out different designs. Producing products that are uniquely suited to each customer has proven challenging for manufacturers because of the significant expenses associated with doing so. Additional expenses incurred during product development are now lower thanks to 3D printing. Though 3D printing has been around for a while, it is only lately that technology has seen widespread application in a broad range of industries, producing anything from prototypes to final products. But AM can 3D print small quantities of customized items at a low cost. The biomedical field, which typically requires unique patient-specific commodities, places a premium on this. Wohlers Associates predicts that during 2020, over half of all 3D printing will be used for commercial product production [7], with the majority of this time spent on producing unique functional objects. Medical professionals are intrigued by this technique because it might be used to manufacture a wide range of medical devices based on CT-scanned tissue replicas [8]. 3D printing is gaining popularity over traditional methods because of its ability to manufacture complicated geometries with more precision, greater material savings, greater creative freedom, and greater potential for personalization. Metals, polymers, ceramics, and even concrete may all be used in modern 3D printing. Polylactic acid (PLA) and acrylonitrile butadiene styrene (ABS) are two common polymers used in 3D-printed composites. Traditional methods are too time-consuming, labour-intensive, and costly for widespread usage in the aerospace sector, hence high-tech metals and alloys are often used instead. Additive manufacturing in construction often uses concrete, although ceramics are commonly utilized for 3D-printed scaffolding. Several advantages of 3D printing over conventional techniques contribute to its growing popularity. These include the ability to precisely manufacture complex geometries, to save materials to the maximum extent possible, to have complete creative control over the design process, and to add a unique, personalized touch to products. Traditional manufacturing procedures, such as casting, are difficult to apply to complex geometries like lattice systems and need extra equipment and post-processing time, while AM eliminates these barriers to mass production. Conversely, improved machine design is required to increase production rates while simultaneously decreasing costs. Another issue that prevents AM from being used for mass production is the high cost and lengthy nature of the process. The aim of this chapter is to offer a high-level introduction to 3D printing by discussing its fundamental principles, common materials, current state, and wide range of applications. The research limitations and implementation difficulties of this technique will also be discussed in this chapter.

12.2 AM CLASSIFICATION

Additive manufacturing (AM) is a technique for constructing complicated geometries and structures using 3D model data. It is also known as digital fabrication technology, rapid prototyping, layer manufacturing, and solid-free form fabrication, among others [9]. This methodology was developed to speed up the process from the initial idea to the finished product [10]. Furthermore, low-volume custom gadgets and goods may be 3D printed using AM technology. Multiple authors have used various additive manufacturing (AM) techniques, including selective laser sintering (SLS), stereolithography (SLA), electron beam melting (EBM), fusion filament fabrication (FFF), selective laser melting (SLM), and inkjet printing, in the

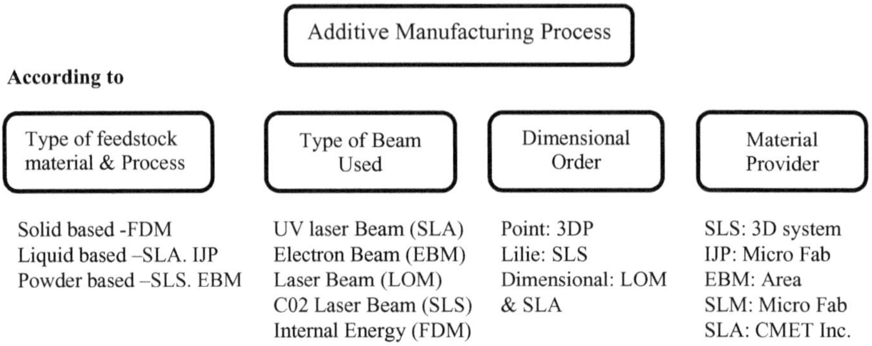

Figure 12.1 Classification of additive manufacturing [136].

course of product development (IJP). As the material is deposited in successive layers, the desired object is gradually built up. This is true of all of these technologies. Some of the materials now in use in AM are: ceramics, wax, metals, concrete, polymer powers, acrylonitrile butadiene styrene (ABS), adhesive-coated sheets, and polylactic acid (PLA). Scaffolding used in 3D printing is often made of ceramics or concrete. Mechanical quality and anisotropic properties of additive-produced components continue to limit AM's viability on a large scale. An improved AM pattern is needed for controlling anisotropic behaviour and susceptibility to defects. Changes to the printing conditions have an influence on the produced goods' quality [11]. In the near future, a wide range of AM tactics will be implemented in today's enterprises. Polymer filament is the primary material used in FDM, the most popular AM method. SLS, SLA, LOM, EBM, SLM, and IJP are just a few of the many AM techniques available. There are merits and drawbacks to every possible strategy. When deciding on the best AM process for a given application, it is crucial to take into account the fabrication cost, time, and accuracy [12]. Generally speaking, AM technologies may be categorized as indicated in Figure 12.1.

12.2.1 Stereolithography

In 1986 [13], the rapid prototyping technique known as stereolithography (SLA) gained widespread use. It uses ultraviolet (UV) light and an electron beam to start off the chain reaction of the monomer solution on the resin layers (EB). Epoxy and acrylic are the two examples of UV-activated monomers that undergo a dramatic transformation into polymer chains after activation. After polymerization, the resin layer hardens to secure subsequent layers. This method is utilized to create high-tech nanocomposites [14] since it just requires one step, is less costly, produces high-quality components that are slow, and generates no waste. A schematic diagram is shown in Figure 12.2.

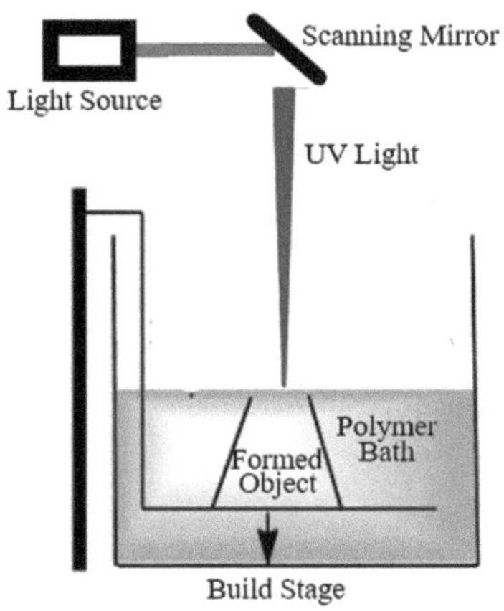

Figure 12.2 Stereolithography.

12.2.2 Fusion deposition modelling

Scott Crump developed fusion deposition modelling (FDM) in the 1800s, and a company in the United States (Stratasys) popularized it in the 1990s [15]. Advanced techniques in thermal energy, surface chemistry, and layer creation form the basis of this method. The filament material, which is obtained as (uncoiled/unwound) metal wire, is heated to a semi-liquid condition, produced, and sprayed on a substrate using a specially designed moving head on the base (platform). Once the print job is complete, the melted filaments are fused together and allowed to cool to room temperature. Layer thicknesses may be anywhere between 50 and 500 mm when using FDM filaments with diameters ranging from 1.75 to 2.85 mm. Mechanical properties of printed components are profoundly affected by processing parameters such as layer width, layer thickness, orientation, and others. Another reason mechanical inadequacy exists is because of interlayer distortion. The key advantages of FDM are its speed, ease of use, minimal equipment cost, and ability to create customized goods. Poor mechanical characteristics, poor surface quality, longer production time, a layer-by-layer appearance, and so on are only some of the problems with this approach [16]. A schematic design of the FDM approach is shown in Figure 12.3.

12.2.3 Inkjet printing

For additive manufacturing (AM) of ceramics, composites, and metals, inkjet printing (IJP) is a significant non-contact approach [17]. Typically, more complex ceramic structures, such as scaffolds for tissue engineering purposes [18], are printed using the IJP approach. Wax support materials and thermoplastic building materials are maintained in a molten state in two heated reservoirs, a setup developed by Solidscape Inc. The IJP head (X, Y planes) takes these materials as input and fires droplets in the right

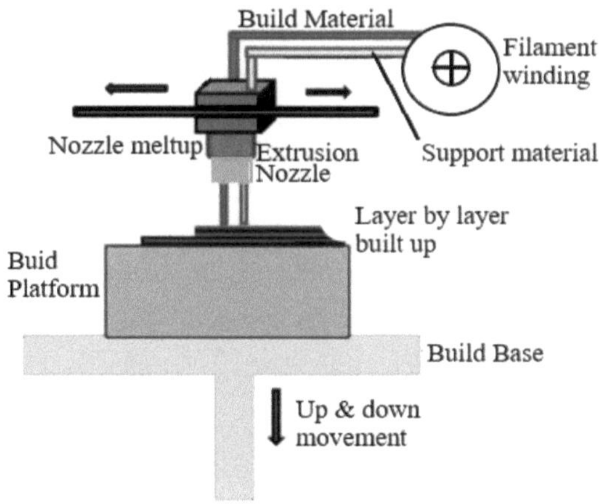

Figure 12.3 Fusion deposition modelling.

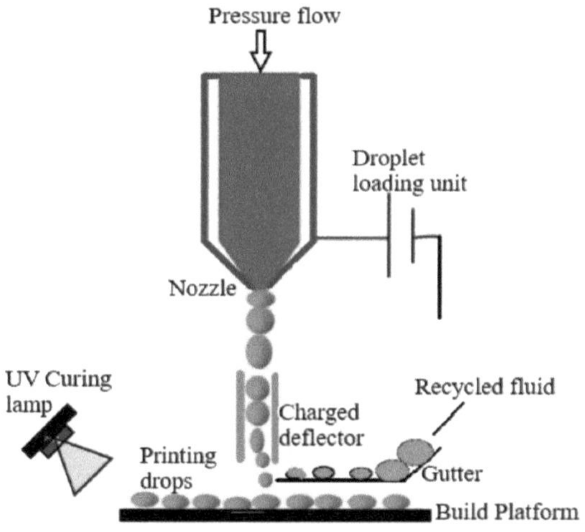

Figure 12.4 Inkjet printing.

direction, creating a uniform layer of chemicals. At the same time as both substances cool and solidify, a layer is formed. Therefore, for each layer, the same process is carried out once more. Once a part is removed, the wax shims supporting it are melted [19]. Inkjet printing (IJP) is efficient, rapid, flexible, and capable of generating high-quality printing and intricate structural designs at low cost. Major limitations of this method include low-quality printing, poor durability in the printing head, poor adherence between layers, and an inability to print in large quantities. In Figure 12.4 we see a simplified diagram of the inkjet printing method.

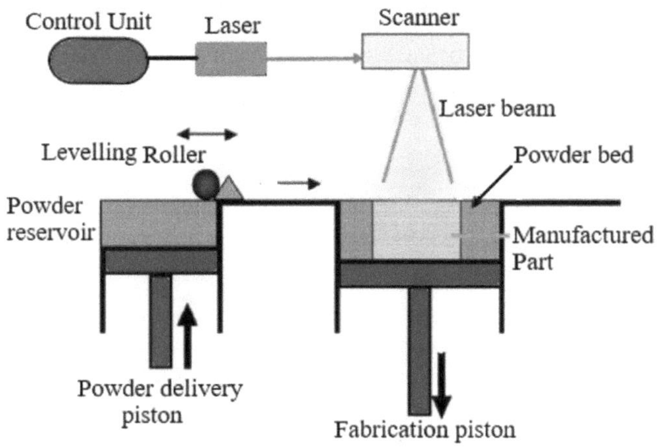

Figure 12.5 Selective laser sintering.

12.2.4 SLS (selective laser sintering)

To create a form using selective laser sintering (SLS), material such as ceramics, metals, polymers, hybrids, and composites in powdered form is heated without melting and then dispersed uniformly, layer by layer, over the model's supporting structure. After being sliced into a cross-section of a CAD model by a laser system, powder particles sinter together to form the final product. This procedure is repeated until a finished model is achieved [20]. Solid-Liquid-Sintering (SLS) offers excellent mechanical properties, is fast, does not need support, and has good layer adhesion. The primary problem of this method is the need for post-processing due to the inner porosity of the structure. Additionally, large, flat surfaces and microscopic holes cannot be printed consistently [21]. Particle sizes in SLS processes typically fall in the range of 10–50 μm [22]. Figure 12.5 presents a simplified layout for the SLS (Selective Laser Sintering) method.

12.2.5 Selective laser melting

Bridges and crowns for teeth may now be made with remarkable precision using a technique called selective laser melting (SLM). Materials suitable for the powder-based fusion method include metals, polycarbonate (PC), polymers such as PLA, ABS, wax, and powdered plastics [23]. Furthermore, this technique employs powder layers that are just nanometres thick and are densely packed on top of a base. As the laser beam moves away from the melt pool, the liquid substance solidifies. Because of this, a substantial structure takes shape. An additional layer is applied after the powdered material layer has melted and fused the layer underneath it. Figure 12.3 shows that the process is repeated to create the whole structure. Important aspects of SLM include the chemistry of the binder, the particle shape and size, the interaction between the binder and powder, the rate of deposition, and the processing that comes afterwards [24]. However, the superior print quality and high resolution of SLM make it a fantastic option for producing elaborate graphics. Figure 12.6 depicts a simplified layout of the Selective Laser Melting method.

12.2.6 Electron beam melting (EBM)

It's an electron gun and a computer-controlled process for making highly thick 3D components from metal powder [25]. First introduced to the market was the EBM printing machine by Arcamin Sweden [26]. The powder feedstock is fed to the construction container from hoppers next to it, creating a nearly vacuum environment in which the parts are fabricated. Figure 12.7 depicts the overall framework of the electron beam melting method.

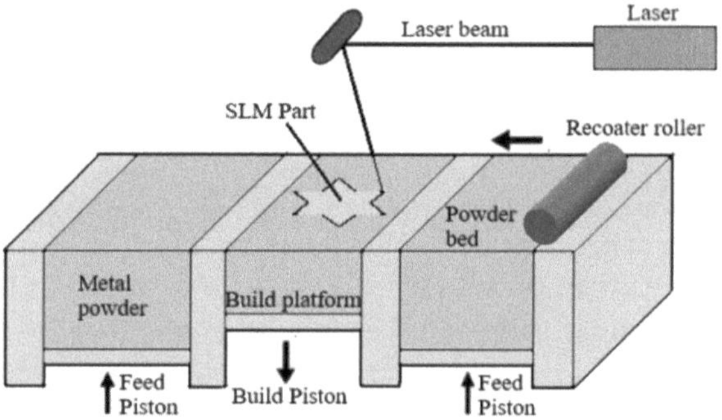

Figure 12.6 Selective laser melting.

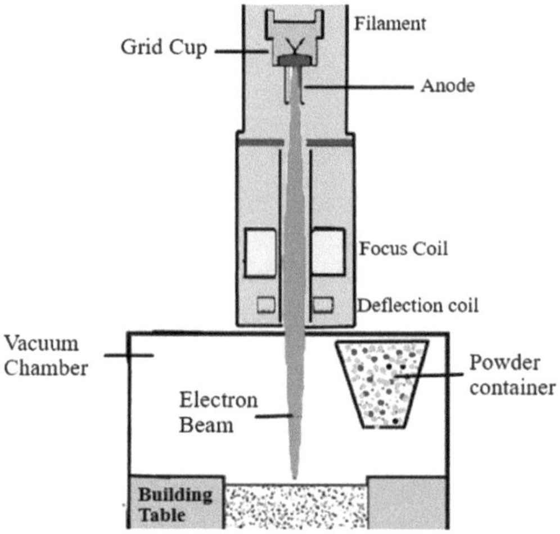

Figure 12.7 Electron beam melting.

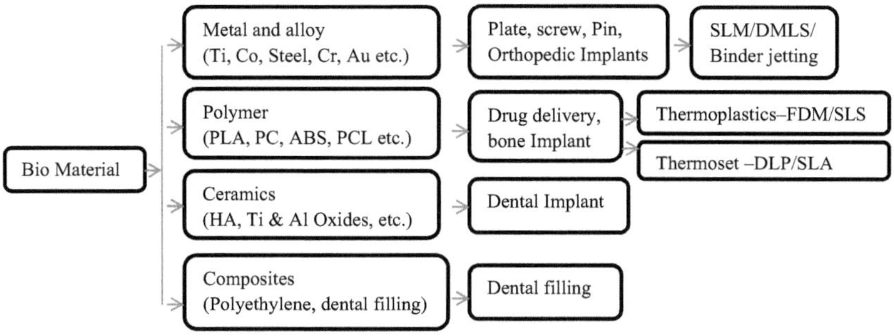

Figure 12.8 Various classifications of AM materials.

12.3 3D PRINTING MATERIALS

Additive manufacturing today utilizes a wide variety of materials, including ceramic, polymers, metals, plastics, and their hybrids. However, improvements are continually being made to advanced materials [27]. A variety of 3D-printable materials are available for use in creating a wide range of parts. The several categories of AM materials are shown in Figure 12.8.

12.3.1 Metal and alloy

Among its many impressive properties, the metal utilized in AM has exceptional physical properties, including high compressive strength and resilience to fatigue, which allow for the creation of intricate designs. Not only is it used to manufacture aeroplane parts, but also human organs. Essential metals include those based on cobalt, aluminium, nickel, stainless steel, titanium, and many more. In particular, cobalt-based alloys are recommended due to their high specific stiffness, elongation, durability, and recovery capacities [28]. Thus, it finds widespread use in the dental industry.

12.3.2 Ceramics

Creating high-end ceramics for use in tissue engineering and biomaterials like dental and bone scaffolds is becoming more common thanks to AM technology [29]. To date, AM has been utilized to make fracture- and pore-free 3D prints out of materials like ceramics and concrete. Ceramics are fireproof and endure a long time. It may be used in the aerospace industry, in hospitals, and in dental practices. Ceramic materials include alumina, zirconia, and bioactive glasses. In contrast, alumina finds widespread use in a number of industries and fields [30].

12.3.3 Polymer

Prototypes of fully functional buildings and polymers are often employed in AM's component creation process [31]. It has a lot of advantages, including portability, affordability, durability against chemicals, adaptability in manufacturing, and so on [32]. Polymers are used in the creation of biomaterials, medical devices, and orthopaedic implants [33]. Tissue engineering, cartilage healing, as well as other applications continue to use polymers [34].

12.3.4 Composite

For a number of years, a wide variety of industrial applications have been researching the potential for additive manufacturing to be used in composite materials (construction of buildings, the medical field, the toy industry, and aerospace) [34]. Traditional materials are heavier and less versatile than composite materials, which are lighter and more adaptable. Composite materials, for instance, have transformed industries for the purpose of achieving increased performance owing to the exceptionally low weight and adaptability of these materials. The most prevalent types of composites are carbon fibre-reinforced polymer composites and glass fibre-reinforced polymer composites [35]. As a result of its remarkable properties, carbon fibre composites find the majority of their applications in the aircraft industry (high strength, corrosion resistance, and superior fatigue performance).

12.3.5 Biomedical materials

To diagnose and cure medical conditions, biomedical engineers work on a wide range of tools and technologies. Among the many biomedical applications are replacement organs such as kidneys, heart, teeth, knees, eyes, ears, noses, and hips. Printing for these medical uses paves the way for individualized, complex implant geometries and on-demand fabrication, both of which have the potential to reduce both inventory and production costs. Further, because no specialized equipment is needed for 3D printing, the unit cost is uniform across all products. This kind of cost analysis has proven useful in the field of biomedical orthopaedics. Biological uses for this technology are on the rise as researchers discover its many advantages. In the field of medicine, applications include implants and tissue engineering [36]. Applications for 3D printing in medicine may be constructed out of a variety of different materials. In addition to its other desirable qualities, the perfect biomaterial would be simple to print, biocompatible with living organisms, and morphologically indistinguishable from real tissue. Metals and alloys are used for orthopaedic implants, plates, screws, and other components that need to have a high level of strength. Ceramics have the potential to be used in the production of bioactive orthopaedic implants. Porous orthopaedic implants often make use of composite, which is a biological material that has outstanding mechanical qualities. On the other hand, polymers are the kind of chemicals that are used in the medical field the most often. Biomaterials made from polymers are the most popular kind, about 86% [37].

Different implant applications call for different materials to be used for biological objectives. To choose the best material for medical AM applications such as personalized implants, surgical equipment, prosthetic limbs, tissue scaffolds, and so on, one must consider the final product's function and desired characteristics. Nylon and acrylonitrile butadiene styrene (ABS) are great materials for use in medical devices because of their high strength, long life, and versatility. Polylactic acid (PLA) is, nevertheless, a biodegradable polymer with applications in medicine. An ideal biomaterial would have several desirable properties, such as being easy to print, biocompatible, having a comparable morphology to genuine tissue, being nontoxic, etc. Metals and alloys are used in the construction of orthopaedic implants, plates, screws, and other structural components. Exceptional mechanical qualities of ceramics utilized in porous orthopaedic implants may allow for their usage in the creation of bioactive orthopaedic implants. On the other hand, polymers are by far the most common substance utilized in healthcare. Over 86% of all biomaterials now in use are polymer based. However, tissue engineering makes use of biomimetic materials. Scaffolds are the focus of a number of ongoing efforts to enhance them utilizing time-tested techniques including salt leaching and phase separation. However, bone tissue engineering plays a crucial role in the treatment of wounded tissues and harvested tissues, and has been the subject of a number of process techniques, including rapid prototyping (RP), electrospinning, and others. It is common practice to use polymers (polycaprolactone, poly (lactic-co-glycolic) acid) and ceramics (wollastonite bredigite) in bone tissue engineering [38,39].

12.3.6 Smart materials

Smart materials are capable of changing shape and structure when subjected to external forces (water and heat). A self-evolving structure and soft robotics system are the most prominent examples of additive-produced items developed with the help of smart materials. There is a subgroup of smart materials known as 4DP materials [40]. Biomedical prostheses and micro-electro-mechanical systems both make use of shape memory polymers and alloys (Ni–Ti) [41].

12.3.7 Plastics

FDM, SLS, and SLA may all use plastics as their material of choice. Acrylonitrile butadiene styrene (ABS) filament is the most common kind of plastic used to make car, phone, and electronic appliance housings. However, the polybutadiene-based thermoplastic component utilized in elastomer production gives it resilience to stress and pliability. ABS plastic, which is often used in additive manufacturing, is melted at temperatures between 230°C and 260°C. ABS may be made in a chemical process and has high strength and toughness and can be recycled several times. However, it is not biodegradable and it loses volume when exposed to air. Therefore, heating the printing base is necessary to prevent warping. The ABS plus material, however, is used in all state-of-the-art Stratasys FDM machines. If you're looking for a transparent material with the same impact resistance as regular ABS, you have a few options. You may use ABS and PC together in certain FDM machines. Polylactic acid (PLA) is used in AM because of its many desirable properties. These include low printing temperatures, biodegradability, and decreased shrinkage (190°C–230°C). However, PLA is more challenging to deal with because of how quickly it cools and solidifies after being handled. Polyethylene terephthalate (PET), polycarbonate (PC), and other polymers are also utilized in 3DP [42].

12.4 MECHANICAL PROPERTIES

Large-scale printing is still restricted because of the poor mechanical qualities and anisotropic behaviour of the component created by AM. Therefore, an optimal AM component arrangement is necessary for regulating anisotropic behaviour and defect sensitivity. On the other hand, AM has the capability of producing components in a diverse array of dimensions (micro to macro scale). Saroha et al. [43] used an electrodeposition method to deposit Ni coating on ABS material. This was done in order to compare their results to those of other researchers who use metallization. Ni-coated ABS improved its microhardness by 27.1% and reduced its wear rate by 95.5% compared to pure ABS. Numerous studies have shown that applying a metal coating on AM components is one approach to improve their mechanical qualities.

12.4.1 Metals

The manufacturing of metal additives has significant potential for expansion. The number of firms supplying additive manufacturing systems expanded from 49 in 2014 to 97 in 2016, with 49% of those companies active in the production of metal AM products [44]. Research, prototyping, or other specific applications, such as Boeing's fabrication of the F-15 pylon Rib [45], were the most common uses for this technology in the aerospace sector[45]. It has found applications in the fields of both medicine and transportation. Metal additive manufacturing (AM) offers a great deal more freedom than conventional manufacturing methods when it comes to the design of intricate geometries and one-of-a-kind connections. In order to provide both structural and protective solutions, multi-functional components may be developed. In metal 3D printing, one of the first steps involves melting powder particles or wire using a laser or an electron beam. This is an essential phase. The molten substance is cooled and solidified layer by layer, starting at the bottom and working its way up. Powder bed fusion (PBF) and direct energy deposition (DED) are the most common techniques for printing metals. However, in recent years, other methods have been developed, including binder jetting [46], cold spraying [47], friction stir welding [48], direct metal writing [48], and diode-based processes [49]. These methods have the potential to enhance precision and velocity. Inconel 625, Inconel 718, and Hastelloy X are Ni-based superalloys that can withstand high temperatures. Cobalt-chromium (Co28Cr6Mo) alloys are resistant to extreme heat. High-temperature-resistant (TRL) aluminium alloys include AlSi12, AlSi10Mg, AlSi7Mg0.6, 1050A, 2017A, 2219, 6061, 7020, 7050, 7075, and 5083. The only way to bridge the gap between what can be accomplished with additive manufacturing (AM) and what could be accomplished with conventional casting and forging processes is to [50] increase the maturity of available resources to promote the fabrication of critical components or large-scale production, and expand the wide range of compatible materials. Metals with structures, microstructures, and three-dimensional multiscale designs that are not observed in cast and wrought materials may be created by rapidly cooling the material and guiding the crystallization process. A few microstructural characteristics of AM-processed metals are fine grains, anisotropic with elongated grains, non-equilibrium (metastable phases, solute trapping, non-equilibrium compositions) microstructure, and metallurgical flaws like porosity due to partially and non-melted powders and gas entrapment [51]. Surface characteristics common to AM-fabricated metal components include weld cracks, projecting partially melted powder particles, ejected molten droplets, and recesses. Due to the possibility of bulk and surface structural changes during AM, the mechanical behaviour of AM metals may significantly deviate from that of traditionally treated materials [52]. Therefore, it is crucial to have a deeper understanding of the numerous interactions between powder's metallurgical characteristics, processing, microstructure, and mechanical properties.AM has been found to be effective with a variety of metals and metal alloys, including titanium and its alloys, steel and its alloys, aluminium and its alloys, nickel and its alloys, and even cobalt-based and magnesium alloys. High-performance materials like titanium and its alloys have many applications across several sectors [53]. Significant progress has been made in the field of titanium and titanium alloys, and both Ti [54] and Ti6Al4V [55] are currently employed in commercial aerospace and medical applications.

AM makes use of a number of distinct kinds of steel, such as austenitic stainless steels, maraging steels [56], precipitation hardenable stainless steels [57], and tool steels [58]. The high hardness and strength of these alloys makes them perfect for use in tool and mould production. As austenitic steels and precipitation hardenable stainless steels increasingly rely on AM characteristics, this is especially true for precipitation hardening. Dense metallic components made using AM are on par with, or even somewhat better than, those made using more conventional methods. This effect can only be achieved by careful manipulation of the material's porosity and microstructure. Porosity [59] is the primary defect that causes fracture propagation and may be controlled by adjusting the applied volume energy and the feedstock quality [60]. A little amount of energy is all that's needed to create a vacuum of an unpredictable form in the material [61]. When too much force is applied, however, the pores become round. It is possible to improve material quality by boosting flowability and homogeneity [60] by using denser powder beds and smaller, homogeneous spherical particles. Preserving the alloy's purity necessitates avoiding exposure to any contaminants. CoCr alloys have actually been examined for prospective use in the medical and dental areas, and the development of nickel-based superalloys (Inconel 625 [62] and 718 [63]) was motivated by the demand from these industries. Studies using Mg alloys (for biological resorbable applications), Au, and Cu have also been performed. Improvements in our understanding of metals and metal alloys are broadening the range of materials that may be employed in AM. In a number of important performance categories, including as strength-to-weight ratios, fracture resistance, tensile strength, corrosion resistance, and oxidation resistance, these alloys surpass more conventional alloys.

12.4.2 Multi-use functional materials

Both additive manufacturing and magnetoelectric composite materials are growing as individual disciplines, and they are poised to be merged to power the coming fourth industrial revolution. The interplay between the manufacturing process, the resulting material's structure, and its characteristics is highlighted as a defining characteristic of this new category of materials. Even though the idea of multiferroic materials was initially presented around the turn of the century [64], it has only recently caught the attention of the scientific and technological communities, particularly in the case of those materials that display intrinsic or extrinsic magnetoelectric coupling. The effect of an applied electric (or magnetic) field on a material and its consequent change in magnetization is described by magnetoelectric coupling [65].

12.4.3 Polymer

Polymers are the most often used material in 3D printing because of their adaptability and versatility across a wide range of 3D printing processes. Because of their intrinsic lack of strength and usability, pure polymer goods printed using 3D printers are often only used for conceptual prototypes.

Continued investigation into the mechanical properties of 3D-printed polymers has led to the creation of novel methods and materials for fabricating high-performance advanced polymer composites. Both polycarbonate (PC) and polylactic acid (PLA) [66] may be worked with using 3D printing techniques. Although it may be difficult to keep PLA resin at the right viscosity when temperatures are low, there are two options worth exploring: raise the temperature of the whole production process and use enough plasticizer. The second strategy is more resistant to heat deterioration. Senatov et al. [67] are now employing PLA-based composite mixes to create 3D-printed scaffolds for tissue engineering. A variety of 3D porous biocompatible scaffolds made from PLA and bioactive CaP glass have been produced for use in tissue engineering [68]. Polymeric cellular structures are created with the use of chemical or physical blowing agents. Abueidda et al. [69] used 3D printing to create new polymeric cellular materials and analysed the mechanical properties of three triply periodic minimum surfaces (TPMS). PLA has less environmental effect but less mechanical capabilities than ABS, which has stronger mechanical strength characteristics but produces an unpleasant odour when processed. Better mechanical characteristics, including elastoplastic and orthotropic behaviour with substantial asymmetry in compression and tension, were found in 3D-printed PLA compared to injection-moulded PLA by Song et al. [70].

12.4.4 Composites

The evolution of 3D printing of polymer composites has allowed for the investigation and creation of novel materials. Industrial production could benefit greatly from combining 3D printing with polymer matrix composites, especially in terms of functionality and mechanical performance [71]. However, a fundamental obstacle remains: the lack of printable materials that can scale up the industrial use of 3D printing for high-performance composites. AM composites cannot compete with older technologies in terms of speed or reproducibility. Interest in moving beyond rapid prototyping into mass customization of functional components, however, has been rising steadily. As a result of this change, it may be possible to synthesize new materials with superior mechanical properties by using other matrix materials. Therefore, there has to be more study done to determine which materials and applications are best for 3D printing polymer composites. Fibre-reinforced hydrogel composites were fabricated swiftly by Bakarich et al. [72] by merging 3D printing with a UV-curable glue. Advanced hydrogel-based composites may be made with the help of the right ink resolutions. Utilizing AM technology, cement mortar reinforcing parts were 3D printed using photopolymer and titanium alloy fibres. Using this technique, we were able to simultaneously create a novel surface form and improve the matrix's toughness. Reinforcing fibres' shear strength, flexural toughness, and fracture resistance were highlighted. Optimizing the fibre surface design may increase the energy-absorbing capacity of fibre-reinforced mortars [73]. An alumina particle-reinforced polymer matrix was developed for usage as a filament in FDM [74] or as nanofillers in SLM.

Fibre reinforcing might be used in the future to enhance the mechanical properties of polymers for use in 3D printing [75]. Improving the mechanical characteristics of 3D-printed polymer composites by using continuous fibre reinforcement has proven challenging in recent years [76]. Tekinalp et al. [77] researched the load-bearing capability of composite products made using carbon fibre and ABS resin feedstock, and they also looked into the challenges of 3D-printed fibre-reinforced composites. The strength and rigidity of samples made using FDM and compression moulding (CM) have been greatly enhanced. Furthermore, the increased tensile strength of the CM samples indicated that porosity was a more important factor in determining tensile characteristics than fibre orientation. By combining CF with polymer feedstock, the final 3D printed goods will be stronger and stiffer with less deformation [78].

12.4.5 Ceramics

An increasingly important tactic [79] in the field of biomaterials and tissue engineering is the use of ceramic scaffolds for bone and teeth. Although 3D printing ceramics is very accurate, there are still certain drawbacks, such as the material limitations and the layer-by-layer visual appearance [80]. Sintered ceramic items need expensive and time-consuming post-processing to get the correct shape. The ability to 3D-print intricate designs and then fuse the layers together to form strong ceramics has become more attractive. Additional advantages from 3D printing of porous ceramics or lattices include the production of high-quality, lightweight materials that may be tailored to meet specific requirements. Ceramic scaffolds for tissue engineering have advanced to the point where they can replace time-consuming and labour-intensive methods like casting and sintering. Common approaches to 3D printing ceramics include powder bed fusion, paste extrusion, stereolithography, and inkjet printing (suspension). The most common method for making thick, unaltered ceramic samples is inkjet printing. The rheology of the fluid used in 3D inkjet printing must be controlled so that it flows easily, does not clog the nozzle, and dries rapidly. The viscoelastic behaviour of inks governs inkjet printing of ceramic suspensions [81]. When printing ceramic suspensions using an inkjet printer, the filament is often broken and thinned. Subtractive layer sintering (SLS) is another popular method for 3D printing ceramics (selective laser sintering). However, ceramics may crack due to the sudden temperature change between fusion heating and ambient temperature. Additionally, ceramic matrix composites produced using selective laser sintering (SLS) and the sol-gel method (SLG) have been proven [82]. Binders with low melting temperatures are required for usage with ceramic powders because they do not readily fuse or melt at the temperatures reached by low-power laser heating. The ceramic momentarily trapped by the laser-activated binder is sintered at high temperature in the green section to solidify it permanently. After sintering, the auxiliary binder may be washed away. When combining ceramics with glass or polymers, indirect SLS is often used. Better printing resolution, mechanical performance, and faster settings were achieved by Vorndran et al. [83] when they used phosphoric acid as a binder for their tricalcium phosphate ceramics. Binder removal from the powder bed was also simplified. Without a sintering post-treatment, SLS-made ceramics tend to be less dense and more porous than cast ceramics. Object flowability, density, and shrinkage during the printing process are all influenced by the particle size distribution of the ceramic material. According to the findings of a glass-ceramic system, a larger proportion of microscopic ones lowered flowability, which led to worse printing resolution and increased contraction. In addition, during the sintering process, glass-ceramic powder experiences a greater degree of contraction than other powders do as a result of its lower bulk density (poor packing) [84]. For the purpose of 3D printing ceramics, high-viscosity slurries are not appropriate since the extrusion of ceramic paste requires rheology [85]. After 3D printing, the characteristics of the finished ceramic may be affected by a number of parameters, including the liquid-to-solid ratio, the amount of air that is trapped in the material, the temperature, the drying and de-binding process, the solidification kinetics, and the interlayer adhesion. In direct ceramic stereolithography (also known as CSL), a heat-curing binder that contains a significant amount of ceramic fillers is used to generate the green body of the ceramic. By using ceramic particles of smaller grain size that have increased light scattering capabilities, one may accomplish improved light curing of the binder. In addition, a nano-ceramic resin that can be printed with a laser has been created [86]. Liu et al. [87] studied how zirconium or magnesium solution infiltration and subsequent in situ precipitation affected alumina ceramic stereolithography. Although fracture toughness was much lower than in uninfiltrated samples, significant strength gains were seen. Unlike powder printing employing SLS technology, the process produces 3D-printed objects unaffected by the particle size distribution of ceramic powder in the paste. When comparing a mono-sized alumina particle distribution to one with a bimodal particle distribution, Wu et al. [87] discovered that the latter resulted in higher density.

12.4.6 Concrete

The use of concrete additive manufacturing in construction has grown in recent years.

Additive manufacture of buildings has become more popular, with the approach of contour crafting taking the lead [88]. High pressure is applied to bigger nozzles, and the concrete paste is extruded. To avoid the appearance of individual layers, a trowel-like device has been designed and attached to printers [88]. In the building sector, 3D printing is still in its infancy. Consequently, the future of the technology is uncertain. Perhaps most important to effective contour shaping is the concrete's natural freshness. To 3D-print intricate patterns, a material must have both the buildability (the capacity to support further layers) and the extrudability (the ability to be pushed through a nozzle) of concrete. Le et al. [89] developed a high-performance polypropylene fibre-reinforced mortar that was 3D printed. The largest aggregate size that they used was 2 mm. The key elements that went into the mortar were Portland cement, fly ash, silica fume, and sand (maximum aggregate size of 2 mm).

We used superplasticizers and a retarder to make the material more malleable while extruding it via a nozzle with a 9 mm diameter and a 100 ms open time. A total of 61 layers may make it into the final product (about 400 mm). When extruding, materials must be developed to allow for longer workability before setting, and they must also have strong early strength to prevent following layers from collapsing. In the printing technique devised by Gosselin et al. [90], the accelerator and premix mortar are pumped via individual tubes and mixed just before being extruded. This allows the printed layers' early strength to be retained and the premix mortar's rheology to be conserved for a longer period of time, facilitating the accurate building of subsequent layers. This method might eliminate the need for temporary supports in the production of larger structures with complex geometries by using a six-axis robotic arm and by maintaining control of the material's behaviour during and after the extrusion process. Paul et al. [91] studied several concrete recipes and found that the thixotropic behaviour of the concrete is a crucial element in pumping and printing. Researchers found that printing characteristics, such as nozzle shape, were significant in determining the mechanical properties of the specimens. The interlayer adhesion of 3D-printed concrete structures is a particularly difficult aspect. Obtaining suitable external support and then removing the external support after 3D printing is a complicated process and the tall structures is also another significant problem [92].

Interlayer binding strength was studied by Zareiyan and Khoshnevis [93], who looked at how aggregate size, extrusion, and layer thickness all had a role. Increased strength was attributed to improved interlayer bonding, which was shown when the maximum aggregate size was decreased and the cement to aggregate % was raised. Because of the higher layer thickness and longer time between successive layers, the printed structures' compressive strength decreased, despite improved interlayer bonding. Cold joints across layers will be much more likely to occur when curing durations are shortened [93]. Even though cement and concrete slurry are often utilized in 3D printing, powder bed fusion also has been investigated. Regular Portland cement, calcium aluminate cement, and an aqueous solution of lithium carbonate were used to create the powder bed, as described by Shakor et al. The printed layer's resistance to settling and distortion due to succeeding layer printing is what is meant by shape-stability of the printed component. 3D printing cement paste with added silica fume and nano-clay improves form stability significantly, as reported by Kazemian et al. [94]. Inadequate powder-water contact led to low hydration and compressive strength of 8 MPa, while porosity was near 50%. In the process of 3D printing wet concrete, it may be difficult to maintain control over the flowability and open time for extrusion of the material, in addition to the structural characteristics such as stiffness, interfacial bonding, distortion, and build-ability. The durability of 3D-printed buildings is an important factor to think about. Water in a 3D-printed construction may evaporate more quickly than in traditional concrete since there is no formwork to shield it from air exposure. That might make shrinking and cracking more likely. Powder bed AM has to be improved upon so that strong structures may be made with it [94].

12.5 MATERIALS FOR 3D PRINTING: A COMPARISON

Recent developments in additive manufacturing have made it possible to print three-dimensional objects out of a diverse array of materials, from chocolate to complex multi-functional materials. Printing in three dimensions might make use of a wide variety of materials. You may locate them in the shape of filaments, wire, powder, paste, sheets, or inks, among other possible manifestations. Three-dimensional printing makes extensive use of thermoplastic polymers and thermosetting powders. Examples of the former include polystyrene, polyamides, and photopolymer resins; examples of the latter include ABS, polyamide (PA), polycarbonate (PC), and polylactic acid (PLA). Because of their poor mechanical properties, the majority of applications for 3D-printed polymers involve quick prototyping. In recent times, research has been conducted on the practice of dispersing nanoparticles and fibres to polymers with the intention of enhancing the mechanical strength properties of printed products that will be utilized as load-bearing or functional components.

Complex components of varying sizes are being printed using metal and alloy 3D printers in the aerospace, defence, and automotive sectors. The most common form is a powder made of metal (or wires). Traditional 3D printing methods include SLS, SLM, and DED. Currently, 3D printing can only work with a limited variety of metals and alloys; research is being conducted to expand this list, as well as to apply current techniques to a larger range of materials and develop composite structures [95]. Issues with metal 3D printing include defects like porosity and bead accumulations, as well as material properties and morphologies that change depending on the pieces' orientation. Improvements in processing parameters and post-processing therapy are the subject of continuing theoretical and clinical research.

Using a laser or electron beam to sinter or melt powders is a common feature throughout these techniques. With the advent of 3D printing, ceramics with a high strength-to-weight ratio can now be easily customized and fabricated into complex lattices for use in a wide variety of settings. Ceramic scaffolds of varying forms and durability are essential in fields such as tissue engineering. Ceramics are often 3D printed using ceramic powders or dyes. Lasers may be used to fuse powders together or sinter them using a secondary adhesive. Conversely, inkjet printing is often used to create slurry of ceramic particles that are subsequently sintered at extreme temperature. The biggest problem is the poor quality and inaccurate dimensions that come from using 3D printing for ceramics. However, using ceramic 3D printing, you may tailor the materials' microstructure and chemical makeup to your exact specifications. Since there is room for improvement in AM methods and in the range of materials available for 3D printing ceramics, further study is warranted. The ability to alter materials across levels is not shared by all entities. High-pressure jetting procedures can only combine materials with similar rheological properties and curing temperatures, whereas extrusion-based processes like FDM are confined to blending materials with similar melting temperatures [96]. Bioprinting using multi-jet fused deposition modelling (FDM) technology is gaining popularity [97]. These scaffolds made of hydrogel and cells may mimic native tissue structures. A number of additional 3D printing methods, such as direct energy deposition (DED), enable the continuous extrusion of multiple materials by combining metal alloys or ceramics. It is important to take precautions while working with two different materials so that unwanted phases don't form [98]. In powder-based technologies, powders having a broad range of melting points have been employed effectively. When put in close proximity to particles having a higher melting point, materials with a lower melting point may remelt or deteriorate. Improved numerical modelling and optimization techniques in material processing have the potential to enhance the performance of multi-material production processes and the multi-functional responses they provide [99].

12.5.1 Metal matrix composites (MMC)

Multi-Metal Composites (MMCs) have a ductile metal as its matrix. Fibre has the potential to improve the composite's strength, stiffness, creep resistance, abrasion resistance, and thermal conductivity, while the metal matrix offers the composite with high-temperature ductility and thermal stability. Aluminium and its many alloys, copper, titanium, magnesium, and other elements are the MMC metals that are used most often. On the basis of the kind of reinforcing material that they make use of, metal matrix composites may be divided into three categories (either fibres or particles). Powder metallurgy, spray deposition, and squeeze casting are the three most common methods used to produce MMC [100].

In the mature, practical stage of additive manufacturing, Ti-, Ni-, and Fe-based alloy powder are often employed. The ease of process control makes LM or LMD the most essential processing technology for specific alloys. Ti6Al4V is the most used Ti-based alloy because of its many uses in aerospace and the medical fields. Because of their low coefficients of expansion, high strength-to-weight ratios, and resistance to corrosion, creep, oxidation, and damage, Inconel 625 and 718 are the most widely used nickel-based alloys. Components for jet engines and gas turbines could be made of Inconel 625 or 718. However, there are restrictions on the AM fabrication of Fe-based alloys since the density of AM-produced steels is the same as that of completely dense steel. This restriction is mostly attributable to the fact that steel's main constituents all have their own unique chemical characteristics; however, research is being conducted on potential solutions. The use of conventional machining methods on Ti alloys is problematic and costly [101].

Selective laser melting (a powder bed method) is used to examine AM-recommended AlSi10Mg alloy for a wide range of performance parameters, including high cycle fatigue, fracture behaviour, and microstructure. A total of 91 samples were utilized and were heated to varying degrees (30°C and 300°C, and oriented at 0°C, 45°C, and 90°C). For this purpose, we used statistical approaches (design of experiments) and analysed the data using marginal means plots. Post-heat treatment was determined to be the most influential element in fatigue resistance, whereas construction orientation had no role. Platform heating at 300°C and peak-hardening was shown to be the ideal setting for equalizing variations in fatigue life and increasing fatigue resistance. Si particles are merged into spherical forms, and the as-built microstructure is homogeneous and constant at all angles (0°, 45°, and 90°). This is because all dendrites, laser traces, and heat-affected zones have been erased. Surface or subsurface occurrence of breakthrough fractures is defined by the presence of non-melted areas and pores [102]. Selective laser melting using a pulsed-laser source has been used to treat Al12Si alloy, resulting in excellent microstructure optimization and precise temperature control. Densities of 95% or more, hardness of 135 HV or more, and Si refinements of less than 200 nm were all achievable. In contrast to cast alloy and prior SLM studies, an increase in hardness is seen, and the morphology of the microstructure is demonstrated to be dependent on temperature gradients and significant undercooling [103].

12.5.2 Ceramic components and ceramic matrix composites

Traditional methods of producing ceramic matrix composites are troublesome. But AM technology can handle such materials without the requirement for moulds or specialized equipment for each individual component.

Three-dimensional printing, selective laser sintering, stereolithography, and direct inkjet printing are some of the most widely used techniques for creating ceramic components [104]. Completely dense ceramic free-form components were fabricated and designed utilizing AM technology, and high-strength oxide ceramics (ZrO_2-$Al2O_3$ ceramic) with exceptional mechanical qualities were used. In laser beam melting (SLM) experiments, ZrO^2-Al^2O^3 was completely melted. Density of 100% and flexural strength of 500 MPa were measured in specimens that had not undergone any sintering operations or post-treatment. In terms of microstructure, tetragonal zirconia and alpha-alumina were fabricated into two fine-grained phase microstructures, respectively. Dental restorations and complex ceramic prototypes are only two examples of the various uses for this technique, which offers several advantages over laser sintering processes. Despite these advantages, created components have restrictions on their surface quality and mechanical strength [105].

12.5.3 Polymer matrix composites

To a large extent, AM technology influences the processing of effective and promising polymer composites, such as carbon fibre-polymer composites, due to its ability to manage complicated structures with significant design freedom. Samples that were 3D printed outperformed their traditional compression-moulded composite counterparts in tensile strength, tensile modulus, and yielding by around 115%, 700%, and 91.555%, respectively. Microstructure and mechanical performance were examined as a function of fibre orientation, distribution, and void formation. It has been reported that the use of carbon fibre-reinforced material with appropriate orientation, high dispersion capabilities, and surface modification to increase interfacial adhesion among fibres and matrix significantly improves the performance of load-bearing composite materials [106]. There are a few roadblocks that prevent the widespread use of AM for polymer composites, such as [107] constraints on mechanical properties, extremely slow production rate, and tiny dimensions. However, advanced methods have been developed to overcome these challenges, and carbon fibres in particular have benefited from these innovations. The methods mentioned in ref. [107] have been proposed for the development of carbon fibre-reinforced polymers to increase the material's specific strength. Deposition-related material distortion and warping may be minimized by using carbon fibre reinforcements in the production process. Carbon fibre and additive manufacturing work together to make it possible to make complex parts that would not be possible using each technique alone. However, the advancement of polymer composites by AM technique is significantly influenced by the commercialization of quasicrystalline polymers. Properties of the suggested material were compared to those of hard steel using SLS. Porosity, hardness, wear resistance, and friction coefficient were among the characteristics evaluated in this research. Due to its beneficial features, quasicrystalline polymer composites have found use in the automobile sector [108].

By combining powder and wire feeding laser deposition into a single phase, we were able to manufacture composite sandwich functionally graded structures. This gave us the opportunity to investigate the survivability and performance characteristics of these structures. Copper powder and nickel wire were used to coat the H13 tool steel, which was then used to form structures comprised of copper, nickel, and iron. Electron probe microanalysis (EPMA), scanning electron microscopy (SEM), and energy-dispersive X-ray spectroscopy (EDS) were used to detect the microstructures and compositions of materials, and optical microscopy, X-ray diffraction, X-ray spectroscopy, and energy-dispersive X-ray spectroscopy were used to investigate those microstructures and compositions (phases generated). The use of a dual powder feed deposition approach has been shown to be successful when installing functionally graded Cu-Ni-Fe systems. The presence of unmelted Ni powders in the Cu layer offers difficulty [109]. Materials that have been functionally graded allow for more precision in determining the exact chemical and physical makeup of the final product (FGM). Examples of functionally graded materials (FGM) include graded composite electrodes, tungsten-copper composites, and piezo-electrically graded materials. The fabrication of metal or metal-ceramic FGM may be accomplished by the use of Layered Object Manufacturing (SLM), laser cladding, and ultrasonic consolidation, respectively. Inkjet printing, selective laser sintering, and additive manufacturing are three methods that may be used to produce hybrids of different types of materials [110–112]. These hybrids can be polymer-polymer, polymer-ceramic, or ceramic-ceramic. Researchers also investigated spark plasma sintering as a method of fabricating ZrB_2–SiC/ZrO_2 (3Y) FGM (SPS). It was discovered that employing prefabricated graphite dies with fine outline dimensional control may enhance temperature gradient sintering. The effectiveness of the method was evaluated by assessing the hardness and compositional distribution of SPS-sintered FGM. The graded layers are fortified and toughened by the presence of partly stable ZrO_2 in the FGM, where the composition is distributed along a gradient. There was a small reduction in hardness at the borders of neighbouring composition layers, with values ranging from 17.8 to 19.2 GPa. The phase transition of t-ZrO_2 to m-ZrO_2 has been found to improve fracture toughness from the top to the bottom surface [113]. One of the FGM experiments used ultrasonic consolidation (UC) to join stainless steel, aluminium, and copper foils. Samples of 62-inch-wide by 13-inch-long foils were determined to be optimal owing to the use of the aforementioned stacking sequence. The study concluded that machining or a transitional glue layer would be the best way to improve the FGM production process using UC [114]. Titanium powder of varying densities and porosities was used in EBM metal processing to construct structures that suited the cosmetic and functional needs of the patient and surgical community [115].

12.5.4 Nanocomposites

Several studies have shown the usefulness of nanoparticle-based 3D composite structures for a wide range of tasks, including those requiring use in low-temperature environments or those with harsh chemical or biological conditions. Direct laser writing, optical tweezers, electro-kinetic nanomanipulation, and dip-pen lithography (DPN) are all examples of DLM technology (DLM). The minimal feature size (resolution), the velocity of deposition, and the choice of material are the most important features of the ANM process. However, selective laser sintering (SLS) has emerged as a formidable tool for the fabrication of metal composites' functional components, thanks to its speed, efficiency, and flexibility in creating complex or internal cavity geometries [116]. The SLM process was used to generate large-scale shapes out of nanocrystalline titanium carbide (TiC) reinforced with an Inconel 718 matrix. Microstructure, overall product properties, and metallurgical processes have all been investigated as a function of SLM process variables. The densification process was stymied because larger-sized pore chains or interlayer micro-pores could not be produced within the limited amount of laser energy available [117]. New additive methods, including thermo-kinetic deposition procedures and spray-drying, are employed to include nano-ceramic coatings, which meet prerequisites and provide desirable rheological qualities [118]. High solids content and stable processing conditions are essential for the ceramic production procedure. Because of AM technology, TiC/Ti nanocomposite parts may be manufactured using selective laser sintering (SLS). Adjusting the laser process settings allowed us to investigate a wide range of reactions in SLM-processed TiC/Ti, including those related to nanopowder type, densification behaviour, and tribological characteristics. We employed both ball-milled TiC/Ti nanocomposite powder and manually combined nano-TiC/Ti powder in this study. It was determined that the density of the laser energy and the kind of powder used had the biggest effect on the density achieved. For the ball-milled TiC/Ti nanocomposite powder, the best laser energy was 0.33 kJ/m, resulting in more SLM densification. High laser power led to a decrease in frictional resistance to 0.22 and a wear rate of $2.8 \times 10^6 m^3$/min.

12.6 APPLICATIONS

According to the findings, biomedical industry, which now accounts for 11% of the worldwide AM market, will be one of the primary drivers of future advancements in AM innovation and development. The global 3D printing industry is expected to grow from USD 12.6 billion in 2021 to USD 34.8 billion in 2026, a CAGR of 22.5%. Some of the factors driving the growth of the 3D printing market include the convenience with which customized products can be fabricated, the decline in production costs and process downtime, the funding of 3D printing initiatives by governments, and the discovery of novel industrial-grade 3D printing materials. Additive manufacturing has a wide range of potential medical applications. These days, it's most often utilized to create 3D models of skull deformities, porous titanium implants made using electron beam melting (EBM), and models of skulls with implants fitted to them.

12.6.1 Tissue and organ bioprinting

Bioprinting is a technique that employs a computer-guided pipette to layer different types of live cells (called bio-ink) to construct fake living tissues. Due to severe scarcity of donor organs and tissues, bioprinting is being used to make artificial substitutes. This technique may be used to create and transplant a variety of organs and tissues [119]. These include the heart, bone, tracheal splints, cartilaginous structures, vascular grafts, and multilayered skin. Due to their inexpensive cost compared to human organ transplantation and their capacity to mimic tiny size organs, artificial tissue organoids or constructs may be employed for medical research. There are now three distinct types of printing technologies available: laser printing, inkjet printing (sometimes known as "drop-on-demand printers"), and extrusion printing.

12.6.1.1 Medical devices and tools

Three-dimensional printing technology is increasingly being used in the production of medical equipment. It's particularly helpful for making complex pieces of medical devices that are difficult to manufacture in a conventional way. Affordable equipment tailored to a certain patient's anatomy is also developed. Therefore, this is the primary approach used to produce personalized hearing aids, such as those that need to be tailored to the patient's ear anatomy [120].

Orthopaedic devices, dental surgical tools, and surgical guides are just some of the many medical products that have benefited from the advent of additive manufacturing (implant supporting bar and abutments). Using this technology, surgeons may make custom surgical guides that account for each individual's anatomy and accurately label their instruments. These tools may be used to ensure the safety of patients even while working in confined locations.

This 3D technology may find applications in a variety of medical equipment, including hearing aids, stethoscopes, and spectacles for the visually handicapped [121]. Hearing aids that can print information are helpful for several reasons, including saving money, protecting the health of the user, and improving the effectiveness of the device. Several fields are expected to see an uptick in 3D printer use in the near future. Actually, the term "medical devices" covers a wide range of applications, from those used in procedures involving the face and skull to those involving prostheses, implants, and even human organs. The aircraft industry is well known for its expertise at producing relatively few units at a time. AM is more cost-effective for low-volume production since it does not need the use of costly equipment like moulds or dies.

12.6.1.2 Customized implant and prosthetics

Furthermore, AM technique is used to create prosthetic limbs that are unique to each individual patient. Accelerating production and reducing the price of goods for patients while maintaining or improving their usefulness is a major benefit of additive manufacturing. Several orthopaedic conditions may be treated, and 3D-printed implants and prostheses might play a role in this. Implants for the maxilla, skull, jawbone, and lower jaw are all part of the custom implant spectrum [122]. Scaffoldings, tissue regeneration, bone implants, forensic, clinical application, healthcare device design, and personalized implants are a few of the numerous medical uses of 3D printing [123]. One study compared 3D printing models to generated photos and found that using the printers gives a significant advantage when analysing patient anatomy. Surgeons may evaluate the patient's anatomy in 3D before surgery, perform virtual procedures, and put new tools through their paces with the use of 3D models.

12.6.1.3 Tissue engineering

Musculoskeletal and organ dysfunction may be a consequence of bone deformities, traumas, acute illnesses, and incidents that cause tissue damage. Restoration and replacement of the implant are often required to stabilize joint function after injury [124]. Gas foaming, freeze-drying, electrospinning, solvent casting, and emulsification are just a few examples of the technologies that may be used to create scaffolds for tissue engineering. The capacity to mass-generate unique and specific components for patients at cheap cost and with more efficiency and transparency is a major boon of medical 3D printing [125]. Bone, skin, nerves, arteries, veins, tendons, ligaments, menisci, and urethra are just some of the tissues that 3D printing may be used to repair.

12.6.1.4 Dentistry

The field of dentistry, which encompasses procedures such as oral and maxillofacial surgery, orthogenetic, endodontics, and orthodontics, is only one of the many areas of medicine that have benefited from AM process technology. Both the patient's comfort and the dentist's ease of use benefited from this technological advancement. Additionally, dental restorations done using AM are more long-lasting than those made by dental technicians. Printing bridges, orthodontic appliances, crowns, and other dental components are possible using various additive manufacturing (AM) techniques (FDM, SLS, LOM) [126].

12.6.2 Construction

Manufacturing components and even entire buildings are now possible with the help of AM in the construction industry. As a result of AM, the construction industry can save money, time, and resources without sacrificing quality. Academics from all over the world are currently working to perfect printing facilities [127]. The construction sector might benefit from additive manufacturing since it could streamline the distance between the supplier and the end-user [128]. Providing affordable housing to underdeveloped locations is made easier by AM's ability to print out home components. According to Buchanan and Gardner [129], directed energy deposition and powder bed fusion are the most promising additive manufacturing technologies for usage in the construction industry. Both methods have the potential to produce impressive structures, but each has substantial downsides in terms of cost, complexity, and size in general. On the other hand, wire and arc AM direct energy deposition may make it feasible to produce components of practically unlimited size and speed while also decreasing the cost of manufacturing. However, this approach has limitations in terms of surface polish and dimensional precision. This technique is used to build bridges in the MX3D format. It has potential aeronautical and automotive uses, among others, such as turbine engine maintenance.

The contour crafting (CC) method was invented by Khoshnevis [130] for use in the robotic construction of earth buildings and public works, as well as in aerospace. It can utilize local resources, therefore it can be used to construct inexpensive homes and other structures on the moon. The first ever house to be 3D printed was built in Amsterdam in 2014 utilizing fused deposition modelling (FDM) technology [131]. The goal of DUS Architects' effort was to showcase the printer's portability and help pave the way for it to become widely used in the construction sector by cutting down on wastage and transportation expenses. According to Roodman and Lenssen's [132] calculations, the building uses up more than 40% of the world's raw materials. Using contour crafting, the typical 7-tonne trash produced during the construction of a single-family home may be reduced to practically nothing, and the average building time can be cut from 48 to 24 hours. Although this technique is not currently suitable for elaborate or luxurious buildings, it has the potential to be utilized for projects such as emergency housing and affordable residences.

12.6.3 Consumer product (electronics)

In this industry, additive manufacturing (AM) technology is used to make prototypes of various household items, toys, and sporting goods. Materials created using 3D printing technology have the potential to be used in the manufacture of consumer goods due to their low weight and high flexibility (such as footwear and clothing). This strategy is quick and cheap. When talking about rapid prototyping in electronics, the word "prototype" may be safely ignored. Models/prototypes of the housing and other parts of the testing apparatus may be manufactured rapidly using AM [133].

12.6.4 The aviation industry

Wohlers predicts that by 2025, aerospace will account for 18.2% of the global AM industry [44], making it one of the most promising industries for the future. Additive manufacturing is essential in the aerospace industry because it improves the thermal and mechanical performance of components composed of metallic and polymeric materials, especially nickel and titanium alloys, while also reducing their weight. Powdered material SLS has become a viable solution for the repair, production, and maintenance of long-lasting components including turbine blades and aeronautical tools. AM is also used to create and repair aerospace parts (such as nozzles, combustion chambers, fittings, brackets, moulds, dies, and so on). Complicated geometry: For structural, thermal, and circulatory purposes, intricate shapes are required. For instance, GE Aviation is developing redesigned blade tips to increase thrust. Moreover, it may be simplified by integrating various components like GE fuel nozzles [134]. Last but not least, printable or additively manufactured (AM) functional circuits are possible [135].

12.7 CONCLUSION

3D printing has several advantages, including the capacity to quickly iterate designs, mass-produce unique items, and produce intricate structures with little waste. For better understanding the potential of 3D printing in a variety of fields, a thorough analysis of the processes involved, the materials used, and the present status of popular applications was conducted. Additionally, the difficulties inherent in 3D printing were studied. Because of its cheap price, ease of usage, and fast processing speed, fused deposition modelling (FDM) is the most popular 3D printing method. Fused deposition modelling, also known as FDM, is primarily used for rapid prototyping but produces lower quality printed products and mechanical qualities when compared to powder bed technologies such as selective laser sintering (SLS) and selective laser melting (SLM). Powder bed technologies, which make use of an additional adhesive to fuse, melt, or glue powders together, are able to reach finer resolutions than other technologies. Direct energy deposition (DED) is a technique that melts metal particles by directly delivering energy in the form of a laser or electron beam, in contrast to selective laser melting (SLM), which uses a laser to do the melting. Similar to fused deposition modelling (FDM), but with sufficient heat to melt metals, the feedstock is heated layer by layer before deposition. Inkjet printing is a convenient method for producing 3D ceramic suspensions, albeit the finished products will need to be heat treated. Extrusion is used to print larger objects, such as houses, using a technique called "contour crafting" (concrete). Pioneering in the field of 3D printing, stereolithography is best known for its use with photopolymers and its ability to create objects with high precision. However, due to resource limitations, the procedure is lengthy and challenging. Last but not least, laminated object manufacture (LOM) is a process that involves cutting and laminating sheets or rolls of materials one by one. Filaments, wire, powder, paste, sheets, and inks are only some of the materials that may be utilized in 3D printing.

The mechanical characteristics of the 3D-printed composite were improved with the use of fibres and nanoparticles to strengthen polymers. Metals are often produced using powders (or wires), while SLS, SLM, and DED are the most common 3D printing techniques. The number of metals and alloys that may be used in 3D printing is few, thus there has been a push to expand the scope of existing methods to include a greater variety of composite structures. Tissue engineering ceramic scaffolds are one example of the many uses for ceramics, which have created a trend towards customising materials with a high strength-to-weight ratio and made possible the production of complicated ceramic lattices. Materials for 3D printing ceramics with improved microstructure and composition are scarce. The construction industry has been sluggish to adopt 3D printing concrete despite its many benefits, including mass customization, the elimination of the need for formwork, and the possibility of automating formerly labour-intensive tasks. Recent efforts in the field have centred on developing a concrete mixture that exhibits all the desirable qualities: high flowability, processability, mechanical performance, and aesthetics. A broad variety of composite materials and alloys, including those with a metal matrix, ceramic matrix, polymer matrix, and a mix of metals and polymers, have all been the subject of research. To demonstrate that additive manufacturing is better to more traditional production processes, it has been recommended that a broad variety of performance indicators be used. In practically all of the research, the mechanical qualities, the physical properties (such as relative density), the tribological properties, and the sort of generated microstructure are considered to be major performance variables. Based on research into the processes and attributes of additive manufacturing, it has also been stated that nanocomposites may be additively manufactured and that fabrication of nanocomposites can take place.

Functionally graded materials, which have been discussed in this work, are one of the most successful AM-proposed composite materials because they allow for adjustment of the composition and optimize the attributes of the constructed object. This makes functionally graded materials one of the most successful AM-proposed composite materials.

REFERENCES

[1] Wang X, Jiang M, Zhou Z, Gou J, Hui D. 3D printing of polymer matrix composites: A review and prospective. *Composites Part B: Engineering.* 2017;110:442–58.

[2] Horn TJ, Harrysson OL. Overview of current additive manufacturing technologies and selected applications. *Science Progress.* 2012;95:255–82.

[3] Wickramasinghe S, Do T, Tran P. FDM-based 3D printing of polymer and associated composite: A review on mechanical properties, defects and treatments. *Polymers.* 2020;12(7):1529. https://doi.org/10.3390/polym12071529.

[4] Kamaal M, Anas M, Rastogi H, Bhardwaj N, Rahaman A. Effect of FDM process parameters on mechanical properties of 3D-printed carbon fibre–PLA composite. *Progress in Additive Manufacturing.* 2020;6:63–69. https://doi.org/10.1007/s40964-020-00145-3

[5] Boschetto A, Bottini L. Accuracy prediction in fused deposition modeling. *The International Journal of Advanced Manufacturing Technology.* 2014;73(5):913–28.

[6] Gibson I, Rosen D, Stucker B (Eds.). *Additive Manufacturing Technologies: 3D Printing, Rapid Prototyping, and Direct Digital Manufacturing.* Springer, New York, 2015, pp. 147–73.

[7] Berman B. 3-D printing: The new industrial revolution. *Business Horizons.* 2012;55(2):155–62.

[8] Stansbury JW, Idacavage MJ. 3D printing with polymers: Challenges among expanding options and opportunities. *Dental Materials.* 2016;32(1):54–64.

[9] Chua CK, Leong KF, An J. *Rapid Prototyping of Biomaterials.* Woodhead Publishing, London, 2014.

[10] Ashley S. *Rapid Prototyping System.* American Society of Mechanical Engineers, New York, vol. 113, 1991, p. 34.

[11] Ivanova O, Williams CB, Campbell T. Additive manufacturing (AM) and nanotechnology: promises and challenges. *Rapid Prototyping Journal.* 2013;19(5):353–364.

[12] Wendel B, Rietzel D, Ku¨hnlein F, Feulner R, Hu¨lder G, Schmachtenberg E. *Mater. Eng.* 2008;293(10):1438–7492. https://doi.org/10.1002/mame.200800121

[13] Ligon S C, Liska R, Stampfl J, Gurr M, Mulhaupt R. Polymers for 3D Printing and Customized Additive Manufacturing. *Chemical Reviews.* 2017;117:10212. https://doi.org/10.1021/acs.chemrev.7b00074

[14] Crump SS. Fused deposition modeling putting rapid back into prototyping. In: *The 2nd International Conference on Rapid Prototyping*, Dayton, OH, vol. 3, 1991, 354.

[15] Jafari MA, Han W, Mohammadi F, Safari A, Danforth SC, Langrana NA. A novel system for fused deposition of advanced multiple ceramics. *Rapid Prototyping Journal.* 2000;6:161–175.https://doi.org/10.1108/13552540010337047.
[16] Khalil S, Nam J, Sun W. Multi-nozzle deposition for construction of 3D biopolymer tissue scaffolds. *Rapid Prototyping Journal.* 2005;11(1):9–17. https://doi.org/10.1108/13552540510573347.
[17] Barbulovic-Nad I, Lucente M, Sun Y, Zhang M J, Wheeler AR, Bussmann M. Bio-microarray fabrication techniques--a review. *Critical Reviews in Biotechnology.* 2006;26(4):237–59. doi: 10.1080/07388550600978358.
[18] Do AV, Khorsand B, Geary SM, Salem AK. 3D printing of scaffolds for tissue regeneration applications. *Advanced Healthcare Materials.* 2015; 4(12):1742–1762. doi: 10.1002/adhm.201500168.
[19] Dou R, Wang T, Guo Y, Derby B. Ink-jet printing of Zirconia: Coffee staining and line stability. *Journal of the American Ceramic Society.* 2011;94(11):3787–3792.
[20] Olakanmi E, Cochrane RF, Dalgarno KW. A review on selective laser sintering/melting (SLS/SLM) of aluminium alloy powders: Processing, microstructure, and properties. *Progress in Materials Science.* 2015; 74:401–477. https://doi.org/10.1016/j.pmatsci.2015.03.002.
[21] Salmoria GV, Klauss P, Paggi RA, Kanis LA, Lago A. The microstructural characterization of PA6/PA12 blend specimens fabricated by selective laser sintering. *Polymer Testing.* 2009;28(7):746–751. https://doi.org/10.1016/j.polymertesting.2009.06.010.
[22] Duan B, Wang M, Zhou WY, Cheung WL, Li ZY, Lu WW. Three-dimensional nanocomposite scaffolds fabricated via selective laser sintering for bone tissue engineering. *Acta Biomaterilia.* 2010;6(12):4495–4505. doi: 10.1016/j.actbio.2010.06.024.
[23] Santos EC, Osakada K, Shiomi M, Kitamura Y, Abe F. Microstructure and mechanical properties of pure titanium models fabricated by selective laser melting. *Proceedings of the Institution of Mechanical Engineers, Part C: Journal of Mechanical Engineering Science.* 2004;218(7):711–719. doi:10.1243/0954406041319545.
[24] Utela B, Storti D, Anderson R, Ganter M. A review of process development steps for new material systems in three dimensional printing (3DP). *Journal of Manufacturing Process.* 2008;10:96–104. doi:10.1016/j.jmapro.2009.03.002.
[25] Al-Ahmari A, Nasr EA, Moiduddin K, Anwar S, Kindi MA, Kamrani A. A comparative study on the customized design of mandibular reconstruction plates using finite element method. *Advances in Mechanical Engineering.* 2015;7(7). doi:10.1177/1687814015593890.
[26] Gong, X, Anderson, T, Chou, K. Review on Powder-Based Electron Beam Additive Manufacturing Technology. *Proceedings of the ASME/ISCIE 2012 International Symposium on Flexible Automation. ASME/ISCIE 2012 International Symposium on Flexible Automation.* St. Louis, Missouri. June 18–20, 2012. pp. 507–515. ASME. https://doi.org/10.1115/ISFA2012-7256.
[27] Osman RB, Swain MV. A critical review of dental implant materials with an emphasis on titanium versus zirconia. *Materials.* 2015;8:932–58.

[28] Uhlmann E, Kersting R, Klein TB, Cruz MF, Borille AV. Additive manufacturing of titanium alloy for aircraft components. *Procedia CIRP.* 2015;35:55–60.
[29] Wen Y, Xun S, Haoye M, Baichuan S, Peng C, Xuejian L, et al. 3D printed porous ceramic scaffolds for bone tissue engineering: A review. *Biomaterials Science.* 2017;5:1690–98.
[30] Travitzky N, Bonet A, Dermeik B, Fey T, Filbert-Demut I, Schlier L, et al. Additive manufacturing of ceramic-based materials. *Advanced Engineering Materials.* 2014;16:729–54.
[31] Caminero MA, Chacón JM, García-Moreno I, Rodríguez GP. Impact damage resistance of 3D printed continuous fibre reinforced thermoplastic composites using fused deposition modelling. *Composites Part B.* 2018;148: 93–103.
[32] Xin W, Man J, Zuowan Z, Jihua G, David H. 3D printing of polymer matrix composites: A review and prospective. *Composites Part B.* 2017;110:442–58.
[33] Bedi TS, Kumar S, Kumar R, Corrosion performance of hydroxyapatite and hydroxyapatite/titanium bond coating for biomedical applications. *Materials Research Express.* 2019;7:1–16.
[34] Regassa Y, Lemu HG, Sirabizuh B. Trends of using polymer composite materials in additive manufacturing. *IOP Conference Series: Materials Science and Engineering.* 2019;659:012021.
[35] Sathishkumar TP, Satheeshkumar S, Naveen J. Glass fiber-reinforced polymer composites—A review. *Journal of Reinforced Plastics and Composites.* 2014;33(13):1258–75.
[36] Yahya BEK. 3D printing technology; methods, biomedical applications, future opportunities and trends, *Journal of Materials Research and Technology.* 2021;14:1430–50.
[37] Culmone C, Smit G, Breedveld P. Additive manufacturing of medical instruments: A state-of-the-art review. *Additive Manufacturing.* 2019;27:461–473, https://doi.org/10.1016/j.addma.2019.03.015.
[38] Eilbagi M, Emadi R, Raeissi K, et al. Mechanical and cytotoxicity evaluation of nanostructured hydroxyapatite-bredigite scaffolds for bone regeneration. *Materials Science and Engineering: C.* 2016;68:603–12.
[39] Gentile P, Chiono V, Carmagnola I, Hatton P, et al. An overview of poly (lacticcoglycolic) acid (PLGA)-based biomaterials for bone tissue engineering. *International Journal of Molecular Sciences.* 2014;15(3):3640–59.
[40] Demir AG, Previtali B. Additive manufacturing of cardiovascular CoCr stents by selective laser melting. *Materials & Design.* 2017;119:338–50.
[41] Gieseke M, Noelke C, Kaierle S, Wesling V, Haferkamp H. Selective laser melting of magnesium and magnesium alloys. In: Hort N., Mathaudhu SN, Neelameggham NR, Alderman M (Eds.), *Magnesium Technology 2013*, The Minerals, Metals & Materials Series (MMMS), Springer, 2013, pp. 65–8.
[42] Ramirez D, Murr L, Martinez E, Hernandez D, Martinez J, Machado B, Medina F, Frigola P, Wicker R. Novel precipitate–microstructural architecture developed in the fabrication of solid copper components by additive manufacturing using electron beam melting. *Acta Materialia.* 2011;59(10):4088–99.
[43] Saroha V, Pabla BS, Bhogal SS. *International Journal of Innovative Technology and Exploring Engineering.* 2019;8(10):2164–2167.

[44] Wohlers T. 3D printing and additive manufacturing state of the industry. Annual Worldwide Progress Report, Wohlers Report, 2017.
[45] National Research Council. *Accelerating Technology Transition: Bridging the Valley of Death for Materials and Processes in Defense Systems*, Washington, DC: National Academies Press, 2004. https://doi.org/10.17226/11108
[46] Bai Y, Williams CB. An exploration of binder jetting of copper. *Rapid Prototyping Journal.* 2015;21(2):177–85.
[47] Sova A, Grigoriev S, Okunkova A, Smurov I. Potential of cold gas dynamic spray as additive manufacturing technology. *The International Journal of Advanced Manufacturing Technology.* 2013;69(9–12):2269–78.
[48] Sharma A, Bandari V, Ito K, Kohama K, Ramji M, BV HS. A new process for design and manufacture of tailor-made functionally graded composites through friction stir additive manufacturing. *Journal of Manufacturing Processes.* 2017;26:122–30.
[49] Matthews MJ, Guss G, Drachenberg DR, Demuth JA, Heebner JE, Duoss EB, Kuntz JD, Spadaccini CM. Diode-based additive manufacturing of metals using an optically-addressable light valve. *Optics Express.* 2017;25(10):11788–800.
[50] Gibson I, Rosen DW, Stucker B. *Additive Manufacturing Technologies: Rapid Prototyping to Direct Digital Manufacturing*, Springer, New York, 2009.
[51] Seifi M, Salem A, Beuth J, et al. Overview of materials qualification needs for metal additive manufacturing. *JOM.* 2016;68(3):747–64.
[52] Delgado J, Ciurana J, Sereno L. Comparison of forming manufacturing processes and selective laser melting technology based on the mechanical properties of products. *Virtual and Physical Prototyping.* 2011;6:167–78.
[53] Sheydaeian E, Toyserkani E. A new approach for fabrication of titanium-titanium boride periodic composite via additive manufacturing and pressure-less sintering. *Composites Part B: Engineering.* 2018;138:140–8.
[54] Attar H, Calin M, Zhang L, Scudino S, Eckert J. Manufacture by selective laser melting and mechanical behavior of commercially pure titanium. *Materials Science and Engineering: A.* 2014;593:170–7.
[55] Vaithilingam J, Kilsby S, Goodridge RD, Christie SD, Edmondson S, Hague RJ. Functionalisation of Ti6Al4V components fabricated using selective laser melting with a bioactive compound. *Materials Science and Engineering: C.* 2015;46:52–61.
[56] Casalino G, Campanelli S, Contuzzi N, Ludovico A. Experimental investigation and statistical optimisation of the selective laser melting process of a maraging steel. *Optics & Laser Technology.* 2015;65:151–8.
[57] Murr LE, Martinez E, Hernandez J, Collins S, Amato KN, Gaytan SM, Shindo PW. Microstructures and properties of 17-4 PH stainless steel fabricated by selective laser melting. *Journal of Materials Research and Technology.* 2012;1(3):167–77.
[58] Mazumder J, Choi J, Nagarathnam K, Koch J, Hetzner D. The direct metal deposition of tool steel for 3-D components. *JOM.* 1997;49(5):55–60.
[59] Maskery I, Aboulkhair N, Corfield M, Tuck C, Clare A, Leach RK, Wildman RD, Ashcroft I, Hague RJ. Quantification and characterisation of porosity in selectively laser melted Al–Si10–Mg using X-ray computed tomography. *Materials Characterization.* 2016;111:193–204.

[60] Sutton AT, Kriewall CS, Leu MC, Newkirk JW. Powder characterisation techniques and effects of powder characteristics on part properties in powder-bed fusion processes. *Virtual and Physical Prototyping*. 2017;12(1):3–29.
[61] Vilaro T, Colin C, Bartout J-D. As-fabricated and heat-treated microstructures of the Ti6Al-4V alloy processed by selective laser melting. *Metallurgical and Materials Transactions A*. 2011;42(10):3190–9.
[62] Yadroitsev I, Thivillon L, Bertrand P, Smurov I. Strategy of manufacturing components with designed internal structure by selective laser melting of metallic powder. *Applied Surface Science*. 2007;254(4):980–3.
[63] Körner C, Helmer H, Bauereiß A, Singer RF. Tailoring the grain structure of IN718 during selective electron beam melting. *MATEC Web of Conferences: EDP Sciences*, 2014, p. 08001.
[64] Spaldin NA, Fiebig M. The renaissance of magneto electric multiferroics. *Science*. 2005;309:391–92.
[65] Liang X, et al. A review of thin-film magnetoelastic materials for magnetoelectric applications. *Sensors*. 2020;20:1532.
[66] Zhuang Y, Song W, Ning G, Sun X, Sun Z, Xu G, Zhang B, Chen Y, Tao S. 3D-printing of materials with anisotropic heat distribution using conductive polylactic acid composites. *Materials & Design*. 2017;126:135–40.
[67] Senatov FS, Niaza KV, Zadorozhnyy MY, Maksimkin AV, Kaloshkin SD, Estrin YZ. Mechanical properties and shape memory effect of 3D-printed PLA-based porous scaffolds. *Journal of the Mechanical Behavior of Biomedical Materials*. 2016;57:139–48.
[68] Serra T, Planell JA, Navarro M. High-resolution PLA-based composite scaffolds via 3-D printing technology. *Acta Biomaterialia*. 2013;9(3):5521–30.
[69] Abueidda DW, Bakir M, Abu Al-Rub RK, Bergström JS, Sobh NA, Jasiuk I. Mechanical properties of 3D printed polymeric cellular materials with triply periodic minimal surface architectures. *Materials & Design*. 2017;122:255–67.
[70] Song Y, Li Y, Song W, Yee K, Lee KY, Tagarielli VL. Measurements of the mechanical response of unidirectional 3D-printed PLA. *Materials & Design*. 2017;123:154–64.
[71] Alaimo G, Marconi S, Costato L, Auricchio F. Influence of meso-structure and chemical composition on FDM 3D-printed parts. *Composites Part B: Engineering*. 2017;113:371–80.
[72] Bakarich SE, Gorkin III R, Panhuis M, Spinks GM. Three-dimensional printing fiber reinforced hydrogel composites. *ACS Applied Materials and Interfaces*. 2014;6(18):15998–6006.
[73] Farina I, Fabbrocino F, Carpentieri G, Modano M, Amendola A, Goodall R, Feo L, Fraternali F. On the reinforcement of cement mortars through 3D printed polymeric and metallic fibers. *Composites Part B: Engineering*. 2016;90:76–85.
[74] Singh R, Singh N, Amendola A, Fraternali F. On the wear properties of Nylon6-SiCAl2O3 based fused deposition modelling feed stock filament. *Composites Part B: Engineering*. 2017;119:125–31.
[75] Hou Z, Tian X, Zhang J, Li D. 3D printed continuous fibre reinforced composite corrugated structure. *Composite Structures*. 2018;184:1005–10.
[76] Parandoush P, Lin D. A review on additive manufacturing of polymer-fiber composites. *Composite Structures*. 2017;182:36–53.

[77] Tekinalp HL, Kunc V, Velez-Garcia GM, Duty CE, Love LJ, Naskar AK, Blue CA, Ozcan S. Highly oriented carbon fiber–polymer composites via additive manufacturing. *Composites Science and Technology*. 2014;105:144–50.
[78] Love LJ, Kunc V, Rios O, Duty CE, Elliott AM, Post BK, Smith RJ, Blue CA. The importance of carbon fiber to polymer additive manufacturing. *Journal of Materials Research*. 2014;29(17):1893–8.
[79] Wen Y, Xun S, Haoye M, Baichuan S, Peng C, Xuejian L, Kaihong Z, Xuan Y, Jiang P, Shibi L. 3D printed porous ceramic scaffolds for bone tissue engineering: A review. *Biomaterials Science*. 2017;5(9):1690–8.
[80] Travitzky N, Bonet A, Dermeik B, Fey T, Filbert-Demut I, Schlier L, Schlordt T, Greil P. Additive manufacturing of ceramic-based materials. *Advanced Engineering Materials*. 2014;16(6):729–54.
[81] Bienia M, Lejeune M, Chambon M, Baco-Carles V, Dossou-Yovo C, Noguera R, Rossignol F. Inkjet printing of ceramic colloidal suspensions: Filament growth and breakup. *Chemical Engineering Science*. 2016;149:1–13.
[82] Liu F-H, Shen Y-K, Liao Y-S. Selective laser gelation of ceramic–matrix composites. *Composites Part B: Engineering*. 2011;42(1):57–61.
[83] Vorndran E, Klarner M, Klammert U, Grover LM, Patel S, Barralet JE, Gbureck U. 3D Powder printing of β-tricalcium phosphate ceramics using different strategies. *Advanced Engineering Materials*. 2008;10(12):B67–B71.
[84] Sun C, Tian X, Wang L, Liu Y, Wirth CM, Günster J, Li D, Jin Z. Effect of particle size gradation on the performance of glass-ceramic 3D printing process. *Ceramics International*. 2017;43(1):578–84.
[85] Dehurtevent M, Robberecht L, Hornez J-C, Thuault A, Deveaux E, Béhin P. Stereolithography: A new method for processing dental ceramics by additive computer-aided manufacturing. *Dental Materials*. 2017;33(5):477–85.
[86] Henriques B, Pinto P, Silva FS, Fredel MC, Fabris D, Souza JCM, Carvalho O. On the mechanical properties of monolithic and laminated nano-ceramic resin structures obtained by laser printing. *Composites Part B: Engineering*. 2018;141:76–83.
[87] Liu W, Wu H, Zhou M, He R, Jiang Q, Wu Z, Cheng Y, Song X, Chen Y, Wu S. Fabrication of fine-grained alumina ceramics by a novel process integrating stereolithography, and liquid precursor infiltration processing. *Ceramics International*. 2016;42(15):17736–41.
[88] Wu H, Cheng Y, Liu W, He R, Zhou M, Wu S, Song X, Chen Y. Effect of the particle size and the debinding process on the density of alumina ceramics fabricated by 3D printing based on stereolithography. *Ceramics International*. 2016;42(15):17290–4.
[88] Khoshnevis B. Automated construction by contour crafting—Related robotics and information technologies. *Automation in Construction*. 2004;13(1):5–19.
[89] Le TT, Austin SA, Lim S, Buswell RA, Gibb AGF, Thorpe T. Mix design and fresh properties for high-performance printing concrete. *Materials and Structures*. 2012;45(8):1221–32.
[90] Gosselin C, Duballet R, Roux P, Gaudillière N, Dirrenberger J, Morel P. Large-scale 3D printing of ultra-high performance concrete—A new processing route for architects and builders. *Materials & Design*. 2016;100:102–9.
[91] Paul SC, Tay YWD, Panda B, Tan MJ. Fresh and hardened properties of 3D printable cementitious materials for building and construction. *Archives of Civil and Mechanical Engineering*. 2018;18(1):311–9.

[92] Duballet R, Baverel O, Dirrenberger J. Classification of building systems for concrete 3D printing. *Automation in Construction.* 2017;83:247–58.
[93] Zareiyan B, Khoshnevis B. Interlayer adhesion and strength of structures in Contour Crafting—Effects of aggregate size, extrusion rate, and layer thickness. *Automation in Construction.* 2017;81:112–121. https://doi.org/10.1016/j.autcon.2017.06.013.
[94] Kazemian A, Yuan X, Cochran E, Khoshnevis B. Cementitious materials for construction-scale 3D printing: Laboratory testing of fresh printing mixture. *Construction and Building Materials.* 2017;145:639–47.
[95] Ryder MA, Lados DA, Iannacchione GS, Peterson AM. Fabrication and properties of novel polymer-metal composites using fused deposition modeling. *Composites Science and Technology.* 2018;158:43–50.https://doi.org/10.1016/j.compscitech.2018.01.049
[96] Vaezi M, Chianrabutra S, Mellor B, Yang S. Multiple material additive manufacturing–Part 1: A review: This review paper covers a decade of research on multiple material additive manufacturing technologies which can produce complex geometry parts with different materials. *Virtual and Physical Prototyping.* 2013;8(1):19–50.
[97] Jung JW, Lee J-S, Cho D-W. Computer-aided multiple-head 3D printing system for printing of heterogeneous organ/tissue constructs. *Scientific Reports.* 2016;6: 21685.
[98] Hofmann DC, Kolodziejska J, Roberts S, Otis R, Dillon RP, Suh J-O, Liu Z-K, Borgonia J-P. Compositionally graded metals: A new frontier of additive manufacturing. *Journal of Materials Research.* 2014;29(17):1899–910.
[98] Chivel Y. New approach to multi-material processing in selective laser melting. *Physics Procedia.* 2016;83:891–8.
[99] Gaynor AT, Meisel NA, Williams CB, Guest JK. Multiple-material topology optimization of compliant mechanisms created via PolyJet three-dimensional printing. *Journal of Manufacturing Science and Engineering.* 2014;136(6):061015.
[100] Callister WD, Rethwisch DG, *Materials Science and Engineering: An Introduction.* Vol. 7, Wiley, New York, 2007.
[101] Fang N, Wu Q, A comparative study of the cutting forces in high speed machining of Ti–6Al–4V and Inconel 718 with a round cutting edge tool. *Journal of Materials Processing Technology* 2009;209:4385–89.
[102] Brandl E, et al. Additive manufactured AlSi10Mg samples using selective laser melting (SLM): Microstructure, high cycle fatigue, and fracture behavior. *Materials & Design.* 2012;34:159–69.
[103] Chou R. et al., Additive manufacturing of Al-12Si alloy via pulsed selective laser melting. *JOM.* 2015;67:590–96.
[104] Bellini A, Shor L, Guceri SI. New developments in fused deposition modeling of ceramics. *Rapid Prototyping Journal.* 2005;11:214–20.
[105] Wilkes J, et al. Additive manufacturing of ZrO2-Al2O3 ceramic components by selective laser melting. *Rapid Prototyping Journal.* 2013;19:51–7.
[106] Tekinalp HL, et al. Highly oriented carbon fibre-polymer composites via additive manufacturing. *Composites Science and Technology.* 2014;105:144–50.
[107] Jones JB, Wimpenny DI, Gibbons GJ. Additive manufacturing under pressure. *Rapid Prototyping Journal.* 2015;21:89–97.

[108] Kenzari S, et al. Quasicrystal-polymer composites for additive manufacturing technology. *Acta Physica Polonica A*. 2014;126:449–52.
[109] Li L, Syed W, Pinkerton A, Rapid additive manufacturing of functionally graded structures using simultaneous wire and powder laser deposition. *Virtual and Physical Prototyping*. 2006;1:217–25.
[110] Beal V, et al. Evaluating the use of functionally graded materials inserts produced by selective laser melting on the injection moulding of plastics parts. *Proceedings of the Institution of Mechanical Engineers, Part B: Journal of Engineering Manufacture*. 2007;221:945–54.
[111] Domack M, Baughman J, Development of nickel-titanium graded composition components. *Rapid Prototyping Journal*. 2005;11:41–51.
[112] Müller E, et al. Functionally graded materials for sensor and energy applications. *Materials Science and Engineering: A*. 2003;362:17–39.
[113] Hong C-Q, et al. A novel functionally graded material in the ZrB 2–SiC and ZrO 2 system by spark plasma sintering. *Materials Science and Engineering: A*. 2008;498:437–41.
[114] Kumar S. Development of functionally graded materials by ultrasonic consolidation. *CIRP Journal of Manufacturing Science and Technology*. 2010;3:85–7.
[115] Parthasarathy J, Starly B, Raman S, A design for the additive manufacture of functionally graded porous structures with tailored mechanical properties for biomedical applications. *Journal of Manufacturing Processes* 2011;13:160–70.
[116] Gorynin IV, et al. Additive technologies based on composite powder nanomaterials. *Metal Science and Heat Treatment*. 2015;56:519–24 [CrossRef] [Google Scholar].
[117] Zheng B, et al. Microstructure and properties of laser-deposited Ti6Al4V metal matrix composites using Ni-coated powder. *Metallurgical and Materials Transactions A*. 2008;39:1196–1205 [CrossRef] [Google Scholar].
[118] Gadow R, Kern F, Killinger A. Manufacturing technologies for nanocomposite ceramic structural materials and coatings. *Materials Science and Engineering: B*. 2008;148:58–64.
[119] Sun Z, Lee SY. A systematic review of 3-D printing in cardiovascular and cerebrovascular diseases. *The Anatolian Journal of Cardiology*. 2017;17:423–35.
[120] Dodziuk H. Applications of 3D printing in healthcare. *Kardiochirurgia i Torakochirurgia Polska/Polish Journal of Thoracic and Cardiovascular Surgery*. 2016;13(3):283–93.
[121] Choonara YE, du Toit LC, Kumar P, Kondiah PPD, Pillay V. 3D-printing and the effect on medical costs: A new era? *Expert Review of Pharmacoeconomics & Outcomes Research*. 2016;16:23–32.
[122] Dana A, Mektepbayeva D. Patient specific in situ 3D printing, in Kalaskar DM (ed.), *3D Printing in Medicine. Patient Specific In Situ 3D Printing*, Woodhead Publishing, 2017, pp. 91–113. https://doi.org/10.1016/B978-0-08-100717-4.00004-1
[123] Milovanovic J, Trajanovic M. Medical applications of rapid prototyping. *Mechanical Engineering*. 2007;5:79–85.
[124] Turnbull G, Clarke J, Picard F, Riches P, Jia L, Han F, et al. 3D bioactive composite scaffolds for bone tissue engineering. *Bioactive Materials*. 2018;3:278–314.

[125] Tofail SAM, Koumoulos EP, et al. Additive manufacturing: Scientific and technological challenges, market uptake and opportunities. *Mater Today.* 2017;21:22–37.
[126] Hao W, Liu Y, Zhou H, et al. Preparation and characterization of 3D printed continuous carbon fiber reinforced thermosetting composites. *Polymer Testing.* 2018;65:29–34.
[127] Bogue R. 3D printing: The dawn of a new era in manufacturing? *Assembly Automation.* 2013;33:307–11.
[128] Huang Y, Leu MC, Mazumder J, Donmez A. Additive manufacturing: Current state, future potential, gaps and needs, and recommendations. *Journal of Manufacturing Science and Engineering.* 2015;137(1): 014001–10.
[129] Buchanan C, Gardner L. Metal 3D printing in construction: A review of methods, research, applications, opportunities and challenges. *Engineering Structures.* 2019;180:332–48.
[130] Khoshnevis B. Automated construction by contour crafting—Related robotics and information technologies. *Automation in Construction.* 2004;13(1):5–19.
[131] 3D Print Canal House – DUS Architects. https://houseofdus.com/project/3d-print-canal-house.
[132] David MR, Nicholas L. *Worldwatch Paper 124: A Building Revolution: How Ecology and Health Concerns Are Transforming Construction.* Worldwatch Institute, 1995, p. 67.
[133] Giff CA, Gangula B, Illinda P. *3D Opportunity in the Automotive Industry: Additive Manufacturing Hits the Road.* Deloitte University Press, UK, 2014, p. 28.
[134] How 3D Printing Will Change Manufacturing - GE Reports. https://www.ge.com/reports/epiphany-disruption-ge-additive-chief-explains-3d-printing-willupend-manufacturing/.
[135] 3D Printing Is Merged with Printed Electronics (NASDAQ:SSYS). http://investors.stratasys.com/releasedetail.cfm?ReleaseID=659142.
[136] R Kumar, M Kumar, JS Chohan. Material-specific properties and applications of additive manufacturing techniques: a comprehensive review. *Bulletin of Materials Science* 2021;44:181. https://doi.org/10.1007/s12034-021-02364-y

Chapter 13

Fused deposition modeling in knee arthroplasty: review with the current and novel materials

Prashant Veer and Bahadur Singh Pabla
National Institute of Technical Teachers Training and Research (NITTTR)

Jatinder Madan and Vettivel Singaravel Chidambaranathan
Chandigarh College of Engineering and Technology (Degree Wing)

CONTENTS

13.1 INTRODUCTION

13.1.1 Joint arthroplasty

Continuous pain in knee joints, difficulty to perform daily chores, and a limited range of mobility are a few causes that lead people to undertake knee replacement surgery [1]. Joint arthroplasty is a frequent and typically

DOI: 10.1201/9781003430186-13

effective surgical treatment for individuals with arthritic impairments commonly caused by aging or sports-related accidents. There are two broad categories of joint replacement surgeries, namely:

a. Total knee arthroplasty (TKR)
b. Partial knee arthroplasty (PKR)

Table 13.1 draws a broad comparison between total knee arthroplasty and partial knee arthroplasty.

The earliest metal-on-polyethylene TKR implant was popularized in the early 1970s [2].

13.1.2 3D printing

Additive Manufacturing (AM) is a new era in production, which is nearly devoid of complexity and is extremely customizable [3]. Figure 13.1 depicts the stages required in fabrication using a 3D printing machine.

The most common and recognizable 3D printing technology is Fused Deposition Modeling (FDM). It can quickly produce end-use functional items and prototypes with no geometrical limits [4]. The process factors mentioned in Table 13.2 influence the part characteristics like surface quality and dimensional accuracy. Various researchers have focused on exploring the link between part and process parameters to obtain functionally dependable and tailored components with optimal quality [5,6].

Table 13.1 Types of knee replacement surgery

Total knee arthroplasty	*Partial knee arthroplasty*
Both sides of the knee are replaced.	One side of the knee is replaced.
The complete knee is replaced.	Only the damaged region of the knee is replaced.
The incision produced is larger but lasts longer than the partial replacement.	The incision produced is smaller but does not last as long as the total replacement.
It might take anywhere from 1 to 3 hours, depending on the treatment.	The patient's hospital stay and recovery period are reduced, and they have a higher chance of moving more normally.
The risk of infection, blood loss, and clots is relatively higher, making the posttreatment rehabilitation process time-consuming.	The risk of infection, blood loss, and clots is minimized, making the posttreatment rehabilitation process easier.

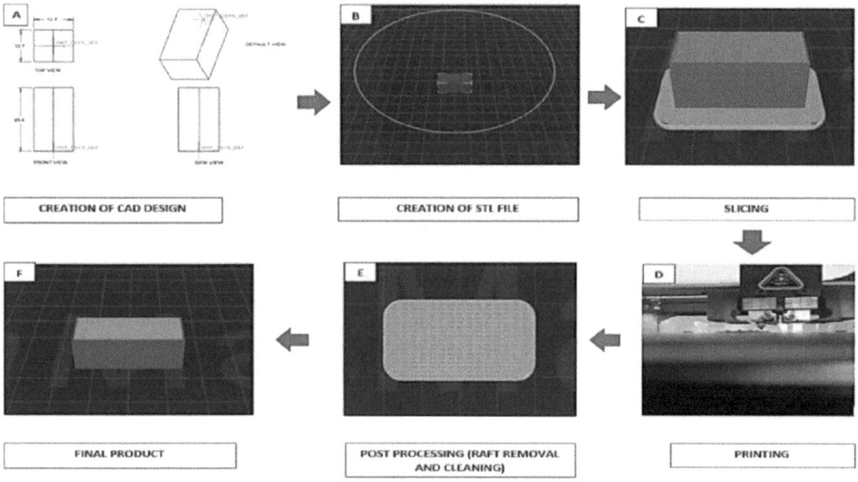

Figure 13.1 Fabrication steps in 3D printing.

Table 13.2 Process parameters [6]

Process parameter	*Information*
Build orientation	Build orientation refers to the alignment of the specimen body with the platform (bed). It includes the coordinate axes of the specimen and its orientation with the bed (Figure 13.2).
Raster orientation	The alignment of the raster with the reference axis is raster orientation. The most commonly used raster orientations are shown in Figure 13.3.
Extrusion temperature	It is the temperature at which the material flows and extrudes out of the nozzle. The material in filament form is fed to the heating element attached to the nozzle, which raises its temperature above the glass transition temperature and makes it flow. The heating element of a Raise 3D RXP-2200 series (Raise 3D Technologies, Inc.) is shown in Figure 13.4.
Infill density	Infill percentage or density, shown in Figure 13.5, refers to the proportion of infill volume filled with filament material inside a 3D component. It significantly impacts the strength, adaptability, and lightness of the components.
Layer thickness	It is the nozzle tip-measured vertical axis height of the deposited layers. The diameter of the nozzle and the filament material employed are usually the determining factors.

(Continued)

Table 13.2 (Continued) Process parameters [6]

Process parameter	*Information*
Infill pattern	It is the pattern in which individual layers are printed one at a time. Various infill designs are deployed to construct a robust and long-lasting inner structure. Figure 13.6 illustrates different infill patterns used to make FDM objects.
Print speed	Print speed is the relative speed of the printer nozzle with respect to the print bed. It may also be approximated as the redeposition rate of layers on the bed.

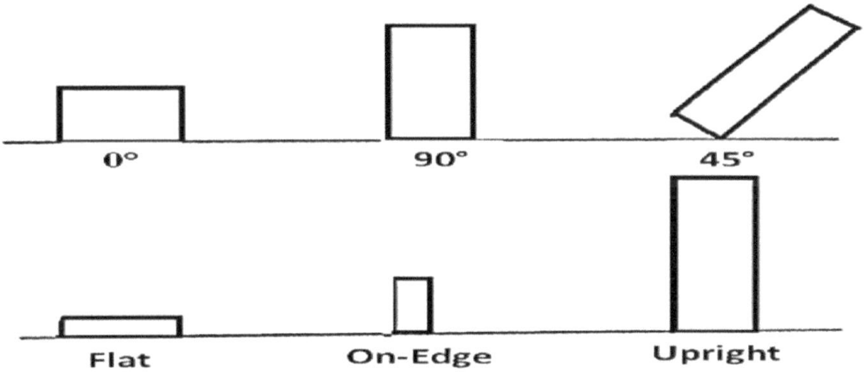

Figure 13.2 Build orientation.

Figure 13.3 Heating element.

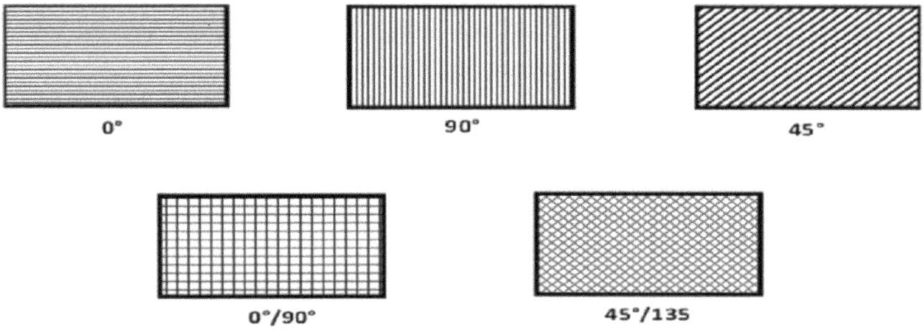

Figure 13.4 Raster orientation.

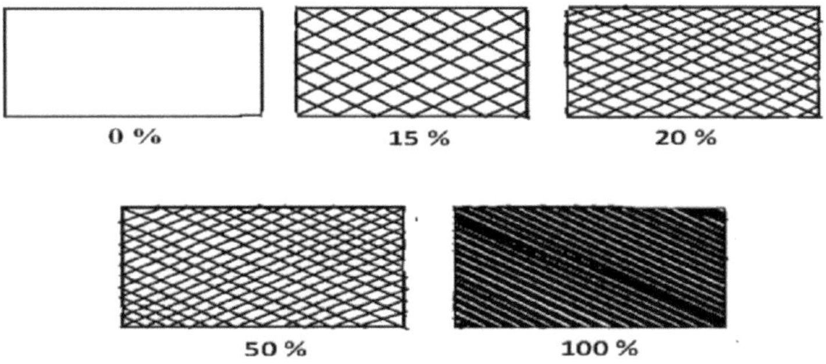

Figure 13.5 Infill densities.

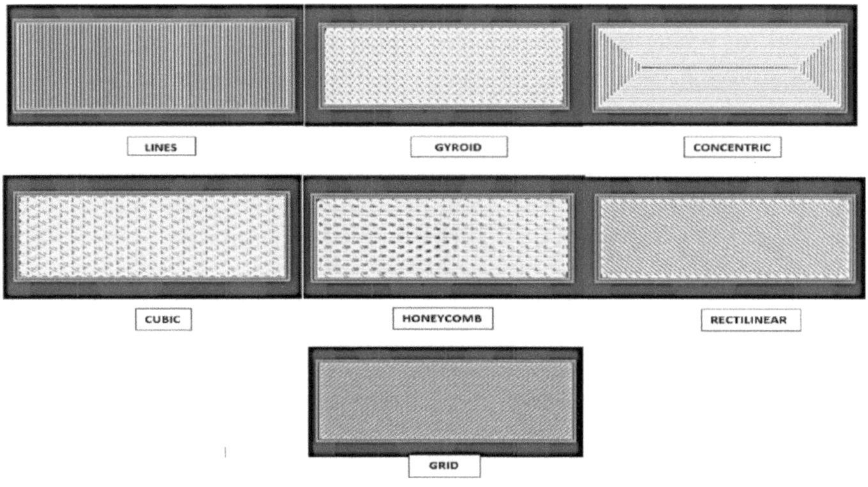

Figure 13.6 Infill patterns.

13.2 MATERIALS FOR KNEE SPACERS

Long-term implant insertion in the human body necessitates exceedingly cautious material and design choices. Thermoplastics for implants must have structural and mechanical properties similar to the tissue they are replacing. First and foremost, the embedded polymer components must continue functioning smoothly within the body's biological system [7]. Secondly, biostability is important when selecting a polymer for biomedical applications [8]. Thirdly, the materials must be inert and depict biocompatibility from a biological standpoint [9,10]. The materials that brush against one another as the knee is flexed are used to classify the various types, as depicted in Table 13.3.

Each patient's choice of design and materials for a knee replacement prosthesis is unique. Metals (commonly titanium or chrome-cobalt alloys) constitute the tibial and femoral components. These implants are either cemented or osteo-integrated, which involves a porous metallic stem extending into the tibial component upon which the patient's natural bone grows. A plastic spacer is placed between the tibial and femoral implant surfaces, on which the femoral component slides. It prevents the two components from coming in contact and causes excessive rubbing between them, leading to their premature failure. Usually, polyethylene is used to make the spacer. However, Thermoplastic Polyurethane (TPU) is also gaining popularity as the new potential material for spacers.

13.2.1 Ultra-high-molecular-weight polyethylene (UHMWPE)

The form and microstructure determine the qualities of any material and so is critical for a finer comprehension of UHMWPE. It consists of polyolefin fiber predominantly made up of olefins. It is a subset of the thermosetting plastic. $[C_2H_4]_n$ is the repeating unit, with n indicating the degree of polymerization. The polymerization degree of UHMWPE utilized in orthopedic implants ranges from 71,000 to 214,000 [11]. Zeigler's process is used to polymerize ethane monomers into high-molecular-weight polyethylene [12]. Bracco et al. [13] reported that the microstructure of UHMWPE is

Table 13.3 Types of knee implants

Name	*Material type for the component*		
	Femoral	*Tibial*	*Spacers*
Metal on plastic	Metal	Metal	Plastic
Ceramic on plastic	Ceramic	Ceramic	Plastic
Ceramic on ceramic	Ceramic	Ceramic	Ceramic
Metal on metal	Metal	Metal	Metal

more important than the molecular mass in determining its characteristics. Chakrabarty et al. [14] reported that UHMWPE is deployed for implants of knee joints, with advances made over time. Its outstanding self-lubricating property, biocompatibility, and low moisture absorption make it one of the top biomaterial implant materials [15].

Despite the attempts, it eventually releases wear debris, leading the implant to fail prematurely. Despite being the greatest option to date, UHMWPE has the severe drawback of accumulating wear debris over the years of extensive use, thus reducing its lifetime [16]. One of the main causes of UHMWPE failure has been identified as oxidative degradation. It causes a drop in resistance to abrasive wear, which generates higher debris while sliding, leading to osteolysis [17–20]. Oonishi et al. [21] reported that despite depicting biocompatibility in bulk form, the debris due to wear from UHMWPE causes osteolysis in the surrounding tissues, ultimately loosening the implant and leading to another replacement surgery.

13.2.2 Thermoplastic polyurethane (TPU)

Thermoplastic elastomers (TPE) comprise a rigid thermoplastic and a soft elastomer component. They are distinguished by their flexibility and ease of processing [22]. The thermoplastic component is recyclable, has a high processing temperature, is inexpensive, has a high melting viscosity, and has good mechanical qualities [23]. TPE is used in AM for bioengineering applications, preparing prototypes, and complete goods [24]. TPEs are classified into several families. Thermoplastic polyurethane (TPU) is the most often deployed thermoplastic elastomer in the 3D printing process, and it is utilized in various industries including medical, automobile production, and sports equipment. TPU has recently been offered as a viable contender for replacing UHMWPE as a bearing material in arthroplasty.

Hugle et al. [25] observed that TPU had interested the biomedical industry to create biocompatible materials to replace soft load-bearing tissues such as articular cartilage, meniscus, and the intervertebral disc. Schwartz et al. [26] reported that in a modeled in vivo wear scenario, the elastomeric polyurethane Pellethane® 236380A was reported to have lower wear rates than UHMWPE.

Agnieszka [27] described the synthesis, manufacturing, mechanical, and tribological characterization of a novel uncatalyzed polyurethane for FDM 3D printing technology. The tribological reaction of PU was examined by Luo et al. [28]. A friction simulator (Prosim pendulum) with physiologically realistic loads and movements was used to evaluate bovine medial compartmental knees. For the negative control, a robust medial segment was studied, while for the positive control, a stainless steel hemiarthroplasty was studied; and three different polyurethane hemiarthroplasty slabs were also examined. Lower-modulus PU resulted in lower contact pressure and frictional shear forces, resulting in lower levels of opposed cartilage attrition. PU with a higher modulus (22 MPa) produced higher amounts of frictional shear stress, resulting in wear of opposite cartilage, albeit not as severe as that produced by stainless steel.

According to Jianhua Xiao [29], the best break elongation and yield strength were obtained at a 45° orientation angle and 215°C temperature.

Yahiaoui [30] concluded that

- The Amontons-Coulomb rule governs the friction of TPU/steel contact.
- At room temperature, TPU wear kinetics follow Archard's law.
- TPU's tribo-rheological behavior causes stick-slip issues when critical load and velocity values are reached.
- The mean friction coefficient rises as the contact temperature rises.

Lee et al. [31] reported that the compression strength increased with the infill rate regardless of thickness. This rise was more pronounced when the infill rate was 100% rather than 10% or 50%. To protect the part against excessive compression, the infill rate had to be 100%. Most knee spacers are fabricated per the patients' requirement by compression or injection molding. However, to ease the customization of the fabrication process, 3D printing techniques like FDM are gaining popularity. The fabrication of different biocompatible materials through the FDM process is discussed in the following sections.

13.3 FABRICATION BY FDM PROCESS

Ning et al. [32] concluded that FDM has traditionally used thermoplastics, with polycarbonate (PC), acrylonitrile butadiene styrene (ABS), polyamide (PA), polylactide (PLA), as well as any two thermoplastics combined being the most popular. FDM uses layer-wise addition to make complicated objects with little tooling and waste [33]. Various parameters like infill density and pattern, the thickness of a layer, the speed of printing, the temperature of the nozzle, the direction of build, the raster angle, and the layer's orientation affect the mechanical qualities and quality of FDM-printed parts [34].

13.3.1 Ultra-high-molecular-weight polyethylene (UHMWPE)

As discussed, ultra-high-molecular-weight polyethylene (UHMWPE) is the most commonly used material for knee spacers owing to its superior creep, wear resistance, and high strength and stiffness [35,36]. The very high viscosity of UHMWPE detriments its easy melting and flowability and makes its processing through FDM a difficult task [35,37–40].

Reinforcement of UHMWPE is a promising way to enhance its properties [36]. The blending of UHMWPE with other additives can enhance its flow properties [35]. Ramli et al. [35] observed that direct blending of PEG with UHMWPE is unsuitable, even if it offers good internal lubrication. Adding up to 60% high-density polyethylene by weight improved the processability of the blend and its melt flow index (MFI) performance. It also depicted thermal stability in thermogravimetric analysis (TGA) [35]. Dontsov et al. [41] attempted the manufacture of UHMWPE mixed with 17% HDPE grafted with maleic anhydride and 12% polypropylene; 170°C NT, 100°C, and 20 mm/s PS offered maximum physical and tribo-mechanical properties of the composite. The wear performance of FDM-printed specimens was slightly lower than that of compression-sintered specimens. Panin et al. [42] deployed a UHMWPE-PP blend prepared by FDM and hot isostatic pressing (HIP). The shore D hardness increased with the addition of PP to UHMWPE. Also, the value of elastic modulus for the composite increased compared to neat UHMWPE when prepared by HIP, while this increase was not as significant as that for FDM specimens. This is depicted in Figure 13.7. Under severe loading conditions, the composite's wear performance is 15% higher than that of neat UHMWPE.

Panin et al. [43] reported that UHMWPE with 17% HDPE-g-VTMS could easily bind 5% glass fibers (GF) with different aspect ratios (AR). An increase in glass filler induced the adhesion of matrix and filler and deteriorated the tribological performance. UHMWPE loaded with milled glass fibers (MGF) had an elastic modulus 1.3–1.6 times higher than neat UHMWPE, depicted in Figure 13.7. The wear rate decreases with an increase in MGF up to 5%. An increase in chopped glass fibers (CGF) increased specific gravity, hardness, and elastic modulus compared to the composite (Figures 13.8 and 13.9). A brief summary of the work reviewed is displayed in Table 13.4.

13.3.2 Thermoplastic polyurethane (TPU)

As discussed earlier, the biocompatible nature and superior mechanical behavior of TPU make it suitable as a potential alternative to UHMWPE.

Miller et al. [47] recently completed 3D printing of TPU and examined the mechanical characteristics of printed specimens, concentrating on fatigue and cyclic efforts. TPU displayed greater energy dissipation capabilities in cyclic compression tests than softer polymers such as acrylate and silicone. TPU was able to overcome and endure fatigue breakage owing to the mechanism. Furthermore, the monotonic compression, shearing, and tensile fatigue of 3D-printed samples equaled or surpassed injection-molded counterparts. They also discovered that, in comparison to stiffer materials, the incorporation of alternative printing internal structure topologies (or print infill geometries) has no significant effect

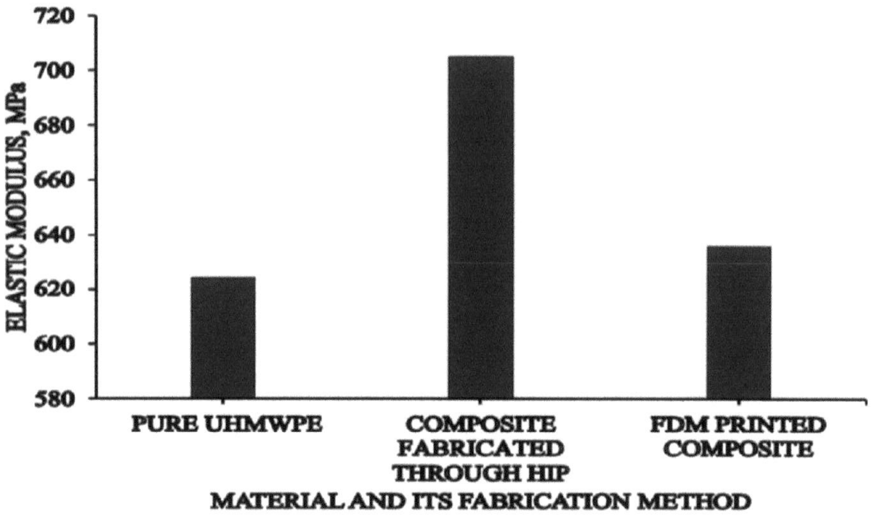

Figure 13.7 Variation of elastic modulus for pure UHMWPE, its composite fabricated through the HIP, and the composite printed on FDM [43].

Table 13.4 Summary of UHMWPE composites' performance based on FDM parameters

Ref material	*Findings*
[44] UHMWPE with polypropylene (10:90) composite	The impact strength was studied. The LT had a major effect on the response, followed by PS and the number of shells. Other parameters had a moderate effect.
[45] UHMWPE with 17% HDPE-SMA and 12% PP	0.5% carbon nanofibers and hydroxyapatite nanoparticles decreased wear and friction coefficient by 1.2–1.6 times. This composite prepared by FDM surpassed the mechanical properties of hot-pressed powder mix and extruded granules.
[46] UHMWPE with 20% PP	The maximum density was found in specimens made by twin screw compounded hot compression, followed by powder technology hot compression, and finally, twin screw compounded FDM. The maximum elastic modulus was obtained by twin screw compounded FDM specimens, followed by twin screw compounded hot compressed specimens and powder technology hot compressed specimens.

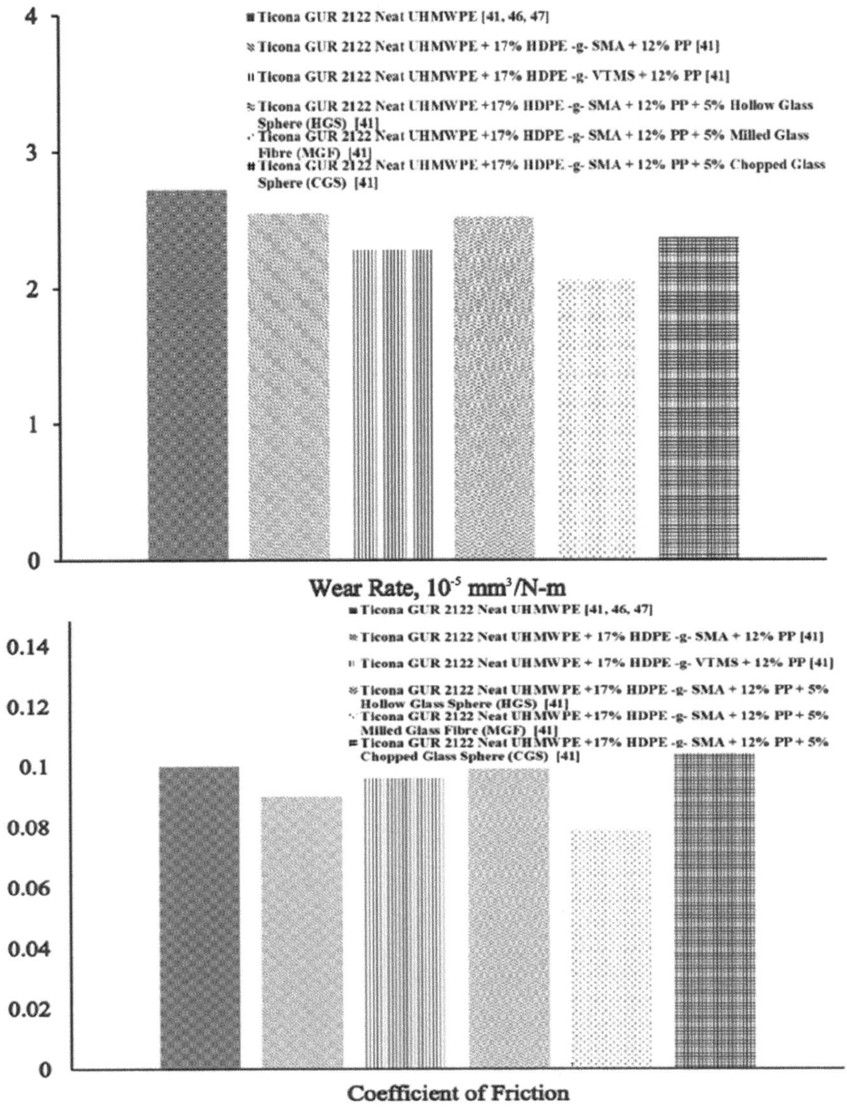

Figure 13.8 Variation of tribological properties with addition of different materials in UHMWPE [42,45,46].

on the fatigue performance of TPU. This is due to soft polyurethanes' adaptability and compliance. These properties highlight the advantages of 3D-printed TPU for artificial meniscus implants (Figure 13.10). The research conducted for TPU with FDM as the fabrication method has been discussed in detail in Table 13.5.

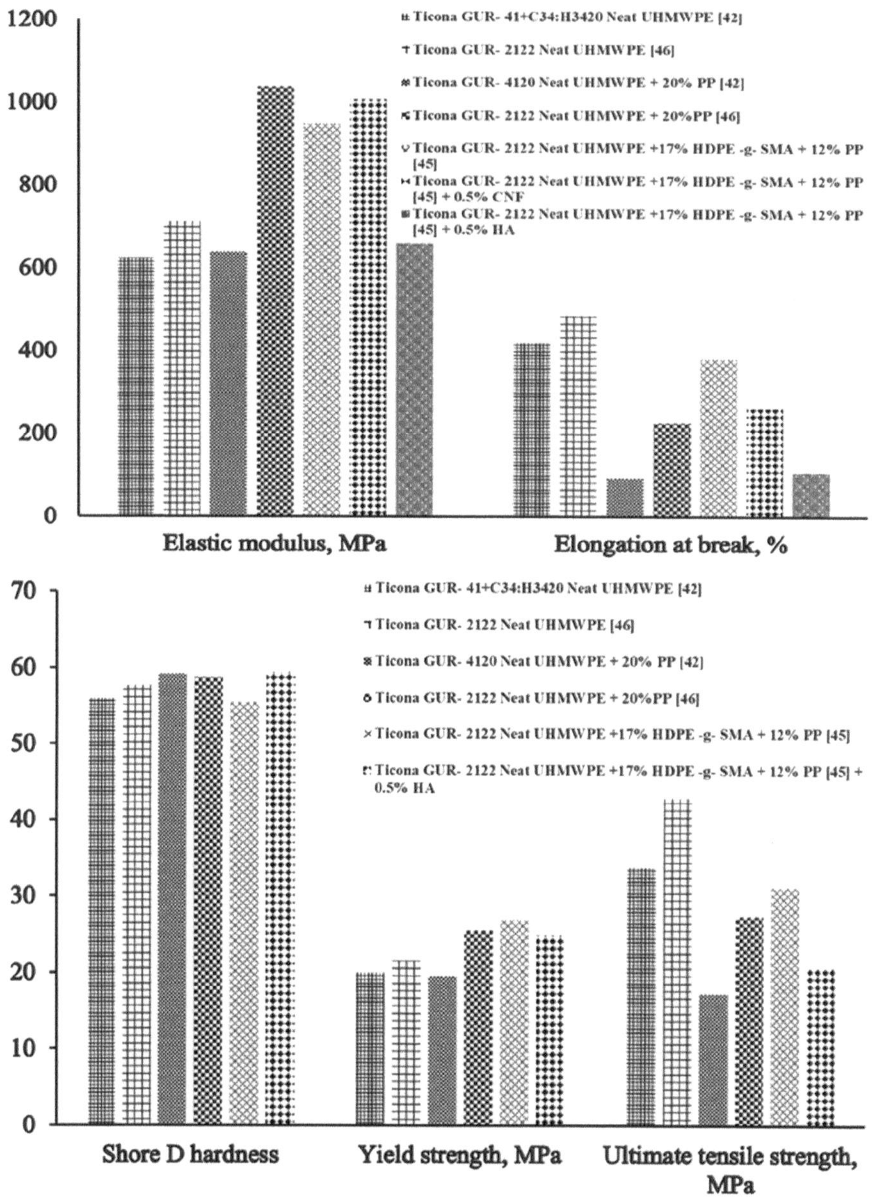

Figure 13.9 Variation of mechanical properties with addition of different materials in UHMWPE [42,45,46].

13.3.3 Polyvinylidene fluoride (PVDF)

Piezoelectric polymers, particularly polyvinylidene fluoride (PVDF), are gaining popularity as smart biomaterials [54] in tissue engineering [55] and

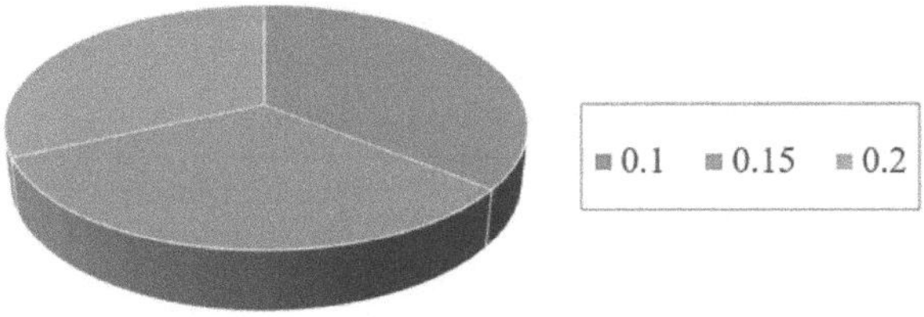

Figure 13.10 Tensile strength variation with LT [48].

Table 13.5 Summary of TPU's performance based on FDM parameters

Ref. parameters	*Findings*
[48] LT, BO	The X orientation provided the best flexural and tensile strength, but some fluctuation in the impact strength was observed. The optimum operating parametric values for TPU were 0.2 mm LT, printed in the X orientation.
[49] LT, RO, ET, BT	With the increase in LT, the tensile strength decreased for RO normal to the length of the specimen. For RO along the length of the specimen, the tensile strength displayed no significant variation from 0.1 to 0.2 mm LT with a sudden decrease at 0.3 mm.
[50] ET, BT, LT, PS, IFD, RO	With the bed increment and nozzle temperature increment, the elastic modulus and UTS increased, while the porosity and elongation percentage at break decreased. Increasing print speed decreased the elastic modulus, UTS, and elongation % and increased the porosity from 14% to 33%. Increasing the flow rate from 100 to 120 increased elastic modulus from 17 to 22 MPa, UTS from 21 to 25 MPa, and decreased the porosity from 14% to 3%. The best orientation that provided the best mechanical properties was 0° at a 100% flow rate.
[51] PS, ET	Increased nozzle temperature and decreased PS increased the specimen's density, but this increase was not significant beyond 240°C. Also, with an increase in nozzle temperature, the surface roughness increased.
[52,53] IFP, RO	100% IFD and 0° provided the maximum recovery speed, highest elastic modulus, and the ultimate tensile strength.

biosensors [5,56]. Among other fluoropolymers, PVDF has been reported to possess the maximum flexural modulus [57]. With the emergence of additive manufacturing, numerous researchers have endeavored to use it to fabricate piezoelectric PVDF and composites. Some of this research is discussed in Table 13.6.

13.3.4 Polylactic acid (PLA)

Polylactic acid is one of the renewable polymers from plant starch [64]. It is a much more printable material with superior mechanical qualities

Table 13.6 Summary of PVDF and its composites' performance based on FDM parameters

Ref. material parameter	*Findings*
[58] PVDF RO, PS, BT	A 0° orientation offered the best elastic modulus. β phase in the prepared specimens depended on the PS and the BT.
[59] PVDF LT, IFD (70%)	A BT range of 90°C–110°C and a PS range of 10–25 mm/s increased the dimensional tolerance. The elastic modulus values were reported to be less than bulk PVDF (1.7 GPa). This was due to lesser IFD.
[60] PVDF reinforced with barium titanate	The tensile strength increased by 45% with the addition of barium titanate.
[61] PVDF with zirconium tungstate LT (0.15 mm), IFD (70%), PS (15 mm/s)	A 10% zirconium tungstate by weight added in PVDF increased the elastic modulus by 30%.
[62,63] PVDF with barium titanate and graphene PS (50 mm/s), RO (45°), IFD (100%)	An increase of 43.45% and 48.43% in peak and break strength respectively.

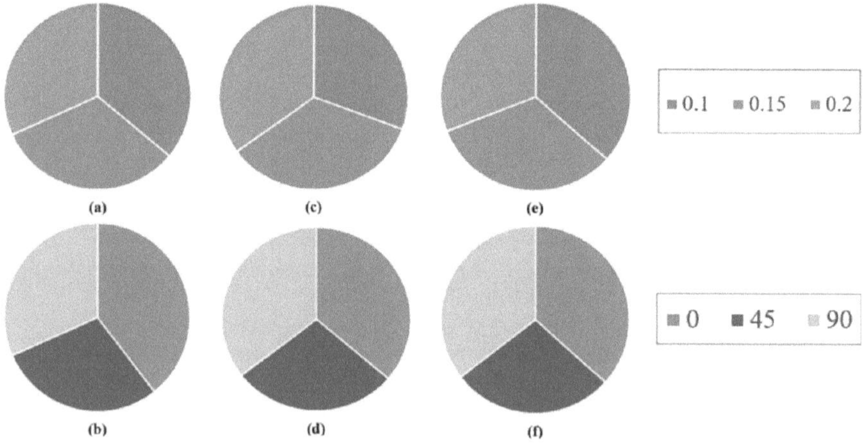

Figure 13.11 Variation of tensile strength with LT and RO [77].

than most other plastics [65,66]. Despite its enormous potential because of its renewability, biodegradable nature, and biocompatibility [67–69], this plastic has received minimal research related to the FDM process (Figure 13.11). Table 13.7 includes a summary of the effects of the process parameters on the properties of PLA.

Table 13.7 Summary of PLA's performance based on FDM parameters

Ref. parameter	*Findings*
[70] IFD, PS, LT	Mechanical properties are enhanced with increasing infill density and degraded with increasing printing speed. An increase in layer thickness negatively affected adhesion between layers, thus decreasing the elastic modulus.
[71] LT, perimeters, IFP	LT (0.1 mm), the number of outermost layers (6), and gyroid IFP were the best parameters for tensile strength. LT (0.15 mm), the number of outermost layers (4), and concentric IFP were the best parameters for flexural strength.
[72] IFD, RO, LT, BO	Elastic modulus and UTS depicted a linear relationship with IFD. For flat and on-edge orientation, 0°/90° gave the best elastic modulus and UTS, while ±45° offered the best elongation.
[73] RO, LT, RW	Higher tensile strength at lower RO, lower LT, and higher RW.
[74] BO	The sample produced in the X direction had the highest tensile strength.
[75] ET, PS, BO	The specimens created in the Y orientation had higher mechanical strength than those formed in the X, Z, and 45° orientations. The mechanical qualities of the pieces weakened with the rise in ET.
[76] LT, N, IFP, IFD	The infill % had the maximum impact on the compressive force, then the number of contours, LT, and IFP. Because of decreased warping, the compressive strength improved as the fraction of shells and infill increased. The diagonal infill design outperformed the linear infill pattern.
[77] LT, RO	90° and 0.1 mm combination gave the best tensile and flexural strength (Figure 13.11).

13.3.5 Acrylonitrile butadiene styrene (ABS)

Acrylonitrile butadiene styrene is an amorphous engineering thermoplastic with high impact resistance. Acrylonitrile, butadiene, and styrene are the three monomers that constitute ABS. It has high stiffness, excellent impact strength at relatively low temperatures, good insulating qualities, better weldability, and wear and strain resistance [78,79]. It is also a potential material for making small implants like mid-ear prostheses [80–82]. Table 13.8 consists of ABS performance based on FDM parameters.

13.3.6 Other materials

Several other materials used in the biomedical field are also reviewed in Table 13.9.

A comparative illustration of tensile strength with variation in layer thickness is shown in Figure 13.12.

Table 13.8 Summary of ABS's performance based on FDM parameters

Ref parameter	*Findings*
[83] LT, RO	±45° and 0.4 mm LT and 0.3 mm and ±45° gave the best tensile and flexural strength.
[84] PS	Surface roughness decreased with an increase in layer thickness.
[85] LT, PO, RW, air gap	Maximum compressive strength at LT 0.254 mm, width 59.44 mm, and 0.422 mm, air gap 0.00026 mm.
[86] IFP	The triangle infill design had a better tensile strength than grid or cubic patterns.
[87] BO, RO	The fabrication orientation (A) and deposition angle (B) significantly influenced the tensile strength. The sample constructed in the X direction had greater tensile properties of 28 MPa at a low raster orientation angle of 0°.
[88] BO, RO	Compared to Z orientation, items constructed with X or Y orientation achieved maximum strength. A component alignment along X and 60° orientation (X60) yielded the highest tensile properties (34.5 MPa), whereas a component alignment along Z and 90° orientation yielded the lowest tensile properties (14.8 MPa) (Z90).
[89] LT, RO, IFD	The specimen with the greatest ultimate strength of 32.649 MPa was created at a 65° raster angle, 0.5 mm layer thickness, and 80% infill percentage. Higher infill percentages and a 65° raster angle displayed superior tensile load resistance.

Table 13.9 Summary of various materials' performance based on FDM parameters

Ref, material, parameters	*Properties*
[90] Nylon IFD, LT, PS	At 0.1 mm LT, 100% IFD, and 70 mm/s PS, the maximum tensile strength and impact strength were 43.5 and 1.746 MPa, respectively. With a 0.3 mm LT, 100% IFD, and 70 mm/s PS, the maximum hardness of 68 was reached. Printing speed had the smallest impact of all the characteristics considered.
[91] PA12 LT, NT, occupancy rate, IFP	Optimum layer height, extrusion temperature, rate of occupancy, and filing pattern were 0.25 mm, 250°C, 50%, rectilinear (for the highest tensile strength); 0.20 mm, 260°C, 25%, and full honeycomb (for the highest elongation %); 0.25 mm, 250°C, 50%, rectilinear (for the highest impact strength).
[92] ULTEM 9085 thermoplastic RO, BO	The edge-printed sample had superior strength among all three orientations. With 0° RO, an average ultimate tensile strength of 89 MPa was attained. Because of perpendicular raster deposition to tension load application, the sample constructed with 90° orientation with an edge direction had the least average tensile strength (61 MPa).

(*Continued*)

Table 13.9 (*Continued*) Summary of various materials' performance based on FDM parameters

Ref, material, parameters	*Properties*
[93] PEEK 5% wt.CF/PEEK 5% wt. GF/PEEK NT, BT, PS, LT	The tensile, flexural, and impact strength increased with nozzle temperatures and platform temperature. 440° NT and 280° BT enhanced the mechanical properties of 5 wt.% CF/PEEK and 5 wt.% GF/PEEK. The tensile, flexural, and impact strength decreased with an increase in printing speed and layer thickness. Optimum mechanical properties were obtained at 5 mm/s printing speed and 0.1 mm layer thickness, overlap interval, and extrusion flow (100%).
[94] PETG LT, OR	Higher overlap ratios and thinner layers enhanced Young's modulus of elasticity and UTS while also providing a larger contact area between adjacent fibers and layers. An optimum combination of layer thickness was 0.2 mm at 20% overlap.

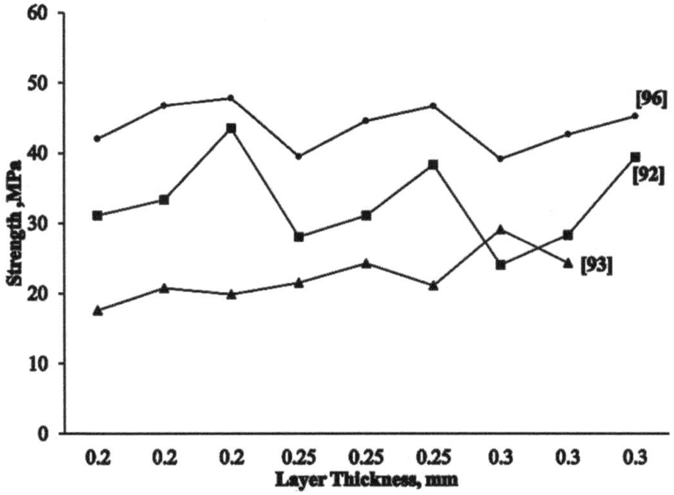

Figure 13.12 UTS [90,94] and tensile strength [91] variation with LT.

3D printing in knee arthroplasty can quickly fabricate tibial, femoral, and spacer components compared to injection molding. By deploying suitable parametric combinations, FDM can be easily used to produce these components at a much lower cost. A brief conclusion of the advantages and disadvantages of variation in various process parameters of FDM is depicted in Table 13.10.

Table 13.10 Conclusions

Parameter	Advantages	Limitations
Layer thickness	An increase in LT decreases print time.	Decreases proper adhesion, surface quality, and mechanical properties.
Temperature	Decreases viscosity, so better filling quality, lower porosity, and better adhesion.	With overflow, geometrical accuracy decreases.
Print speed	Reduces print time.	Turbulent flow, defects in sample poor adhesion.
Flow rate	E increases with flow rate, reduced voids, and porosity.	Lower geometrical accuracy.
Orientation angle	For 0° and load direction X, elongation increases.	Porosity increases.

13.4 CONCLUSIONS

A review of the commonly available materials for biomedical applications has been performed in this study. The most common materials in knee arthroplasty spacers are UHMWPE and TPU. Other materials commonly deployed are PVDF, PLA, and ABS. The properties of these materials have been discussed, and the influence of the various process parameters has also been discussed in detail. The present study shall be useful for researchers to study applying FDM 3D printing for biomedical applications.

ABBREVIATIONS

BO: Bed orientation
BT: Bed temperature
FDM: Fused deposition modeling
IFD: Infill density
IFP: Infill pattern
LT: Layer thickness
NT: Nozzle temperature
OR: Overlap ratio
PS: Print speed
RO: Raster orientation
RW: Raster width
TPU: Thermoplastic polyurethane
UHMWPE: Ultra-high-molecular-weight polyethylene
VTMS: Vinyl-trimethoxy silane

REFERENCES

[1] R. Trebše. Joint replacement: Historical overview. In: *Infected Total Joint Arthroplasty*, 2012, pp. 7–11. Springer, London. https://doi.org/10.1007/978-1-4471-2482-5_2.

[2] D.H. Williams, D.S. Garbuz, B.A. Masri. Total knee arthroplasty: Techniques and results. *British Columbia Medical Journal*, 2 (2010) 447–454. https://bcmj.org/articles/total-knee-arthroplasty-techniques-and-results.

[3] T. Huang, S. Wang, K. He, Quality control for fused deposition modelling based additive manufacturing: Current research and future trends, *2015 First International Conference on Reliability Systems Engineering (ICRSE)*, Beijing, China, 2015, pp. 1–6. https://doi.org/10.1109/ICRSE.2015.7366500.

[4] A. Peng, X. Xiao, R. Yue, Process parameter optimization for fused deposition modelling using response surface methodology combined with fuzzy inference system. *The International Journal of Advanced Manufacturing Technology*, 73 (1–4) (2014) 87–100.

[5] S. Vyavahare, S. Teraiya, D. Panghal, S. Kumar, Fused deposition modeling: A review. *Rapid Prototyping Journal* 26(1), 176–201. https://doi.org/10.1108/RPJ-04-2019-0106.

[6] A. Dey, N. Yodo, A systematic survey of FDM process parameter optimization and their influence on part characteristics. *Journal of Manufacturing and Materials Processing*, 3 (3) (2019) 64.

[7] A.J.T. Teo, A. Mishra, I. Park, Y.-J. Kim, W.-T. Park, Y.-J. Yoon, Polymeric biomaterials for medical implants and devices. *ACS Biomaterials Science & Engineering*, 2 (2016) 454–472.

[8] P.J. Joseph Francis, K.J. Arun, A.A. Navas, J. Irene, Biomedical applications of polymers—An overview. *Current Trends in Biomedical Engineering & Biosciences - Mini Review*, 15 (2018) 1–2.

[9] R.G. Kammula, J.M. Morris, Considerations for the biocompatibility evaluation of medical devices. *Medical Device & Diagnostic Industry*, 23 (2001) 82–92.

[10] D.F. Williams, On the mechanisms of biocompatibility. *Biomaterials*, 29 (2008) 2941–2953.

[11] M.C. Sobieraj, C.M. Rimnac, Ultra high molecular weight polyethylene: Mechanics, morphology, and clinical behavior. *Journal of the Mechanical Behavior of Biomedical Materials*, 2 (2009) 433–443.

[12] E. Peltola, V.-M. Tiainen, Y. Takakubo, B. Rajchel, J. Sobiecki, Y.T. Konttinen, M. Takagi, 7 - Materials used for hip and knee implants. In: S Affatato (Ed.), *Woodhead Publishing Series in Biomaterials, Wear of Orthopaedic Implants and Artificial Joints*, Woodhead Publishing, 2013, 178–218. https://doi.org/10.1533/9780857096128.1.178..

[13] P. Bracco, A. Bellare, A. Bistolfi, S. Affatato, Ultra-high molecular weight polyethylene: Influence of the chemical, physical and mechanical properties on the wear behavior. A review, Materials (Basel), 10 (2017) 791. https://doi.org/10.3390/ma10070791.

[14] G. Chakrabarty, M. Vashishtha, D. Leeder, Polyethylene in knee arthroplasty: A review, *Journal of Clinical Orthopaedics and Trauma*, 6 (2015) 108–112. https://doi.org/10.1016/j.jcot.2015.01.096.

[15] H. Wang, Y. Wang, Q. Su et al., Self-lubricating ultrahigh molecular weight polyethylene thin films with excellent wear resistance at light friction loads on glass and silicon. *Journal of Macromolecular Science, Part B Physics*, 58 (2019) 317–329. https://doi.org/10.1080/00222348.2019.1565155.
[16] J.C. Baena, J. Wu, Z. Peng, Wear performance of UHMWPE and reinforced UHMWPE composites in arthroplasty applications: A review. *Lubricants*, 3 (2015) 413–436.
[17] S.M. Kurtz, O.K. Muratoglu, M. Evans, A.A. Edidin, Advances in the processing, sterilization, and crosslinking of ultra-high molecular weight polyethylene for total joint arthroplasty. *Biomaterials*, 20 (18) (1999) 1659–1688.
[18] A.A. Besong, B. Eng, R. Student et al., Quantitative comparison of wear debris from UHMWPE that has and has not been sterilised by gamma irradiation. *Journal of Bone and Joint Surgery. British Volume* 80 (2) (1998) 340–344. https://doi.org/10.1302/0301-620X.80B2.0800340.
[19] H. Mckellop, F.-W. Shen, P. Campbell, R. Salovey, Effect of sterilization method and other modifications on the wear resistance of acetabular cups made of ultra-high molecular weight polyethylene a hip-simulator study. *Journal of Bone and Joint Surgery. American Volume* 82 (12) (2000) 1708–1725. https://doi.org/10.2106/00004623-200012000-00004.
[20] W.H. Harris, Wear and periprosthetic osteolysis: the problem. *Clinical Orthopaedics and Related Research* 393 (2001) 66–70. doi: 10.1097/00003086-200112000-00007.
[21] H. Oonishi, Y. Kadoya, S. Masuda, Gamma-irradiated cross-linked polyethylene in total hip replacements—Analysis of retrieved sockets after long-term implantation, *Journal of Biomedical Materials Research*, 58 (2001) 167–171.
[22] T.A. Osswald, J.P. Hernández-Ortiz, *Polymer Processing*, Tim A. Osswald, Juan P. Hernández-Ortiz (Eds.) West. Carolina Univ. (2006) i–xxvii. https://doi.org/10.3139/9783446412866.fm
[23] S.S. Banerjee, K.D. Kumar, A.K. Sikder, A. K. Bhowmick, Nanomechanics and origin of rubber elasticity of novel nanostructured thermoplastic elastomeric blends using atomic force microscopy, *Macromolecular Chemistry and Physics*, 216 (15) (2015) 1666–1674.
[24] S.S. Banerjee, A.K. Bhowmick, Tailored nanostructured thermoplastic elastomers from polypropylene and fluor elastomer: morphology and functional properties, *Industrial & Engineering Chemistry Research*, 54 (33) (2015) 8137–8146.
[25] T. Hügle, J. Geurts, C. Nüesch, M. Müller-Gerbl, V. Valderrabano, Aging and osteoarthritis: An IneviTable encounter?, *Journal of Aging Research*, 2012 (2012) Article ID 950192, 7 pages. https://doi.org/10.1155/2012/950192.
[26] C.J. Schwartz, S. Bahadur, Development and testing of a novel joint wear simulator and investigation of the viability of an elastomeric polyurethane for total-joint arthroplasty devices, *Wear*, 262 (2007) 331–339. https://doi.org/10.1016/j.wear.2006.05.018.
[27] A. Haryńska, I. Gubanska, J. Kucinska-Lipka, H. Janik, Fabrication and characterization of flexible medical-grade TPU filament for fused deposition modeling 3DP technology. *Polymers*, 10 (12) (2018) 1304. https://doi.org/10.3390/polym10121304.

[28] Y. Luo, L. McCann, E. Ingham, Z.-M. Jin, S. Ge, J. Fisher, Polyurethane as a potential knee hemiarthroplasty biomaterial: an *in-vitro* simulation of its tribological performance. *Proceedings of the Institution of Mechanical Engineers, Part H: Journal of Engineering in Medicine*, 224 (3) (2010) 415–425. https://doi.org/10.1243/09544119JEIM657.
[29] J. Xiao, Y. Gao, The manufacture of 3D printing of medical grade TPU. *Progress in Additive Manufacturing*, 2 (2017) 117–123. https://doi.org/10.1007/s40964-017-0023-1.
[30] M. Yahiaoui, J. Denape, J.-Y. Paris, A.G. Ural, N. Alcalá, F.J. Martínez, Wear dynamics of a TPU/steel contact under reciprocal sliding. *Wear*, 315 (1–2) (2014) 103–114. https://doi.org/10.1016/j.wear/2014/04/005.
[31] H. Lee, R.-I. Eom, Y. Lee, Evaluation of the mechanical properties of porous thermoplastic polyurethane obtained by 3D printing for protective gear. *Advances in Materials Science and Engineering*, 2019 (2019) 5838361. https://doi.org/10.1155/2019/5838361
[32] F. Ning, W. Cong, J. Qiu, J. Wei, S. Wang, Additive manufacturing of carbon fiber reinforced thermoplastic composites using fused deposition modeling. *Composites Part B: Engineering*, 80 (2015) 369–378. https://doi.org/10.1016/j.compositesb.2015.06.013.
[33] D. Popescu, A. Zapciu, C. Amza, F. Baciu, R. Marinescu, FDM process parameters influence over the mechanical properties of polymer specimens: A review. *Polymer Testing* 69 (2018) 157–166. https://doi.org/10.1016/j.polymertesting.2018.05.020.
[34] A. Przybytek, J. Kucinska-Lipka, H. Janik, Thermoplastic elastomer filaments and their application in 3D printing. *Elastomery*, 20 (4) (2016) 32–39.
[35] M.S. Ramli, M.S. Wahab, M. Ahmad, A.S. Bala, FDM preparation of biocompatible UHMWPE polymer for artificial implant. *Journal of Engineering and Applied Sciences*, 11 (2016) 5474–5480.
[36] G. Stephen, J. Daniel, M. Prakash, Wear performance of UHMWPE reinforced with basalt fibre for total disc replacement. *IOP Conference Series: Materials Science and Engineering*, 912 (2020) 052030. https://doi.org/10.1088/1757–899X/912/5/052030.
[37] M. Xie, H. Li, Viscosity reduction and disentanglement in ultrahigh molecular weight polyethylene melt: Effect of blending with polypropylene and poly (ethylene glycol). *European Polymer Journal*, 43 (2007) 3480–3487.
[38] J. Wang, C. Cao, X. Chen, S. Ren, D. Yu, X. Chen, Extensional-shear coupled flow-induced morphology and phase evolution of polypropylene/ultrahigh molecular weight polyethylene blends: Dissipative particle dynamics simulations and experimental studies. *Polymer*, 169 (2019) 36–45.
[39] S.A. Hashmi, S. Neogi, A. Pandey, N. Chand, Sliding wear of PP/UHMWPE blends: Effect of blend composition. *Wear*, 247 (2001) 9–14.
[40] M. Xie, J. Chen, H. Li, M. Li, Influence of poly (ethylene glycol)-containing additives on the sliding wear of ultrahigh molecular weight polyethylene/polypropylene blend. *Wear*, 268 (2010) 730–736.
[41] Y.V. Dontsov, S.V. Panin, D.G. Buslovich, F. Berto, Taguchi optimization of parameters for feedstock fabrication and FDM manufacturing of wear-resistant UHMWPE-based composites. *Materials*, 13 (12) (2020) 2718. https://doi.org/10.3390/ma13122718.

[42] S. V. Panin, D. G. Buslovich, L. A. Kornienko, Y.V. Dontsov, V.O. Alexenko, B.B. Ovechkin, Comparison of Structure and Tribotechnical Properties of Extrudable UHMWPE Composites Fabricated by HIP and FDM Techniques, In *AIP Conference Proceedings [Author(s) Proceedings of the Advanced Materials with Hierarchical Structure for New Technologies and Reliable Structures*, Tomsk, Russia (1–5 October 2018)], vol. 2051, 2018, 020229. https://doi.org/10.1063/1.5083472.

[43] S.V. Panin, D.G. Buslovich, Y.V. Dontsov, S.A. Bochkareva, L.A. Kornienko, F. Berto, UHMWPE-based glass-fiber composites fabricated by FDM. Multiscaling aspects of design, manufacturing and performance. *Materials*, 14 (2021) 1515. https://doi.org/10.3390/ma14061515.

[44] A.S. Bala, S.B. Wahab, Optimal FDM parameters setting in enhancing impact strength of PP/UHMWPE composite using fractal factorial design, *Journal of Academic Research (Applied Sciences)* 17 (2021) 26–33.

[45] S.V. Panin, D.G. Buslovich, L.A. Kornienko, Y.V. Dontsov, S.A. Bochkareva, M.V. Chaikina, Structure and Tribomechanical Properties of Multicomponent Extrudable "UHMWPE-HDPE-g-SMA-PP" Composites Fabricated by Fused Deposition Modeling, In *AIP Conference Proceedings [AIP Publishing Proceedings of the International Conference on Advanced Materials with Hierarchical Structure for New Technologies and Reliable Structures 2019*, Tomsk, Russia (1–5 October 2019), vol. 2167, 2019, 020264. https://doi.org/10.1063/1.5132131.

[46] S.V. Panin, D.G. Buslovich, Y.V. Dontsov, L.A. Kornienko, V.O. Alexenko, S.A. Bochkareva, S.V. Shilko, Two-component feedstock based on ultra-high molecular weight polyethylene for additive manufacturing of medical products. *Advanced Industrial and Engineering Polymer Research*, 4(4) (2021) 235–250. https://doi.org/10.1016/j.aiepr.2021.05.003.

[47] A.T. Miller, D.L. Safranski, K.E. Smith, R.E. Guldberg, K. Gall, Compressive cyclic ratcheting and fatigue of synthetic, soft biomedical polymers in solution, *Journal of the Mechanical Behavior of Biomedical Materials*, 54 (2016) 268–282. https://doi.org/10.1016/j.jmbbm.2015.09.034.

[48] N.V.S.S. Sagar, K.S. Vepa, Experimental investigations for improving the strength of parts manufactured using FDM process. In *Innovative Design, Analysis and Development Practices in Aerospace and Automotive Engineering (I-DAD 2018)*, Springer, Singapore, 2019, pp. 307–313.

[49] J. Leng, J. Wu, The development of a conical screw-based extrusion deposition system and its application in fused deposition modeling with thermoplastic polyurethane. *Rapid Prototyping Journal*, 26/2 (2020) 409–417. https://doi.org/10.1108/RPJ-05-2019-0139.

[50] S. Kasmi, G. Ginoux, E. Labbé, S. Alix, Multi-physics properties of thermoplastic polyurethane at various fused filament fabrication parameters, *Rapid Prototyping Journal* 28 (5) (2021) 895–906. https://doi.org/10.1108/RPJ-08-2021-0214.

[51] Y. Yang, Y. Chen, Y. Wei, Y. Li, 3D printing of shape memory polymer for functional part fabrication. *International Journal of Advanced Manufacturing Technology*, 84 (9–12) (2015) 2079–2095. https://doi.org/10.1007/s00170-015-7843-2.

[52] J. Villacres, D. Nobes, C. Ayranci, Additive manufacturing of shape memory polymers: Effects of print orientation and infill percentage on mechanical

properties. *Rapid Prototyping Journal*, 24 (4) (2018) 744–751. https://doi.org/10.1108/rpj-03-2017-0043.
[53] J. Villacres, D. Nobes, C. Ayranci, Additive manufacturing of shape memory polymers: Effects of print orientation and infill percentage on shape memory recovery properties. *Rapid Prototyping Journal*, 26 (9) (2020) 1593–1602. https://doi.org/10.1108/rpj-09-2019-0239.
[54] P. Sengupta, A. Ghosh, N. Bose, S. Mukherjee, A.R. Chowdhury, P. Datta, A comparative assessment of poly(vinylidene fluoride)/conducting polymer electrospun nanofiber membranes for biomedical applications. *Journal of Applied Polymer Science* 137 (37) (2020) 49115. https://doi.org/10.1002/app.49115.
[55] Y.-S. Lee, G. Collins, T.L. Arinzeh, Neurite extension of primary neurons on electrospun piezoelectric scaffolds. *Acta Biomaterilia*, 7 (11) (2011) 3877–3886.
[56] T. Sharma, S.-S. Je, B. Gill, J.X. Zhang, Patterning piezoelectric thin film PVDF–TrFE based pressure sensor for catheter application. *Sensors Actuators, A*, 177 (2012) 87–92.
[57] D.W. Grainger, 1.3.2C - Fluorinated biomaterials, In: W.R. Wagner, S.E. Sakiyama-Elbert, G. Zhang, M.J. Yaszemski (Eds.), *Biomaterials Science*, fourth ed., Academic Press, 2020, pp. 125–138.
[58] D.A. Porter, T.V. Hoang, T.A. Berfield, Effects of in-situ poling and process parameters on fused filament fabrication printed PVDF sheet mechanical and electrical properties. *Additive Manufacturing*, 13 (2017) 81–92.
[59] N. Momenzadeh, H. Miyanaji, D.A. Porter, T.A. Berfield, Polyvinylidene fluoride (PVDF) as a feedstock for material extrusion additive manufacturing. *Rapid Prototyping Journal* 26 (1) (2020) 156–163.
[60] H. Kim, F. Torres, D. Villagran, C. Stewart, Y. Lin, T.L.B. Tseng, 3D printing of BaTiO3/PVDF composites with electric in situ poling for pressure sensor applications. *Macromolecular Materials and Engineering*, 302 (11) (2017) 1700229.
[61] N. Momenzadeh, H. Miyanaji, T.A. Berfield, Influences of zirconium tungstate additives on characteristics of polyvinylidene fluoride (PVDF) components fabricated via material extrusion additive manufacturing process. *International Journal of Advanced Manufacturing Technology*, 103 (9–12) (2019) 4713–4720.
[62] R. Sharma, R. Singh, A. Batish, Study on barium titanate and graphene reinforced PVDF matrix for 4D applications, *Journal of Thermoplastic Composite Materials* 34 (9) (2019) 1234–1253. doi:10.1177/0892705719865004
[63] R. Sharma, R. Singh, A. Batish, On mechanical and surface properties of electroactive polymer matrix-based 3D printed functionally graded prototypes. *Journal of Thermoplastic Composite Materials* 35 (5) (2020) 615–630. doi:10.1177/0892705720907677.
[64] A. Valerga, M. Batista, J. Salguero, F. Girot, Influence of PLA filament conditions on characteristics of FDM parts. *Materials*, 11 (8) (2018) 1322. https://doi.org/10.3390/ma11081322.
[65] B.M. Tymrak, M. Kreiger, J.M. Pearce, Mechanical properties of components fabricated with open-source 3-D printers under realistic environmental conditions. *Materials & Design*, 58 (2014) 242–246.
[66] A. Lanzotti, M. Grasso, G. Staiano, M. Martorelli, The impact of process parameters on mechanical properties of parts fabricated in PLA with an open-source 3-D printer. *Rapid Prototyping Journal*, 21 (2015) 604–617.

[67] A.P. Mathew, K. Oksman, M. Sain, Mechanical properties of biodegradable composites from poly lactic acid (PLA) and microcrystalline cellulose (MCC). *Journal of Applied Polymer Science*, 97 (2005) 2014–2025.
[68] S. Farah, D.G. Anderson, R. Langer, Physical and mechanical properties of PLA, and their functions in widespread applications—A comprehensive review. *Advanced Drug Delivery Reviews*, 107 (2016) 367–392.
[69] S.L. Messimer, A.E. Patterson, N. Muna, A.P. Deshpande, T. Rocha Pereira, Characterization and processing behavior of heated aluminum-polycarbonate composite build plates for the FDM additive manufacturing process. *Journal of Manufacturing and Materials Processing*, 2 (2018) 12.
[70] A. Farazin, M. Mohammadimehr, Effect of different parameters on the tensile properties of printed polylactic acid samples by FDM: Experimental design tested with MDs simulation. *The International Journal of Advanced Manufacturing Technology*, 118 (2021) 103–118. https://doi.org/10.1007/s00170-021-07330-w
[71] H. Chokshi, D.B. Shah, K.M. Patel, S.J. Joshi, Experimental investigations of process parameters on mechanical properties for PLA during processing in FDM. *Advances in Materials and Processing Technologies* 8 (2021) 1–14.
[72] S.N. Cerda-Avila, H.I. Medellín-Castillo, T. Lim, An experimental methodology to analyse the structural behaviour of FDM parts with variable process parameters. *Rapid Prototyping Journal*, 26 (9) (2020) 1615–1625. https://doi.org/10.1108/RPJ-12-2019-0312.
[73] R. Anitha, S. Arunachalam, P. Radhakrishnan, Critical parameters influencing the quality of prototypes in fused deposition modeling. *Journal of Materials Processing Technology*, 118 (1–3) (2001) 385–388. https://doi.org/10.1016/s0924-0136(01)00980-3.
[74] S. Kramer, J. Jordan, H. Jin, J. Carroll, A. Beese (eds), *Mechanics of Additive and Advanced Manufacturing*, vol. 8. https://doi.org/10.1007/978-3-319-95083-9_8.
[75] U. Khaleeq uz Zaman, E. Boesch, A. Siadat, M. Rivette, A.A. Baqai, Impact of fused deposition modeling (FDM) process parameters on strength of built parts using Taguchi's design of experiments. *International Journal of Advanced Manufacturing Technology*, 101 (2018) 1215–1226. https://doi.org/10.1007/s00170-018-3014-6.
[76] U. Chandrasekhar et al. (eds), *Innovative Design, Analysis and Development Practices in Aerospace and Automotive Engineering (I-DAD 2018), Lecture Notes in Mechanical Engineering*, Springer, Singapore. https://doi.org/10.1007/978-981-13-2697-4_34.
[77] N.V.S.S. Sagar, K.S. Vepa, Experimental investigations for improving the strength of parts manufactured using FDM process. In: U. Chandrasekhar, L.J. Yang, S. Gowthaman (eds), *Innovative Design, Analysis and Development Practices in Aerospace and Automotive Engineering (I-DAD 2018). Lecture Notes in Mechanical Engineering*, Springer, Singapore, 2019. https://doi.org/10.1007/978-981-13-2697-4_34.
[78] J. Feng, C. Carpanese, A. Fina, Thermal decomposition investigation of ABS containing Lewis-acid type metal salts. *Polymer Degradation and Stability*, 129 (2016) 319–327.
[79] L.W. McKeen, *Fatigue and Tribological Properties of Plastics and Elastomers*, 3rd ed., William Andrew, Norwich, NY, 2016.
[80] M. Ziąbka, A review of materials used in middle ear prosthetics. *Ceramic Materials*, 70 (2018) 65–85.

[81] M. Ziąbka, M. Dziadek, E. Menaszek, R. Banasiuk, A. Królicka, Middle ear prosthesis with bactericidal efficacy—In Vitro investigation. *Molecules*, 22 (2017) 1681.
[82] M. Ziąbka, E. Menaszek, J. Tarasiuk, S. Wroński, Biocompatible nanocomposite implant with silver nanoparticles for otology—In vivo evaluation. *Nanomaterials*, 8 (2018) 764.
[83] S.R. Rajpurohit, H.K. Dave, Effect of process parameters on tensile strength of FDM printed PLA part. *Rapid Prototyping Journal* 24 (8) (2018) 1317–1324. https://doi.org/10.1108/RPJ-06-2017-0134.
[84] B.H. Lee, J. Abdullah, Z.A. Khan, Optimization of rapid prototyping parameters for production of flexible ABS object. *Journal of Materials Processing Technology*, 169 (1) (2005) 54–61. https://doi.org/10.1016/j.jmatprotec.2005.02.259.
[85] S. Abid, R. Messadi, T. Hassine, Optimization of mechanical properties of printed acrylonitrile butadiene styrene using RSM design. *International Journal of Advanced Manufacturing Technology*, 100 (2019) 1363–1372. https://doi.org/10.1007/s00170-018-2710-6.
[86] A. Garg, A. Bhattacharya, A. Batish, Chemical vapor treatment of ABS parts built by FDM: Analysis of surface finish and mechanical strength. *International Journal of Advanced Manufacturing Technology*, 89 (2016) 2175–2191. https://doi.org/10.1007/s00170-016-9257-1.
[87] M. Samykano, S.K. Selvamani, K. Kadirgama, W.K. Ngui, G. Kanagaraj, K. Sudhakar, Mechanical property of FDM printed ABS: Influence of printing parameters. *International Journal of Advanced Manufacturing Technology*, 102 (2019) 2779–2796. https://doi.org/10.1007/s00170-019-03313-0.
[88] F. Afrose, S.H. Masood, M. Nikzad, P. Iovenitti, Effects of build orientations on tensile properties of PLA material processed by FDM. *Advanced Materials Research*, 1044 (2014) 31.
[89] S. Attoye, E. Malekipour, H. El-Mounayri, Correlation between process parameters and mechanical properties in parts printed by the fused deposition modeling process In: Kramer, S., Jordan, J., Jin, H., Carroll, J., Beese, A. (eds.), *Mechanics of Additive and Advanced Manufacturing, Volume 8. Conference Proceedings of the Society for Experimental Mechanics Series*. Springer, Cham (2019). https://doi.org/10.1007/978-3-319-95083-9_8
[90] M. Ramesh, K. Panneerselvam, Mechanical investigation and optimization of parameter selection for Nylon material processed by FDM. *Materials Today: Proceedings* 46 (2020) 9303–9307. https://doi.org/10.1016/j.matpr..2020.02.697.
[91] M. Kam, A. İpekçi, Ö. Şengül, Investigation of the effect of FDM process parameters on mechanical properties of 3D printed PA12 samples using Taguchi method. *Journal of Thermoplastic Composite Materials* 36 (1) (2021) 307–325. doi:10.1177/08927057211006459
[92] K.I. Byberg, A.W. Gebisa, H.G. Lemu, Mechanical properties of ULTEM 9085 material processed by fused deposition modeling. *Polymer Testing*, 72 (2018) 335–347. https://doi.org/10.1016/j.polym ertesting.2018.10.040.
[93] W.A.N.G. Peng, Z.O.U. Bin, D.I.N.G. Shouling, L.I. Lei, C. Huang, Effects of FDM-3D printing parameters on mechanical properties and microstructure of CF/PEEK and GF/PEEK. *Chinese Journal of Aeronautics*, 34 (9) (2021) 236–246.
[94] A. Özen, B.E. Abali, C. Völlmecke, J. Gerstel, D. Auhl, Exploring the role of manufacturing parameters on microstructure and mechanical properties in fused deposition modeling (FDM) using PETG. *Applied Composite Materials* 28 (2021) 1799–1828. https://doi.org/10.1007/s10443-021-09940-9

Chapter 14

Wear and surface roughness of additive manufacturing materials and composites

Arunprasath Kanagaraj, Selvakumar Ponnusamy, and Vijayakumar Murugesan
PSN College of Engineering and Technology

CONTENTS

14.1 CONCEPT OF WEAR AND SURFACE ROUGHNESS IN ADDITIVE MANUFACTURING MATERIALS AND COMPOSITES

Production of metals and amalgams by adding substances is right now hard for large-scale manufacturing, due to lack of assembled rate, restricted greatest form size, issues of surface unpleasantness, support structures expulsion, and so forth. The possibility of AM came from the way that a blend of conventional (subtractive) and added substance assembling. AM assists in manufacturing of medium/enormous parts with high mathematical complicity and precision [1,2].

With its undeniable widespread reasonableness and proficiency, it is trying to anticipate how the cycle will progress. Fabrication of multimaterial added substance is venturing out forward by marvellous single material items to multimaterial parts that hold creative guarantee. With every one

DOI: 10.1201/9781003430186-14

of the upsides of 3D printing (material and asset effectiveness, part and creation adaptability, diminished creation lead time, expanded execution), these parts can have various materials with complex calculations and added usefulness [3].

Designing composite materials in assembling has turned into a vital exploration region in further developing rail material mechanical characteristics against wear. A few better measures are through surface designing of essential rail and wheel districts with further developed characteristics to wear opposition [4].

In designing fields, AM has additionally been utilized in numerous modern areas, like aviation, car, common and biomedical designing. Perhaps the most basic difficulties to originators and architect as of late is the way to rapidly create another item from planning models to qualified items, and the response is AM innovation which gives an exceptionally effective cycle to assemble an item with nearly all math intricacy with next to no new tooling [5].

Composites have phenomenal material properties like gentility, unbending nature, and strength with support of particular materials to serve a lengthy field in designing. In the interim, limitations because of the creation interaction lead to unfortunate machinability attributes and show decreased surface quality, exorbitant cutting temperature and instrument wear [6,7].

Composite machining is not quite the same as regular machining. The expansion of supported particles might bring about extreme harm of cutting devices. The hard particles or supported materials might bring about unfortunate surface wrapping up. Be that as it may, with the customary machining, these issues can be handily experienced [8].

Wear, hotness and grating produce issues in working instruments of a machine as a result of the great and rehashed utilization of machines. A colossal expense on machines would be saved in the event that the wear rate on machines is taken into thought as wear decreases life expectancy of material. Adding an exceptionally essential or least measure of amalgam in the material utilized for the mass handling might actually adjust the obligation of bond between the pair [9].

Surface completion is the most extreme fundamental trademark in assembling of parts. This impacts execution of parts and creation cost.

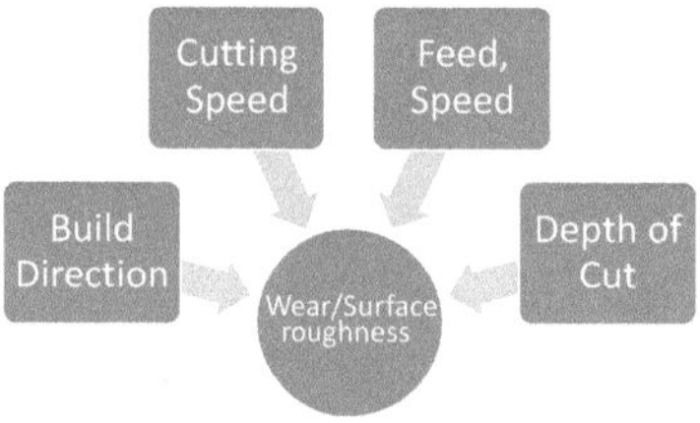

Figure 14.1 Factors responsible for concept of wear and surface roughness.

Present-day ventures need to create great items rapidly with least administrator input. It is exceptionally difficult to consider all elements influencing surface unpleasantness for a specific assembling method [10]. Figure 14.1 shows factors responsible for concept of wear and surface roughness.

14.2 EFFECT OF WEAR AND SURFACE ROUGHNESS ON ADDITIVE MANUFACTURING MATERIALS AND COMPOSITES

Added substance assembling of metallic tribological parts offers extraordinary degrees of opportunity, yet the surface unpleasantness of most as-printed surfaces obstructs the immediate materialness of such designs, and post-processing is fundamental. Rubbing tests were directed in a pin-on-plate design against bearing steel circles under oil with an added substance free mineral base oil for sliding paces among 20 and 170 mm/s [11].

Since many breaks are started from the outer layer of a metal part, surface completion assumes a significant part in weakness execution of AM metals. As a general rule, these materials don't have great surface completion in the as-assembled conditions. As indicated by the unfortunate surface completion of AM metals has three significant causes, the flight of stairs impact from the layer-by-layer producing process balling or adherence of to some degree dissolved powders to the outer layer of the metal, and open pores and zones where liquefying is deficient [12,13].

In light of the holding neck framing process, a hypothetical model of the three-layered surface unpleasantness of manufacturing items is laid out thinking about its real surface conditions. Correlation between the expectations and estimations indicates that the proposed model can be approved and it can give solid expectations in the ME tests' 3DSR SRVF-HB/LB and SRPF-HB/LB), with the inconsistency under 10.52% among anticipated and estimated outcomes. Likewise, the awareness examination shows that rising the expulsion width will diminish SRVF-HB/SRPF-HB, and increment SRVF-LB/SRPF-LB; however, the layer thickness is in the other way around [14].

High requests are put on machine instruments not just as far as required mechanical boundaries and effectiveness yet in addition regarding their dependability. On a basic level, the analytic framework speaks with the machine apparatus framework at such a level that it is conceivable to begin the pre-arranged programme grouping on schedule, limiting or turning away the shortcoming in light of the ongoing indicative outcomes [15].

Added substance producing of extra parts will turn out to be increasingly normal. On account of water powered arrangements, there are additionally a few utilizations of AM innovation connected with topological enhancement, against cavitation upgrades, and so forth.

Subsequently, it was ready with a 100-hour test mission of the AM extra part, which covers the time among harm and supply of the new siphon. The material of the shoe retainer has been recognized and supplanted by another material-accessible as a powder for AM, with comparable properties as the first [16].

Surface unpleasantness is a significant component in streamlined features that can altogether impact the liquid elements and the hotness move. The unpleasantness of the wing skin increment the skin grinding drag, which is one of the parasite drag parts. An inhomogeneous surface harshness circulation on an automated airplane wing after numerous long stretches of not set in stone [17].

Surface unpleasantness typically shifts relying upon the AM strategy. Thus, AM distant from everyone else can't all the while produce parts that meet both mechanical and surface harshness prerequisites. As a rule, the most affecting perspective is the absence of interaction elements understanding. Due to the modern metallurgical what's more, thermo-actual peculiarities, the collaboration systems between the powder bed and the liquefy pool, the powder bed and laser bar, and the softening cycles are trying to make sense of on account of laser added substance fabricating methodology [18].

The surface nature of items is connected with the shaping quality. The cross-part of the material fibre is circular, and holding neck is framed

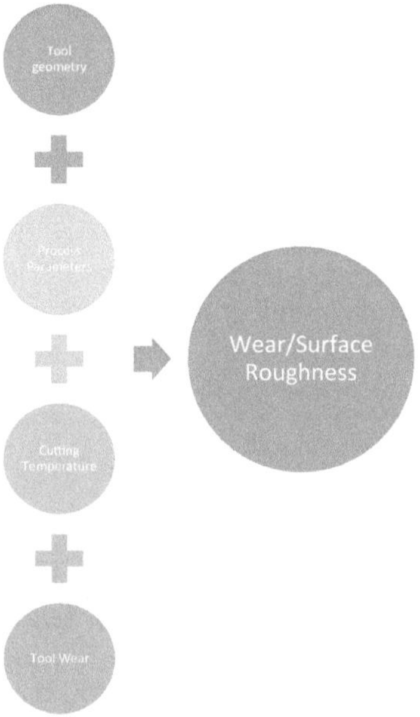

Figure 14.2 Effect of wear and surface roughness on Additive Manufacturing materials and composites.

between adjoining circular areas, which assumes an essential part in the surface nature of items. Along these lines, in light of the holding neck shaping interaction (level and longitudinal holding necks), a three-layered surface unpleasantness hypothetical model of items was laid out in this paper, with the anisotropic qualities and temperature taken into account [19].

Practically Graded Materials offer discrete or consistently evolving properties arrangements over the volume of the parts. The far and wide utilization of FGMs was not quick enough in the past because of limits of the assembling strategies. Critical improvements in fabricating advancements particularly in Additive Manufacturing (AM) empower us these days to make materials with determined changes over the volume/surface of parts [20]. Figure 14.2 represents the effect of wear and surface roughness on Additive Manufacturing Materials and Composites.

14.3 ROLE OF PROCESS PARAMETERS OF WEAR AND SURFACE ROUGHNESS OF AM MATERIALS AND COMPOSITES

Since the time the presentation of 3D printing, enterprises have seen a remarkable development in creation and proficiency. Three-layered printing is the course of added substance producing in which the traditional technique for material evacuation is tested. Layer-on-layer testimony is the fundamental rule of the AM.

Every one of these layers can be seen as a cross-part of the thing that has been softly cut. When contrasted with customary creation techniques, 3D printing permits the making of convoluted shapes with less material. In traditional techniques, the materials go through a few harms because of the instrument workpiece contact making grinding among them and the dispersed hotness that harms the material. Conquering the regular strategy for machining with the assistance of 3D printing is another progression in the businesses [21].

In laser powder bed combination processes, ill tempers of cemented layers in the surface corresponding to the structure plane might slam into the recoating edge causing creation interference or the development of pores that impede the underlying respectability of the last items. Consequently, advancement of LPBF process boundaries and system, alongside post getting done, should be tended to likewise in perspective on unpleasantness decrease. This is especially significant if explicit tribological properties are sought after, as surface unpleasantness influences contact and wear [22].

Surface harshness influences the tasteful nature of parts, and this is a principal concerning the gems business where AM procedures are tracking down expanding application in the last decade, and in the style business that is taking a gander at AM with a developing interest. When mechanical or electro completing is vital, the more awful the as-assembled surface microgeometry, the more material that should be eliminated and this can affect adversely on the costs connected with wrapping up: the reuse of machining chips addresses an extra expense as they should be cleaned from the machining greases and their development addresses a misuse of unrefined substance, hindering the financial seriousness of the added substance innovation [23].

Checking surface quality during machining has impressive pragmatic importance for the presentation of high-esteem items, especially for their gathering points of interaction. Surface harshness is the main measurement of surface quality. The impact of hardware wear minor departure from surface harshness is rarely considered in machining. Also, the decay pattern of surface unpleasantness and instrument wear contrasts under factor cutting boundaries. The expectation models prepared under one bunch of cutting boundaries bomb while cutting boundaries change [24].

On the foundation of the recently evolved 3D surface boundaries, this paper expects to explore the material proportion bends of 3D AM surface geographies, which empowers surface geographical elements to be connected with AM process streamlining [25].

Titanium alloy is broadly utilized in the fundamental construction and primary bearing parts of weapon hardware due to its great mechanical, actual properties and stable substance properties. Versatile twisting under the powerful difference in cutting power and temperature field created by various cycle conditions, which makes it hard to foresee and control the machined surface quality and instrument life [26].

On the reusing of the ABS material since 3D-printed ABS materials and related parts are non-biodegradable, so, there is a huge interest to reuse ABS squander materials and its parts.

Cutting power, and apparatus life demonstrates the viability of LAM in machining earthenware materials. Be that as it may, there has been no deliberate investigation of the impacts of interaction boundaries on surface uprightness in LAM clay materials [27,28].

The ideal arrangement of interaction boundaries that can limit surface unpleasantness, anode wear rate, and amplify material expulsion rate is found in this review. Powder-blended electrical release machining with SiC powder-blended dielectric of solidified 90CrSi steel is led. The pre-owned input boundaries of the advancement cycle are the powder fixation, the beat on schedule, the beat off time, the beat current, and the servo voltage [29].

The higher surface unpleasantness prompts a lower weariness limit, notwithstanding, what is the surface harshness metric that associates to the weakness conduct of as-fabricated AM parts and how to integrate surface profiles into weariness life expectation models stay open inquiries in the examination field of AM [30]. Figure 14.3 shows the role of process parameters of wear and surface roughness of AM materials and composites.

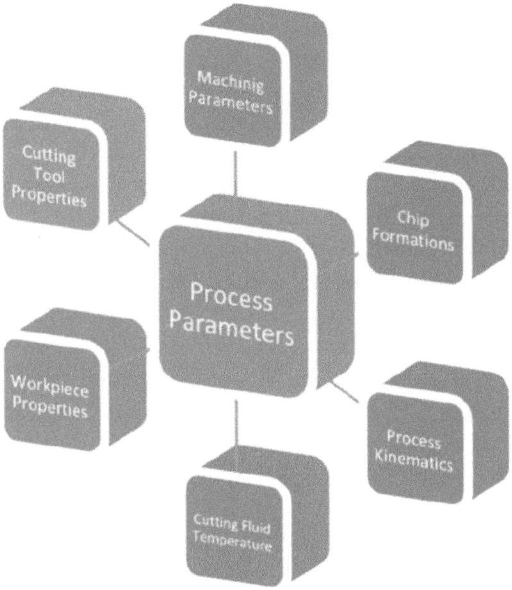

Figure 14.3 Role of process parameters of wear and surface roughness of AM materials and composites.

14.4 INFLUENCE OF SURFACE TOPOLOGY DUE TO WEAR AND SURFACE ROUGHNESS OF AM MATERIALS AND COMPOSITES

The impact of surface treatment methods and conditions on single lap joint strength and interfacial properties has been examined. Valdés et al. examined the multi-objective advancement on glue holding of aluminium-carbon fibre cover, from the multi-objective numerical model perspective [31].

The plan opportunity presented by added substance producing empowers the creation of parts with inner surfaces that are trying to get to post-fabricate. This is of worry, as the surface condition can especially decay exhaustion execution. The most recent couple of years have seen critical examination progresses in these angles. Ti-6Al-4V is the most widely read up combination for AM [32].

Both surface estimation methods reliably demonstrated that all surface medicines prompted enhancements in surface unpleasantness, albeit optical microscopy was affected by surface glares and misjudged the surface harshness upsides of drag-completed examples [33].

The outcomes show that cutting states of higher feed rate and higher wire pressure expanded the feed force by 78% and 20%, individually, since for chip evacuation these circumstances is required a higher power for each grain. On expanding the wire cutting velocity, the feed force diminished by 66% because of a lessening in entrance profundity of the grains [34].

Mechanical strategies, for example, air scraped spot increment the surface region by surface roughening and furnish mechanical interlocking of zirconia with pitch cement. The impact fluctuates as indicated by the sort, size, and infusion tension of the grating particles. Although air scraped area utilizing alumina particles is ordinarily utilized and has shown promising results writers have expressed that the outer layer of Y-TZP ceramics gets harmed, which antagonistically influences its mechanical properties. Silica covering has additionally been accounted for to be a straightforward and compelling technique for holding since it increments hydroxyl bunches for silane coupling and furthermore precisely roughens the surface [35].

Longitudinal torsional ultrasonic-helped crushing is one of the principle strategies to accomplish excellent and high-effectiveness machining of elite execution clay materials. Notwithstanding, it isn't not difficult to precisely describe the three-layered (three-dimensional) surface geography because of numerous irregular elements during LTUG. Focusing on the mind-boggling surface elements brought about by different irregular variables in the LTUG of Si3N4 ceramics, a probabilistic calculation for the level of lingering material on a superficial level in LTUG of Si3N4 earthenware was proposed, and the expectation model for the three-dimensional surface geology and three-dimensional surface unpleasantness boundaries of Si3N4 ceramics in LTUG was laid out by utilizing this calculation [36].

The weariness break ordinarily starts at the surface and afterwards proliferates to the inside, and the break inception process represents the majority of the entire exhaustion life. Thus, it has drawn in various consideration and interests to concentrate on the impact of surface trustworthiness on the weakness life of materials [37].

Titanium grid fibre-built up plastic composites had drawn in expanding interest in marine and different fields in light of their phenomenal mechanical and erosion opposition. Nonetheless, the interfacial holding strength between the titanium lattice and the plastic is a significant variable restricting its wide application. In this chapter, the outer layer of titanium compound TC4 (Ti-6Al-4V) was anodized to shape an oxide layer with a 1–2 μm break structure on the outer layer of titanium amalgam [38].

Ultra-accuracy machining is generally utilized in the assembling of public protection and complex common items as a result of its high accuracy, almost no surface harm and different benefits. The surface unpleasantness of a workpiece is the critical mark of ultraprecision machining innovation, which is impacted by many variables during the time spent machining.

The outcome finished up from the trial is contrasted and the re-enactment result, showing that the mistake between the reproduction and the examination is under 4%, which is of incredible importance to advancing machining process boundaries and improving machining accuracy of

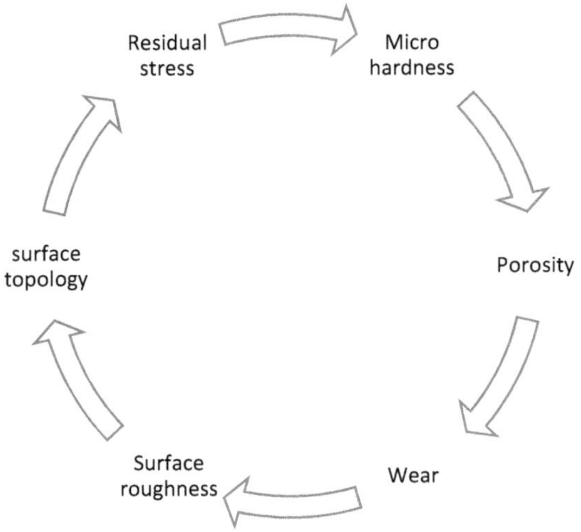

Figure 14.4 Influence of surface topology based on surface roughness and wear on AM and composite materials.

workpiece surfaces for ultraprecision single point precious stone turning [39,40]. Figure 14.4 shows the influence of surface topology based on surface roughness and wear on AM and composite materials.

14.5 IMPROVING SURFACE FINISH AND WEAR RESISTANCE OF ADDITIVE MANUFACTURED COMPONENTS AND COMPOSITES FOR VARIOUS APPLICATIONS

3D printing (3DP) is a way to deal with process parts straightforwardly from its PC helped plan (CAD) record. Added substance producing is changing the scenes of current modern practices. On-request fabricating utilizing added substance producing advances is a recent fad that will essentially impact numerous businesses and item plan conventions. Since there is no requirement for part-explicit tooling, various parts can be assembled utilizing a similar machine [41].

Added substance fabricating is a vanguard creation innovation that has contributed enormously to accelerate supplanting available of complex-moulded parts. A fragile and unavoidable period of added substance innovation is that connecting with the post-handling of the parts, particularly the completing system. Laser cleaning is spreading, truth be told, increasingly more unequivocally, in the field of assembling as a legitimate option in contrast to regular advancements for the surface completing of metallic parts got by added substance processes [42].

The impact of post-handling methods on additively fabricated parts has been talked about. It was observed that laser shock peening (LSP) can cause extreme strain rate age, particularly in more slender parts. LSP have some control over the surface normalities also, nearby grain refinement, accordingly lifting the hardness esteem. Laser cleaning can decrease surface harshness up to 95% and increment hardness, altogether, contrasted with the as-constructed parts. Regular machining processes improve surface quality; notwithstanding, their effect on hardness has not been demonstrated at this point [43].

The progressions of 3D printing innovation have been joined with the utilization of metallic biomaterials to make gadgets and items for the biomedical field. 3D printing has been a progressive cycle that makes the manufacture of metallic biomedical gadgets profoundly unambiguous and all the while simpler than other creation strategies. The reason and generally capacity of every clinical gadget made is reliant upon the kind of metal utilized alongside its manufacture strategy. The significant qualities of metallic biomaterials, including iron, magnesium, zinc, titanium, cobalt, and treated steel, will be talked about. Significant contemplations of these metallic biomaterials incorporate corruption rate, biocompatibility, and mechanical properties [44].

The most famous added substance fabricating advances to create titanium compound parts are electron shaft dissolving, specific laser softening and coordinated energy testimony. It was found that the microstructure is impacted by the most elevated temperature produced and the cooling rate which can be custom-made by the information factors of the AM processes.

AM manufactured parts have lower weakness life contrasted with those of from conventional strategies. Since fast item improvement utilizing 3D printers prompts significant cost decreases all the more as of late, it is normal that more open doors may before long be accessible for the AM of titanium amalgams with fresher AM cycles like virus splash added substance fabricating and added substance contact mix affidavit [45].

Dental rots are a seriously repeating issue in the present occupied existence of individuals all over the planet; consequently, its utilization is exceptionally standard practice. While dental embed improvement, materials, and methods have fundamentally extended and are anticipated, it would additionally expand over the next few decades [46].

The synergistic impacts of wear and erosion on NiTi combination in examination with Ti-6Al-4V compound, the most broadly utilized titanium combination in biomedical applications. Both combinations were handled by an added substance producing laser bar coordinated energy affidavit (LB-DED) method, in particular laser designed net moulding, and investigated by means of tribo corrosion tests by utilizing the ball-on-plate design [47].

Since recent many years, direct printing of parts for different applications by the utilization of 3D-printing innovation has accumulated a ton of consideration. In any case, many issues stay with the far-reaching utilizations

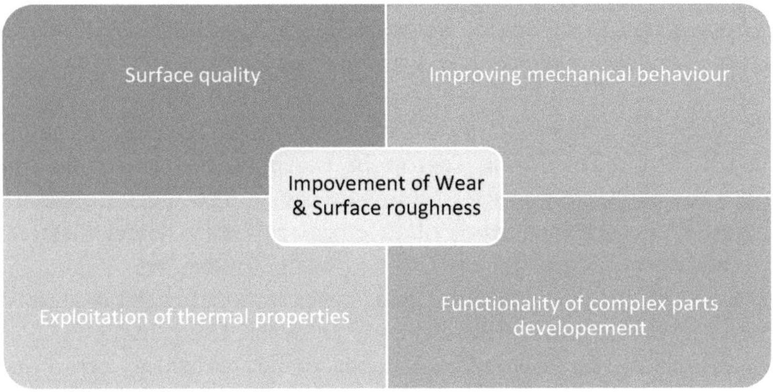

Figure 14.5 Application of improving the surface roughness and wear of AM and composites.

of 3D-printed materials like administrative issues, sterile climate for part creation, and the improvement of material properties with the ideal construction [48].

By re-imagining parts starting from the earliest stage, not restricted to the difficulties introduced by conventional assembling methods, scientists and architects have grown new plan techniques to settle huge scope materials and plan issues around the world. It is imagined that analysts in industry and the scholarly community can use this work to configuration new compounds utilizing metallic AM processes for different current and future applications [49].

Light intensification by animated discharge of radiation as a sound and monochromatic electromagnetic radiation with high power thickness has generally been taken advantage of for different sorts of material handling covering dissolving, machining, joining, storing, and surface designing. The most interesting use of laser currently concerns laser-helped added substance producing that can create a part, instrument, gadget, machine, or a total designing framework beginning from a tiny volume of material through a keen plan of liquid drop reconciliation proceeded with parallel and vertical bearing following a PC supported plan convention [50]. Figure 14.5 shows the application of improving the surface roughness and wear of AM and composites.

14.6 CONCLUSIONS

Wear and surface roughness are two important factors which decide the quality of the complex geometrics of Additive Manufactured components. In the different AM manufacturing conditions, the output surface roughness and wear decide the quality of the product utilized for various applications.

This chapter has discussed the role played by wear and surface roughness of AM produced components in different working conditions. The information drawn from the discussion is summarized below:

- Designing composite materials development in assembling has turned into a vital exploration region in further developing rail material mechanical characteristics against wear. A few better measures are through surface designing of essential rail and wheel districts with further developed characteristics to wear opposition.
- Due to the modern metallurgical what's more, thermo-actual peculiarities, the collaboration systems between the powder bed and the liquefy pool, the powder bed and laser bar, and the softening cycles are trying to make sense of on account of laser added substance fabricating methodology.
- Weariness strength isn't just affected by remaining pressure and porosity, however additionally by surface unpleasantness. Surface imperfections go about as pressure raisers, subsequently surface completing pointed at diminishing surface harshness broadens the weakness life, as tentatively demonstrated and noticed that a proper form filter methodology could work on a superficial level unpleasantness sway on exhaustion life of LPBF Ti6Al4V composite.
- Mechanical strategies, for example, air scraped spot increment the surface region by surface roughening and furnish mechanical interlocking of zirconia with pitch cement. The impact fluctuates as indicated by the sort, size, and infusion tension of the grating particles.
- The progressions of 3D printing innovation have been joined with the utilization of metallic biomaterials to make gadgets and items for the biomedical field. 3D printing has been a progressive cycle that makes the manufacture of metallic biomedical gadgets profoundly unambiguous and all the while simpler than other creation strategies.
- The reason and generally capacity of every clinical gadget made is reliant upon the kind of metal utilized alongside its manufacture strategy. The significant qualities with good wear and surface topology of metallic biomaterials, including iron, magnesium, zinc, titanium, cobalt, and treated steel which are utilized for potential applications in various engineering and non-engineering fields with better outputs.

REFERENCES

1. Popov, V. V., & Fleisher, A. (2020). Hybrid additive manufacturing of steels and alloys. *Manufacturing Review*, 7, 6.
2. Usca, Ü. A., Uzun, M., Kuntoğlu, M., Sap, E., & Gupta, M. K. (2021). Investigations on tool wear, surface roughness, cutting temperature, and chip

formation in machining of Cu-B-CrC composites. *The International Journal of Advanced Manufacturing Technology*, *116*(9), 3011–3025.
3. Bandyopadhyay, A., & Heer, B. (2018). Additive manufacturing of multi-material structures. *Materials Science and Engineering: R: Reports*, *129*, 1–16.
4. Aladesanmi, V. I., Fatoba, O. S., Akinlabi, E. T., & Ikumapayi, O. M. (2021). Analysis of wear properties and surface roughness of laser additive manufactured (LAM) Ti and TiB2 metal matrix composite. *Materials Today: Proceedings*, *44*, 1279–1285.
5. Wang, Y., Zhou, Y., Lin, L., Corker, J., & Fan, M. (2020). Overview of 3D additive manufacturing (AM) and corresponding AM composites. *Composites Part A: Applied Science and Manufacturing*, *139*, 106114.
6. Usca, Ü. A., Uzun, M., Kuntoğlu, M., Sap, E., & Gupta, M. K. (2021). Investigations on tool wear, surface roughness, cutting temperature, and chip formation in machining of Cu-B-CrC composites. *The International Journal of Advanced Manufacturing Technology*, *116*(9), 3011–3025.
7. Aladesanmi, V. I., Fatoba, O. S., Akinlabi, E. T., & Ikumapayi, O. M. (2021). Analysis of wear properties and surface roughness of laser additive manufactured (LAM) Ti and TiB2 metal matrix composite. *Materials Today: Proceedings*, *44*, 1279–1285.
8. Şap, E., Usca, U. A., Gupta, M. K., & Kuntoğlu, M. (2021). Tool wear and machinability investigations in dry turning of Cu/Mo-SiCp hybrid composites. *The International Journal of Advanced Manufacturing Technology*, *114*(1), 379–396.
9. Kumar, A., Choudhary, A., Tiwari, A., James, C., Kumar, H., Arora, P. K., & Khan, S. A. (2021). An investigation on wear characteristics of additive manufacturing materials. *Materials Today: Proceedings*, *47*, 3654–3660.
10. Bhushan, R. K. (2021). Multi-response optimization of parameters during turning of AA7075/SiC composite for minimum surface roughness and maximum tool life. *Silicon*, *13*(9), 2845–2856.
11. Guenther, E., Kahlert, M., Vollmer, M., Niendorf, T., & Greiner, C. (2021). Tribological performance of additively manufactured AISI H13 steel in different surface conditions. *Materials*, *14*(4), 928.
12. Ye, C., Zhang, C., Zhao, J., & Dong, Y. (2021). Effects of post-processing on the surface finish, porosity, residual stresses, and fatigue performance of additive manufactured metals: A review. *Journal of Materials Engineering and Performance*, *30*(9), 6407–6425.
13. Jiang, S., Hu, K., Zhan, Y., Zhao, C., & Li, X. (2022). Theoretical and experimental investigation on the 3D surface roughness of material extrusion additive manufacturing products. *Polymers*, *14*(2), 293.
14. Juneja, S., Chohan, J. S., Kumar, R., Sharma, S., Ilyas, R. A., Asyraf, M. R. M., & Razman, M. R. (2022). Impact of process variables of acetone vapor jet drilling on surface roughness and circularity of 3D-printed ABS parts: Fabrication and studies on thermal, morphological, and chemical characterizations. *Polymers*, *14*(7), 1367.
15. Nahornyi, V., Panda, A., Valíček, J., Harničárová, M., Kušnerová, M., Pandová, I., & Lukáč, O. (2022). Method of using the correlation between the surface roughness of metallic materials and the sound generated during the controlled machining process. *Materials*, *15*(3), 823.

16. Klimek, A., Kluczyński, J., Łuszczek, J., Bartnicki, A., Grzelak, K., & Małek, M. (2022). Wear analysis of additively manufactured slipper-retainer in the axial piston pump. *Materials, 15*(6), 1995.
17. Udroiu, R. (2022). New methodology for evaluating surface quality of experimental aerodynamic models manufactured by polymer jetting additive manufacturing. *Polymers, 14*(3), 371.
18. Mahmood, M. A., Chioibasu, D., Ur Rehman, A., Mihai, S., & Popescu, A. C. (2022). Post-processing techniques to enhance the quality of metallic parts produced by additive manufacturing. *Metals, 12*(1), 77.
19. Jiang, S., Hu, K., Zhan, Y., Zhao, C., & Li, X. (2022). Theoretical and experimental investigation on the 3D surface roughness of material extrusion additive manufacturing products. *Polymers, 14*(2), 293.
20. Nohut, S., &Schwentenwein, M. (2022). Vat photopolymerization additive manufacturing of functionally graded materials: A review. *Journal of Manufacturing and Materials Processing, 6*(1), 17.
21. Juneja, S., Chohan, J. S., Kumar, R., Sharma, S., Ilyas, R. A., Asyraf, M. R. M., & Razman, M. R. (2022). Impact of process variables of acetone vapor jet drilling on surface roughness and circularity of 3D-printed ABS parts: Fabrication and studies on thermal, morphological, and chemical characterizations. *Polymers, 14*(7), 1367.
22. Vishwasrao, P. R., & Nikalje, A. M. Optimization of process parameter in wire electrical discharge machining of H11 tool steel using response surface methodology. *Journal for Research in Applied Science and Engineering Technology. 9*(XII). https://doi.org/10.22214/ijraset.2021.39693
23. Molinari, A., Ancellotti, S., Fontanari, V., Iacob, E., Luchin, V., Zappini, G., & Benedetti, M. (2022). Effect of process parameters on the surface microgeometry of a Ti6Al4V alloy manufactured by laser powder bed fusion: 3D vs. 2D characterization. *Metals, 12*(1), 106.
24. Wang, Y., Wang, Y., Zheng, L., & Zhou, J. (2022). Online surface roughness prediction for assembly interfaces of vertical tail integrating tool wear under variable cutting parameters. *Sensors, 22*(5), 1991.
25. Lou, S., Zhu, Z., Zeng, W., Majewski, C., Scott, P. J., & Jiang, X. (2021). Material ratio curve of 3D surface topography of additively manufactured parts: an attempt to characterise open surface pores. *Surface Topography: Metrology and Properties, 9*(1), 015029.
26. Wang, Z., & Li, L. (2021). Optimization of process parameters for surface roughness and tool wear in milling TC17 alloy using Taguchi with grey relational analysis. *Advances in Mechanical Engineering, 13*(2), doi:10.1177/1687814021996530.
27. Juneja, S., Chohan, J. S., Kumar, R., Sharma, S., Ilyas, R. A., Asyraf, M. R. M., & Razman, M. R. (2022). Impact of process variables of acetone vapor jet drilling on surface roughness and circularity of 3D-printed ABS parts: Fabrication and studies on thermal, morphological, and chemical characterizations. *Polymers, 14*(7), 1367.
28. Pu, Y., Zhao, Y., Meng, J., Zhao, G., Zhang, H., & Liu, Q. (2021). Process parameters optimization using Taguchi-based grey relational analysis in laser-assisted machining of Si3N4. *Materials, 14*(3), 529.

29. Nguyen, A. T., Le, X. H., Nguyen, V. T., Phan, D. P., Tran, Q. H., Nguyen, D. N., & Vu, N. P. (2021). Optimizing main process parameters when conducting powder-mixed electrical discharge machining of hardened 90CrSi. *Machines*, *9*(12), 375.
30. Dinh, T. D., Han, S., Yaghoubi, V., Xiang, H., Erdelyi, H., Craeghs, T., & Van Paepegem, W. (2021). Modeling detrimental effects of high surface roughness on the fatigue behavior of additively manufactured Ti-6Al-4V alloys. *International Journal of Fatigue*, *144*, 106034.
31. Pereira, A., Fenollera, M., Prado, T., & Wieczorowski, M. (2022). Effect of surface texture on the structural adhesive joining properties of aluminum 7075 and TEPEX®. *Materials*, *15*(3), 887.
32. Elambasseril, J., Rogers, J., Wallbrink, C., Munk, D., Leary, M., & Qian, M. (2022). Laser powder bed fusion additive manufacturing (LPBF-AM): The influence of design features and LPBF variables on surface topography and effect on fatigue properties. *Critical Reviews in Solid State and Materials Sciences*, *48*, 1–37.
33. Lee, S., Shao, S., Wells, D. N., Zetek, M., Kepka, M., & Shamsaei, N. (2022). Fatigue behavior and modeling of additively manufactured IN718: The effect of surface treatments and surface measurement techniques. *Journal of Materials Processing Technology*, *302*, 117475.
34. Costa, E. C., Weingaertner, W. L., & Xavier, F. A. (2022). Influence of single diamond wire sawing of photovoltaic monocrystalline silicon on the feed force, surface roughness and micro-crack depth. *Materials Science in Semiconductor Processing*, *143*, 106525.
35. Mohit, K. G., Lakha, T. A., Chinchwade, A., Batul, Q. A., Shaikh, M., & Kheur, S. M. (2022). Effects of surface modification techniques on zirconia substrates and their effect on bonding to dual cure resin cement—An in-vitro study. *The Journal of Indian Prosthodontic Society*, *22*(2), 179.
36. Zhang, Z., Shi, K., Huang, X., Shi, Y., Zhao, T., He, Z., & Song, Y. (2022). Development of a probabilistic algorithm of surface residual materials on Si3N4 ceramics under longitudinal torsional ultrasonic grinding. *Ceramics International*, *48*(9), 12028–12037. https://doi.org/10.1016/j.ceramint.2022.01.060.
37. Chen, S., Zhao, W., Yan, P., Qiu, T., Gu, H., Jiao, L., & Wang, X. (2022). Effect of milling surface topography and texture direction on fatigue behavior of ZK61M magnesium alloy. *International Journal of Fatigue*, *156*, 106669.
38. Du, M., Dong, W., Li, X., Wang, L., Wang, B., & Tang, B. (2022). Effect of surface topography on injection joining Ti alloy for improved bonding strength of metal-polymer. *Surface and Coatings Technology*, *433*, 128132. https://doi.org/10.1016/j.surfcoat.2022.128132.
39. Fu, S., Yang, H., Sun, S., Zhang, M., Liu, Y., Zhang, Y., & Pan, L. (2022). Investigation on the surface roughness modeling and analysis for ultra-precision diamond turning processes constrained by the complex multi-source factors. *Proceedings of the Institution of Mechanical Engineers, Part B: Journal of Engineering* Manufacture, 236(10), 1295–1304. doi:10.1177/09544054221075878.

40. Bailey, C., Morrow, J. A., Stallbaumer-Cyr, E. M., Weeks, C., Derby, M., & Thompson, S. M. (2022). Effects of build angle on additively manufactured aluminum alloy surface roughness and wettability. *Journal of Manufacturing Science and Engineering*, *144*, 1–34. https://doi.org/10.1115/1.4053608
41. Bandyopadhyay, A., & Bose, S. (2022). Additive manufacturing of metal matrix composites for structural and biomedical applications. In: Srivatsan, T.S., Rohatgi, P.K., & Hunyadi Murph, S. (eds.), *Metal-Matrix Composites* (pp. 97–105). Springer, Cham. https://doi.org/10.1007/978-3-030-92567-3_6
42. Annamaria, G., Massimiliano, B., & Francesco, V. (2022). Laser polishing: A review of a constantly growing technology in the surface finishing of components made by additive manufacturing. *The International Journal of Advanced Manufacturing Technology*, *123*, 1401. https://doi.org/10.1007/s00170-022-10265-5
43. Mahmood, M. A., Chioibasu, D., Ur Rehman, A., Mihai, S., & Popescu, A. C. (2022). Post-processing techniques to enhance the quality of metallic parts produced by additive manufacturing. *Metals*, *12*(1), 77.
44. Chua, K., Khan, I., Malhotra, R., & Zhu, D. (2022). Additive manufacturing and 3D printing of metallic biomaterials. *Engineered Regeneration*, *2*, 288-299. https://doi.org/10.1016/j.engreg.2021.11.002.
45. Liu, S., & Shin, Y. C. (2019). Additive manufacturing of Ti6Al4V alloy: A review. *Materials & Design*, *164*, 107552.
46. Selvaraj, S. K., Prasad, S. K., Yasin, S. Y., Subhash, U. S., Verma, P. S., Manikandan, M., & Dev, S. J. (2022). Additive manufacturing of dental material parts via laser melting deposition: A review, technical issues, and future research directions. *Journal of Manufacturing Processes*, *76*, 67–78.
47. Buciumeanu, M., Bagheri, A., Silva, F. S., Henriques, B., Lasagni, A. F., & Shamsaei, N. (2022). Tribocorrosion behavior of NiTi biomedical alloy processed by an additive manufacturing laser beam directed energy deposition technique. *Materials*, *15*(2), 691.
48. Pandey, B., & Chalisgaonkar, R. (2022). A review on post additive manufacturing techniques to improve product quality. In: Dubey, A.K., Sachdeva, A., Mehta, M. (eds.), *Recent Trends in Industrial and Production Engineering* (pp. 11–20). Springer, Singapore.
49. Bandyopadhyay, A., Traxel, K. D., Lang, M., Juhasz, M., Eliaz, N., & Bose, S. (2022). Alloy design via additive manufacturing: Advantages, challenges, applications and perspectives. *Materials Today*, *52*, 207–224, https://doi.org/10.1016/j.mattod.2021.11.026.
50. Kumar, M., Majumdar, J. D., Fecht, H. J., & Manna, I. (2022). Laser-Assisted Additive Manufacturing of Ni-Based Superalloy Components. In: Fecht, H.J., & Mohr, M. (eds.), *Metallurgy in Space . The Minerals, Metals & Materials Series*. Springer, Cham. https://doi.org/10.1007/978-3-030-89784-0_22.

Chapter 15

Electrochemical behaviour of Inconel 718 alloy developed through additive manufacturing process

Babu Raj Anushraj
Chennai Institute of Technology

Jebbas Thangiah Winowlin Jappes and Mahaboob Basha Adam Khan
Kalasalingam Academy of Research & Education

StanlyKochappa Premila Jani
Marri Laxman Reddy Institute of Technology and Management

CONTENTS

DOI: 10.1201/9781003430186-15

15.1 INTRODUCTION

In aviation, medical, automotive, tool, and consumer product enterprises a reliable design and storage of metal parts with complex geometry are of much importance. In order to reduce the market time, the production process, and the production costs of products manufactured by these industries, research has focused on combining multiple production processes into a single machine. The ultimate goal is to reduce production space, time, and staffing needs. Additive manufacturing adoption and implementation in the industry for production could improve the product value of the organization. Additive manufacturing also can produce components in a straight line from CAD demonstration. Additive manufacturing is very similar to CNC machining. In order to produce the finished product in CNC machining the material is removed; in additive manufacturing, the material is being added. The main advantage of additive manufacturing is that the components are created automatically without the requirement of any tool and can produce any type of complex geometry. Thereby, it reduces the need for skilled labour and production time. It is used for creating the prototype models when launching the layer-by-layer manufacturing technique, and due to the sustainable development of the technologies additive manufacturing is shifted to prototype to direct the manufacturing of end products. Various steps involved in additive manufacturing are shown in Figure 15.1.

15.2 ADDITIVE MANUFACTURING FOR GAS TURBINE ENGINE COMPONENTS

Additive manufacturing (AM) technologies produce parts, layer upon layer, by tracing their cross-section directly from a 3D CAD model. As mentioned

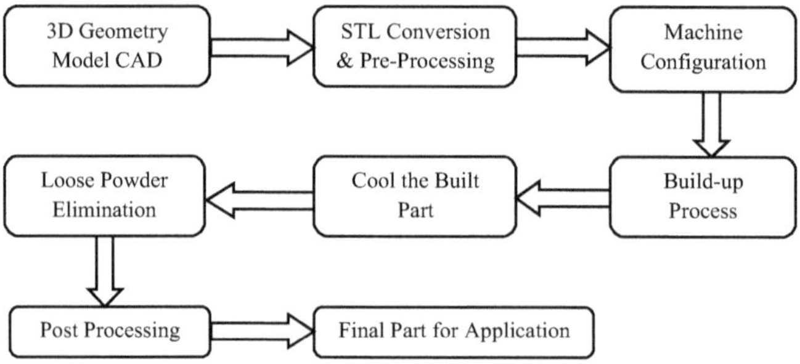

Figure 15.1 Additive manufacturing process chain.

in the introduction section, the technology has plenty of advantages over conventional manufacturing techniques, including production of functionally integrated products, efficient use of materials wherever needed, its synergy with optimal design of complex geometric forms, and the sustainability of the manufacturing technique. This section discusses the advantages of AM technology that make it a promising technique for producing gas turbine engine components. The production of a fully integrated product is among the advantages of AM. Because the method allows for the consolidation and functional integration of pieces in a single construction, completely integrated functioning systems rather than distinct piece parts might be produced [1]. The functional integration of parts reduces the number of pieces, which reduces the obstacles experienced during the assembly process [2]. This might be critical for gas turbine engine components, since it could enable the fabrication of a fully built compressor or turbine at a far reduced cost.

In the aerospace industry, buy-to-fly ratios are approximately 20:1; i.e., 100 kg of input material is needed to produce 5 kg of the final product. This means the remaining 95 kg of material is waste, requiring reprocessing or recycling. This is because many gas turbine engine components must satisfy exceptional requirements. However, AM technology reduces waste or has no waste since the parts are produced by layering the cross-sections of a part, one above the other. Furthermore, since most gas turbine engine components (compressor and turbine blades, nozzles) are very complex in geometry, their production could easily be realized with the implementation of AM techniques [3,4]. The airfoil shape of turbine blades, for instance, requires a special geometry that is designed for aerodynamic performance. Small deviations in the shape can have large consequences on its performance and hence control of the airfoil shape is a critical design step [5].

In addition, when complexity increases, it creates challenges for conventional manufacturing techniques, as complex parts require the assembly of individual parts with nuts, bolts, rivets, and welds, thus reducing the reliability of the component. AM methodology, on the other hand, gives the designer more design flexibility since it allows for the manufacture of complicated part topologies that are impossible to make with conventional manufacturing processes because there are few or no limits on how much complex topology must be generated [6,7]. As a result, design optimization may be utilized to lower the weight of engine parts while preserving their functional requirements and boosting their efficiency by altering their architecture. Production of a sustainable product is another primary requirement in many sectors such as the aerospace and automobile industries. The choice of production technique plays a great role in making the production of components more sustainable. Although only limited research [8–10] has been reported on the assessment of the sustainability of AM technologies, there are indications that the technology is in a promising trend for the production of sustainable products, compared to conventional manufacturing techniques. Therefore, although further investigations are required for the assessment, it seems that AM technology is advancing a preferable production technique for the realization of sustainable aerospace products.

15.2.1 Need for post-processing of AM

Most of the materials always need a heat treatment, in the post-3D printing phase. To relieve stress on components, heat treatment is needed before their removal from the build plate. Among all the post-processing methods available, the heat treatment process is of greater importance and significance. At high heat treatment temperatures, the heat treatment may minimize surface contamination, which may lead to a subsequent enhancement in the mechanical properties of the component. In addition to these heat treatment benefits, the materials listed above are mainly utilized for aeronautical and clinical uses, in which the regulations must be properly followed to prevent damage to the surface and maximize mechanical qualities.

15.2.2 Industrial value of AM

The shipping of complex components has not been easier which increases the production cost of the material. Additive manufacturing made it simple for engineers and designers to produce effective complex designs, cheaper and faster. Industries such as defence, aerospace, automotive, power generating application, space application, and health care are now adapting additive manufacturing to its continuous improvement in technologies [11]. Therefore, there is a large scope for the new applications in the industry on their entry. According to the global market report, based on their technologies the Selective Laser Melting (SLS), Fused Deposition Modelling (FDM), Direct Metal Laser Sintering (DMLS), and Stereolithography (SLA) are most widely used in additive manufacturing. From that DMLS technology can produce high-quality metal parts with the greatest accuracy. Also, it is used to produce complex shapes in automobile and aerospace applications. Due to the cost-efficient production, effortless handling, and detailed resolving power the DMLS technology found growth in the market for additive manufacturing technologies [12].

15.3 3D PRINTING MANUFACTURING TECHNIQUES OF METALS AND ALLOYS

The two techniques depending on the feedstock utilized for production are powder bed fusion (PBF) procedures and Direct Energy Deposition (DED) [13]. Figure 15.2 depicts a schematic of PBF deposition. The PBF technique employs a heat source to fuse the powder particles layer by layer. Different techniques are utilized to disperse the powder layer across the construction plane [14]. By means of melting the metal simultaneously as it is deposited, the Direct Metal Deposition process manufactures the parts [15]. The classification of AM for metals and alloys is shown in Figure 15.3.

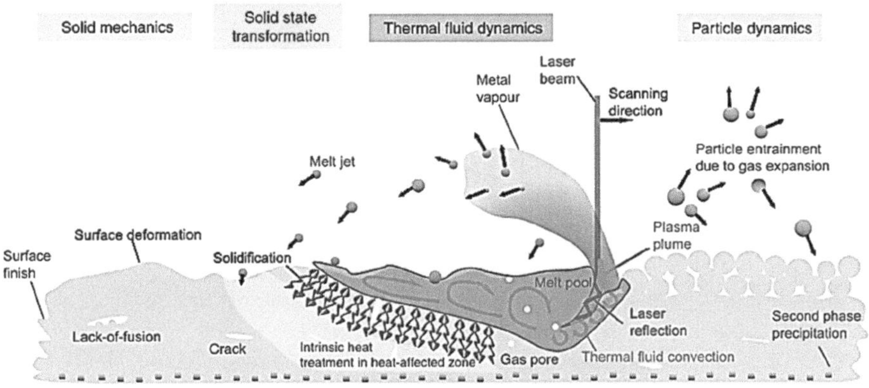

Figure 15.2 Schematic view of PBF used in AM.

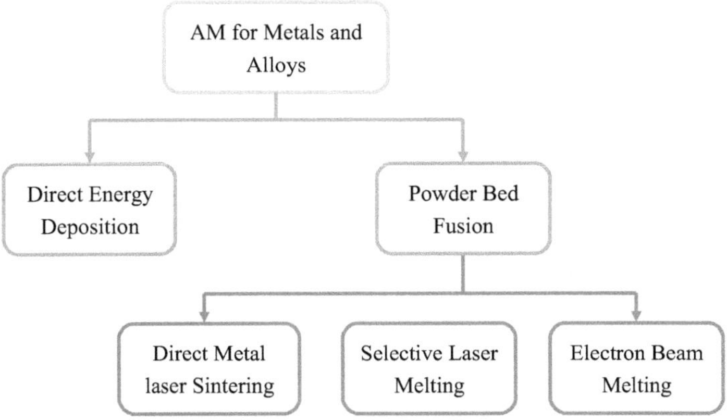

Figure 15.3 Classification of AM for metals and alloys.

15.3.1 Direct energy deposition

Direct Energy Deposition processes use the principle of melting material by thermal energy as they are being deposited on the substrate in a layer-wise sequence [15,16]. The schematic view of DED machine is shown in Figure 15.4.

15.3.2 Powder bed fusion

The PBF is a layer-wise technique that uses a high-energy input such as a laser or electron beam to scan, melt, and freeze particles at a specific location from a powder bed to manufacture high-precision components with generally acceptable surface polish. Figure 15.5 is a schematic of a PBF machine. DMLS, SLM, and EBM are the most promising approaches in the PBF category. These technologies have showed a mature ability and

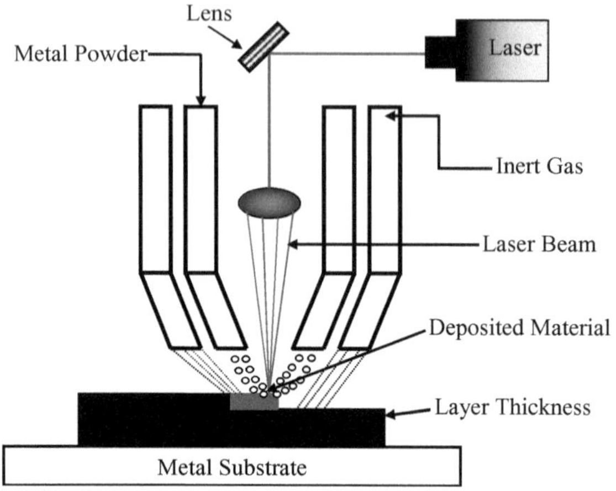

Figure 15.4 Schematic diagram of DED.

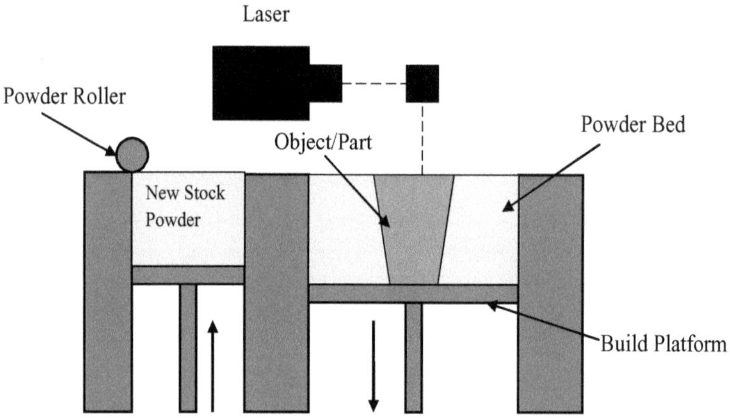

Figure 15.5 Schematic diagram of PBF.

promise for directly manufacturing bespoke metallic components with complicated shapes, high dimensional precision, reduced material waste, and clean structure increases [17].

15.3.3 Direct metal laser sintering

Using a focused CO_2 laser beam, the DMLS technique is a powder bed fusion process that sinters powdered particles to form a layer of part repetitively. Build chambers are heated at an elevated temperature just before the melting temperature of the material used for fabrication in order to limit the amount of laser power input and deformation of the built part. The powder particles are spread layer by layer across the powder bed by a powder levelling scraper [18]. Based on the design specification a focused CO_2 laser beam scans the powdered material and then thermally fuses the material at positions. Allowing for the spread of the loose powder material for a new layer build after each layer build, a piston lowers the completed layer build by the same amount of the layer depth in the z-plane. Accuracy of the part produced via this process is influenced by many elements, e.g., the powder particle size [19].

15.3.4 Selective laser melting

Due to the laser beam fully melting the powders, SLM, a powder bed fusion process, produces close to fully dense structures. SLM technique requires more substantial laser beam power input, which in turn increases the energy cost. Using this technique a moderate to excellent surface finish and feature resolution can be achieved. In this process for part production, polymers, metals, and ceramics can also be utilized [20].

15.3.5 Electron beam melting

Electron beam melting (EBM) processes and the SLS development are alike, as an electron beam melts the powdered particle which is powered by a high voltage of 30 and 60 kV [16]. To prevent oxidation, the printing process occurs at high temperature in a vacuum chamber. With the scan speed being very fast, the surface finish is moderate or poor [21].

15.4 MATERIALS AND METHODOLOGY

15.4.1 Material used

A gas-atomized IN718 powder with a particle size range of 10–50 nm was employed. Table 15.1 shows the IN718 chemical combination that was utilized to produce the sample.

15.4.2 Fabrication process

INTECH DMLS, Pvt Ltd, Bangalore, used an EOS M280 DMLS machine to make the samples. The coupons are printed in the xy-plane (horizontal alignment of the building platform) and in the direction of the building (z-plane). A scoop was used to disseminate the metal powder across the construction bed

Table 15.1 The element distribution of IN718 superalloy

Element	*Ni*	*Fe*	*Cr*	*Nb*	*Mo*	*Ti*	*Al*	*Si*	*C*	*Cu*	*Co*	*W*
Weight (%)	52.44	19.78	17.13	4.77	3.38	1.11	0.59	0.13	0.05	0.16	0.23	0.23

Table 15.2 Process parameter for fabrication of DMLS alloy

Parameter	*Range*
Power	285 W
Scan rate	970 mm/s
Hatching distance	0.15 mm
Layer thickness	40 μm
Beam diameter	80 μm

Electro Chemical	ASTM G5	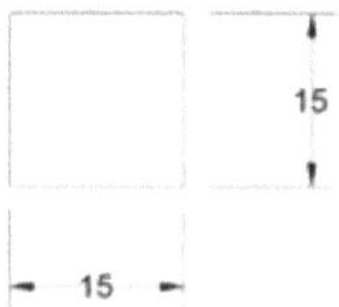	 Electrochemical Sample

Figure 15.6 Details on test samples, ASTM standard, and dimension. *All the dimensions are in mm.

of a powder feeding system. A laser was utilized to fuse the metal powder to the bed. The procedure was repeated until the necessary coupon dimension was attained. The alloy was created in a gradual manner, layer by layer. Table 15.2 shows the optimal process parameters for fabricating the DMLS alloy.

AUTOCAD software was used to design the electrochemical samples developed by DMLS 3D printer. The dimension and ASTM standard is shown in Figure 15.6.

15.4.3 Heat treatment

The heat treatment of the as-built samples was done in a tube furnace. The two separate solutionizing heat treatment settings are depicted in Figure 15.7. The HT 1 thermal treatment for IN718 in turbine blade applications is the traditional heat treatment, whereas the HT 2 heat treatment is the heat treatment for IN718 in turbine blade applications. To ameliorate the microstructure, the heat-treated samples are physically polished with 600-2500 grit abrasive sandpaper and etched with 87 Glyceregia (ASTME 407 STD). A thorough time graph of HT 1 and HT 2 is shown in Figure 15.7.

15.4.4 Electrochemical test

For traditional electrochemical measuring methodologies, the ASTM standard ASTM G3-14 was followed. At room temperature, the exposure area was 1 cm^2. For corrosion testing, a 1.0 M HNO_3 (nitric) acid

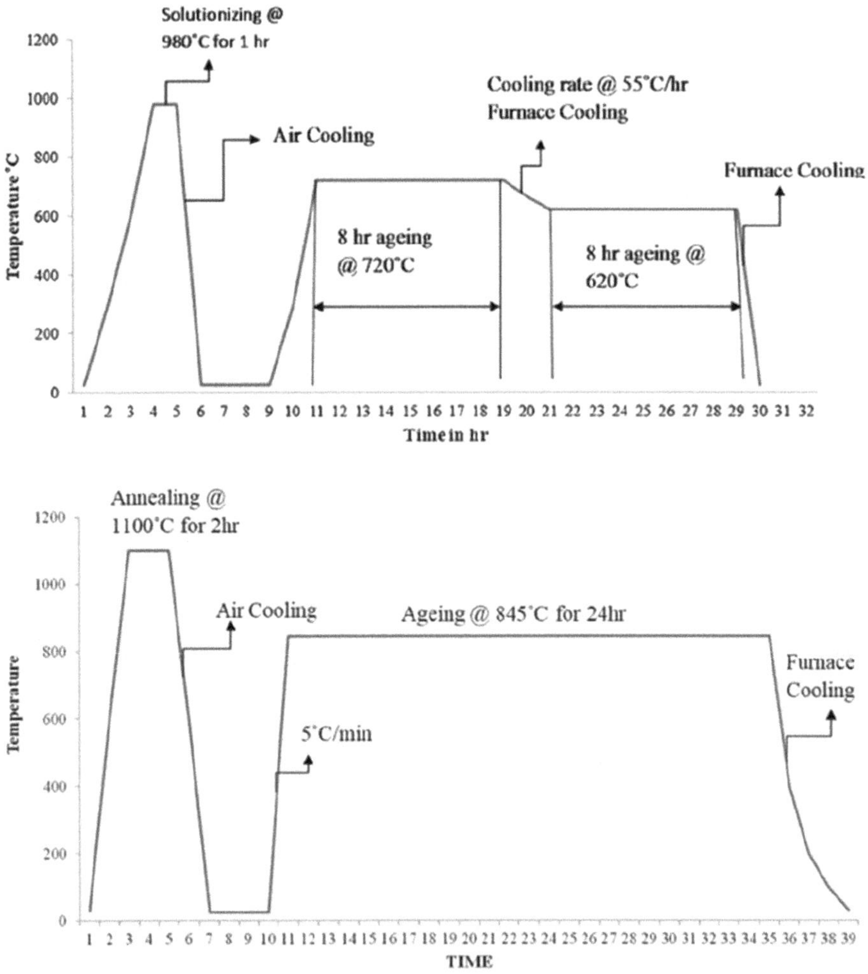

Figure 15.7 (a) Heat treatment time line graph for HT 1. (b) Heat treatment time line graph for HT 2.

solution is used as the expose medium. After achieving a steady open circuit potential, the cruise voltage limits were adjusted at –250 to +250 mV with a 1 mV/s sweep rate. Following the electrochemical polarization observations, Tafel plot was created. After polarization measurement, the exposed materials were characterized using a scanning electron microscope (Make: Zeiss–FE SEM) equipped with energy-dispersive spectroscopy (Make: Brukers EDS). Figure 15.8 depicts a schematic of an electrochemical test setup.

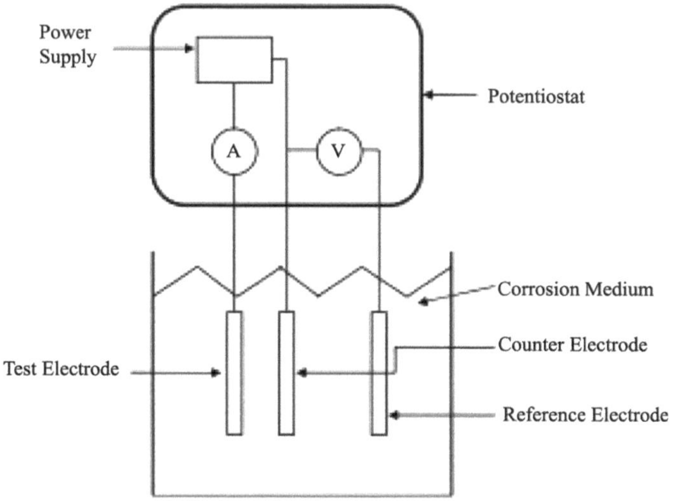

Figure 15.8 Schematic diagram of electrochemical test.

15.5 RESULTS AND DISCUSSION

15.5.1 Heat-treated DMLS

Figure 15.9 illustrates the optical picture of the HT 1 DMLS alloy and commercial alloy morphology. The columnar nature's microstructure and build-up were noticed. As a result, in HT 1, the material was not totally solutionized or recrystallization had not occurred. Grain lengthening had occurred along the construction direction. The sample included columnar dendrites. Although the grains were created, they were in a columnar configuration. In the overlap area, nucleation commenced. The grain was polished at the overlap zone and then stretched outwards from there.

Figure 15.10 depicts the microstructure of the HT 2 DMLS alloy and a commercial alloy. Fine grains with various grain sizes were created on the overlap region for DMLS alloy using the HT 2 technique, as illustrated in Figure 15.10a. Some of the huge grains had columnar structure, and each grain was homogeneous. Figure 15.10b depicts a conventional alloy that has been heat treated to HT 2. The conventional alloy forms big grains, while the DMLS alloy forms an elongated columnar structure.

15.5.2 Electrochemical test

Electrochemical analysis was used to investigate the corrosion kinetics of the cathodic and anodic reactions of IN718 alloy in 1.0 M HNO_3 (nitric) acid solution. Table 15.3 and Figure 15.11 exhibit the corrosive statistics and graph acquired from the integrated programme. HT 2 alloy (–107.54 mV) had the highest corrosion tendency, followed by HT 1 alloy (–134.66 mV) and bare alloy (–143.28 mV). The corrosion density of HT 2

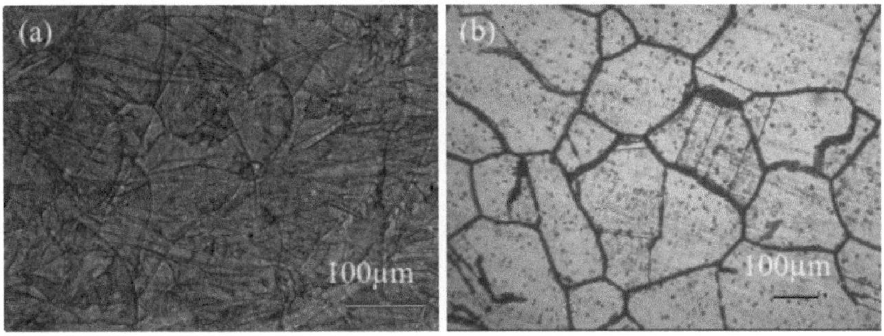

Figure 15.9 Optical image of the HT 1 IN718. (a) DMLS alloy, (b) commercial alloy.

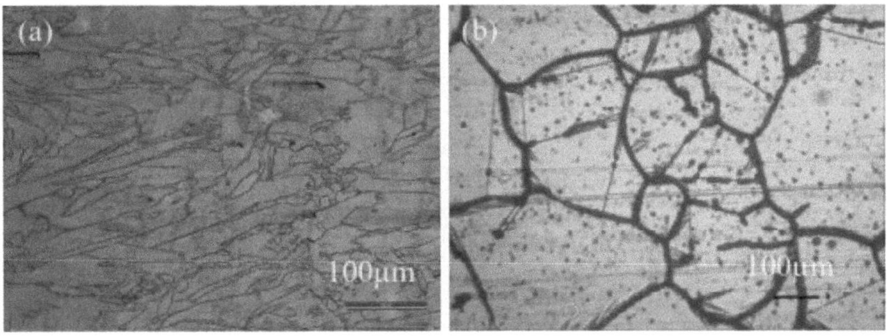

Figure 15.10 Optical image of the HT 2 IN718. (a) DMLS alloy, (b) commercial alloy.

Table 15.3 Polarization data obtained from inbuilt software

Condition	*Corrosion potential* E_{corr} *(mV)*	*Corrosion density* I_{corr} *(mA/cm²)*	*Corrosion rate* C_R *(mm/year)*
Bare	−143.28	2.652E-05	0.0005482
HT 1	−134.66	2.327E-05	0.0004367
HT 2	−107.54	1.931E-05	0.000416

alloy (1.931E-05) is extremely low, followed by HT 1 alloy (2.327E-05) and bare alloy (2.652E-05). Because the corrosion density is low and the wear rate is strong, the HT 2 alloy has a very low rate of corrosion. The corrosion rate is reduced because of the increased corrosion potential and low corrosion concentration. In declining sequence, the corrosion rate of the alloy is HT 2>HT 1>bare, which is directly controlled by the corrosion density.

15.5.3 Corrosion morphology

Scanning electron microscopy (SEM) and energy-dispersive spectrometry are used to analyse the surface analyses of the electrochemical sample (EDS). The corrosion structure of the DMLS coupons exposed in 1.0 M HNO_3 (nitric) acid solution is shown in Figure 15.12. The pitting corrosion mechanism can be seen throughout the sample. The modest quantity of delta phase discovered in the HT 2 DMLS sample may help to strengthen the grain boundary. In comparison to the HT 1 and naked sample, the HT 2 DMLS sample shows high durability.

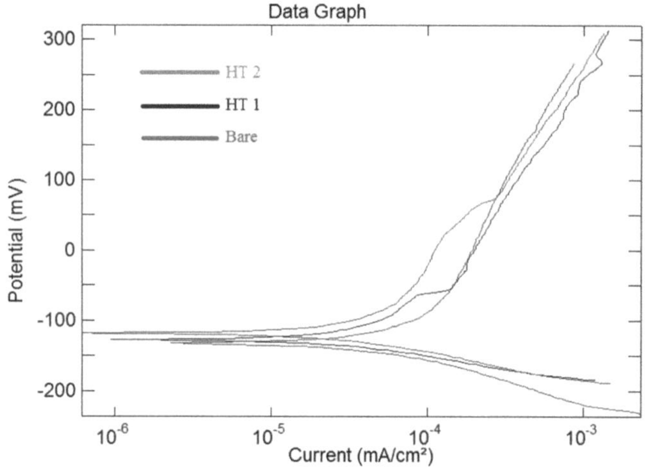

Figure 15.11 Potentiodynamic polarization graph.

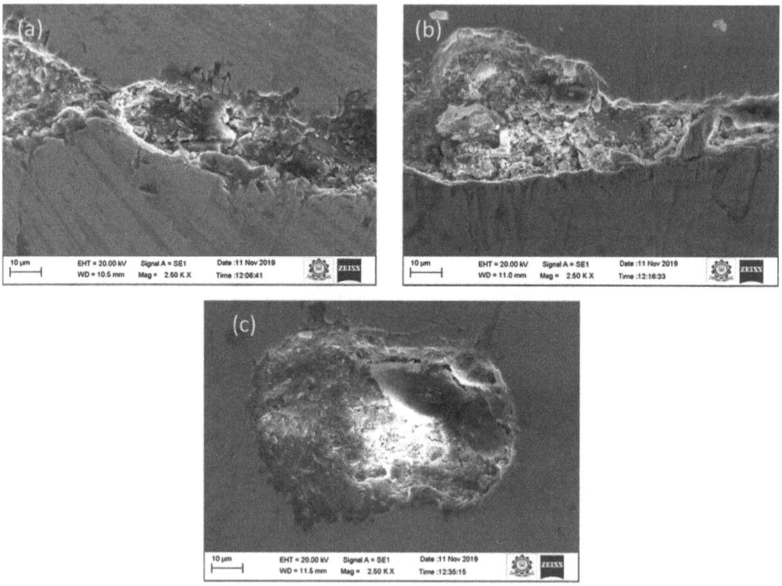

Figure 15.12 SEM image of the electrochemical sample: (a) commercial; (b) heat-treated sample 1; and (c) heat-treated sample 2.

The spectra and EDAX mapping of bare, HT 1, and HT 2 DMLS IN718 alloys are shown in Figures 15.13–15.15. When HT 2 DMLS alloy is compared to HT 1 and bare DMLS alloy, niobium, sodium, and oxygen element precipitation is higher. Chromium, iron, and nickel element precipitate is

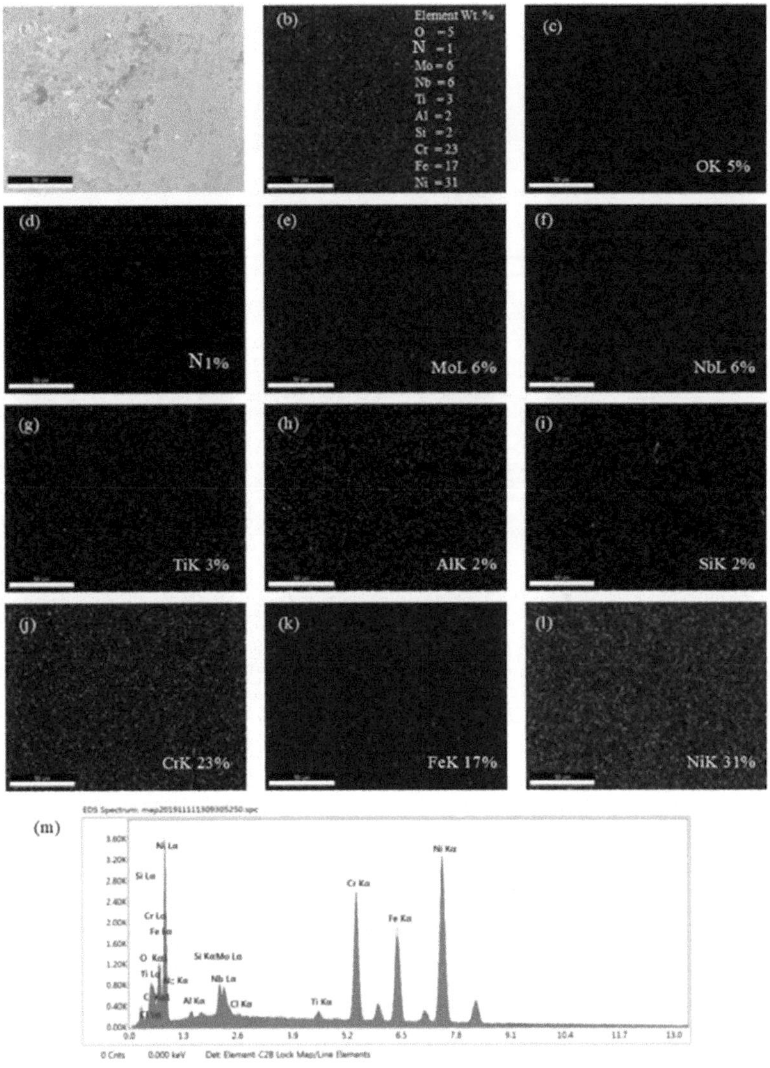

Figure 15.13 EDS mapping of rusted bare sample. (a) SEM image; (b) element composition; (c) distribution of oxygen; (d) distribution of nitrogen; (e) distribution of molybdenum; (f) distribution of niobium; (g) distribution of titanium; (h) distribution of aluminium; (i) distribution of silicon; (j) distribution of chromium; (k) distribution of iron; (l) distribution of nickel; and (m) element spectra.

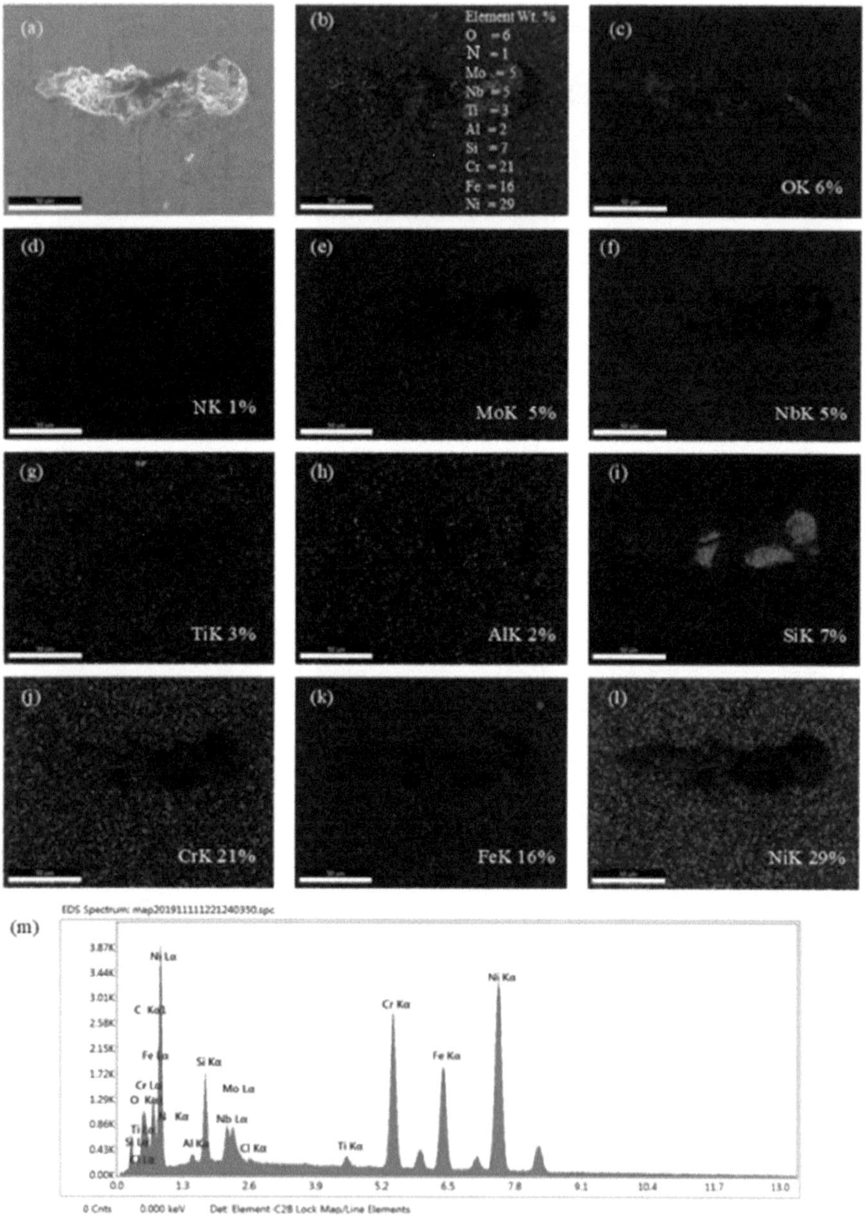

Figure 15.14 EDS mapping of rusted HT I DMLS sample. (a) SEM image; (b) element composition; (c) distribution of oxygen; (d) distribution of potassium; (e) distribution of molybdenum; (f) distribution of niobium; (g) distribution of titanium; (h) distribution of aluminium; (i) distribution of silicon; (j) distribution of chromium; (k) distribution of iron; (l) distribution of nickel; and (m) element spectra.

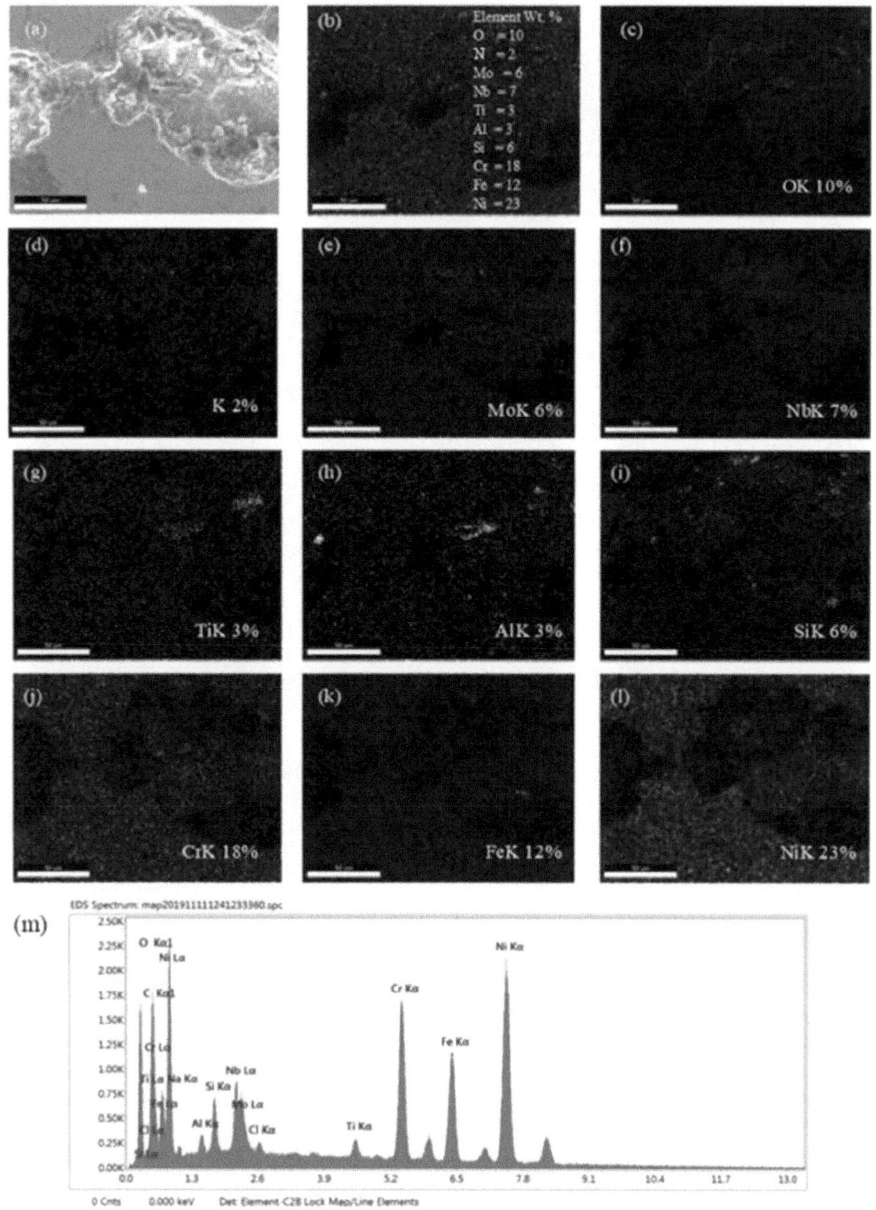

Figure 15.15 EDS mapping of rusted HT 2 DMLS model. (a) SEM image; (b) element composition; (c) distribution of oxygen; (d) distribution of potassium; (e) distribution of molybdenum; (f) distribution of niobium; (g) distribution of titanium; (h) distribution of aluminium; (i) distribution of silicon; (j) distribution of chromium; (k) distribution of iron; (l) distribution of nickel; and (m) element spectra.

high in the bare, HT 1, and HT 2 DMLS alloys. When compared to another heat-treated sample, the naked sample had a higher rate of iron element precipitate. The corrosion resistance of the material may be reduced as a result of the iron element precipitate.

After heat treatment, silicon precipitate is higher in heat-treated DMLS samples than in naked DMLS samples, increasing the DMLS alloy's resistance to corrosion.

15.6 CONCLUSIONS

When compared to traditional materials, the DMLS technique utilized to manufacture IN718 components provides advanced structure of an additive manufacturing process while also ensuring good corrosive behaviour. The resistance to corrosion of heat-treated DMLS IN718 to electrochemical 1.0 M HNO_3 (nitric) acid is attributed to the isolation of high nickel, molybdenum, and oxygen. The corrosion resistance of the DMLS alloy is improved by the precipitation of γ''. The DMLS alloy has a protective oxide covering that helps to prevent corrosion. Dendritic structure and layer-by-layer tracks were discovered in the naked DMLS alloy, and homogeneity happened following the heat treatment procedure.

REFERENCES

1. G. Würtz, H. Lasi, and D. Morar, Additive manufacturing–enabling technology for lifecycle oriented value-increase or value-decrease, *Procedia CIRP*, 33 (2015), 394–399.
2. I. Gibson, D. Rosen, and B. Stucker, Development of additive manufacturing technology, In: *Additive Manufacturing Technologies.* Springer, New York (2015), 19–42. https://doi.org/10.1007/978-1-4939-2113-3
3. M. Gebler, A. J. S. Uiterkamp, and C. Visser, A global sustainability perspective on 3D printing technologies, *Energy Policy*, 74 (2014), 158–167.
4. G. J. Schiller, Additive manufacturing for aerospace, In *Aerospace Conference, 2015 IEEE*, 35 (2015), 1–8.
5. O. Yilmaz, N. Gindy, and J. Gao, A repair and overhaul methodology for aeroengine components, *Robotics and Computer-Integrated Manufacturing*, 26 (2010), 190–201.
6. D. Brackett, I. Ashcroft, and R. Hague, Topology optimization for additive manufacturing, In *Proceedings of the Solid Freeform Fabrication Symposium*, Austin, TX, 1 (2011), 348–362.
7. A. W. Gebisa, and H. G. Lemu, Design for manufacturing to design for additive manufacturing: Analysis of implications for design optimality and product sustainability, *Procedia Manufacturing*, 13 (2017), 724–731.
8. D. Fratila and H. Rotaru, Additive manufacturing–a sustainable manufacturing route, *The 4th International Conference on Computing and Solutions in Manufacturing Engineering, CoSME'16, MATEC Web of Conferences*, Romania, 94 (2017), 03004. doi:10.1051/MATECCONF/20179403004

9. D. G Ahn, Direct metal additive manufacturing processes and their sustainable applications for green technology: A review, *International Journal of Precision Engineering and Manufacturing-Green Technology*, 3 (4) (2016), 381–395.
10. W. R. Morrow, H. Qi, I. Kim, J. Mazumder, and S. J. Skerlos, Environmental aspects of laser-based and conventional tool and die manufacturing, *Journal of Cleaner Production*, 15 (10) (2007), 932–943.
11. R. Sorrentino, P. Martin-Iglesias, O. A. Peverini, and T. M. Weller, Additive manufacturing of radio-frequency components, *Proceedings of the IEEE*, 105 (4) (2017), 589–592.
12. R. Rajesh, and M. V. Kulkarni, Influence of laser scan speed on bronze nickel alloy produced by direct metal laser sintering technique, *Journal of Advanced Microscopy Research*, 13 (2) (2018), 176–180.
13. D. Herzog, V. Seyda, E. Wycisk, and C. Emmelmann, Additive manufacturing of metals, *Acta Materialia*, 117 (2016), 371–392.
14. C. Panwisawas, Y. T. Tang, and R. C. Reed, Metal 3D printing as a disruptive technology for superalloys, *Nature Communications*, 11 (1) (2020), 1–4.
15. B. Graybill, M. Li, D. Malawey, C. Ma, J. M. Alvarado-Orozco, and E. Martinez-Franco, Additive manufacturing of nickel-based superalloys, In *International Manufacturing Science and Engineering Conference*, 51357 (2018), V001T01A015.
16. M. Elahinia, M. Nematollahi, K. S. Baghbaderani, A. Nespoli, and F. Stortiero, Manufacturing of shape memory alloys, In A. Concilio, V. Antonucci, F. Auricchio, L. Lecce, and E. Sacco (eds.), *Shape Memory Alloy Engineering for Aerospace, Structural, and Biomedical Applications*, 2 (2021), 165–193. Butterworth-Heinemann. https://doi.org/10.1016/B978-0-12-819264-1.00006-6
17. L. C. Zhang, J. Wang, Y. Liu, Z. Jia, and S. X. Liang, Additive manufacturing of titanium alloys, *Encyclopedia of Materials: Metals and Alloys*, 1 (2022), 256–274.
18. E. Yasa, O. Poyraz, E. U. Solakoglu, G. Akbulut, and S. Oren, A study on the stair stepping effect in direct metal laser sintering of a nickel-based superalloy, *Procedia CIRP*, 45 (2016), 175–178.
19. M. A. Imran, K. K. Singam, S. P. Jani, and S. Uppalapati, Mechanical properties of carbon particle mixed polylactic acid via fused deposition modeling, *Materials Today: Proceedings*, 46 (2021), 8590–8593.
20. B. Mueller, Additive manufacturing technologies–rapid prototyping to direct digital manufacturing, *Assembly Automation*, 32 (2) (2012), 459–465.
21. G. S. Altug-Peduk, S. Dilibal, O. Harrysson, S. Ozbek, and H. West, Characterization of Ni–Ti alloy powders for use in additive manufacturing, *Russian Journal of Non-Ferrous Metals*, 59 (4) (2018), 433–439.

Chapter 16

Studies on wire arc additive manufacturing of shape memory alloys

Prashanna Rangan R, Gurusamy Visvanathan, Mohandass Muthukrishnan, and Pitchandi K
Sri Venkateswara College of Engineering

CONTENTS

16.1 INTRODUCTION TO ADDITIVE MANUFACTURING

With the emergence of technology, 3D printing has become the buzz word, resulting in affordable Do It Yourself 3D printers for your home to print the stuff right away. Additive Manufacturing (AM) [1,2] is the concept of creating and building material in steps to obtain the finished product, as the name implies, and hence it is also called Rapid Prototyping. In contrast, manufacturing a product by conventional means frequently results in removal of material through milling, machining, carving, sculpting, or other ways. Rapid prototyping employs computer-aided design (CAD) software or 3D object scanners to command machines to deposit material in precise geometric designs, layer upon layer. When used correctly, additive manufacturing [3,4] offers a great plethora of enhanced efficiency, complicated geometry, and easier fabrication. As a result, possibilities for individuals who actively pursue additive manufacturing exist.

In general, for the additive manufacturing process the raw material can be fed in solid, liquid or powder form into the fabricating machine. Rather than the typical subtractive manufacturing methods, the lack of material waste reduces costs for high-value components, and AM has also been found to reduce lead times. AM enables the production of customized parts

DOI: 10.1201/9781003430186-16

with complicated geometries and minimal scrap. The digital technique, which is ideal for rapid prototyping, allows for design changes to be made rapidly and accurately during the production processes. A general data flow of the Additive Manufacturing process is as elucidated in Figure 16.1. The sector-wise AM applications as depicted by the Wohlers Report 2017 is picturized in Figure 16.2.

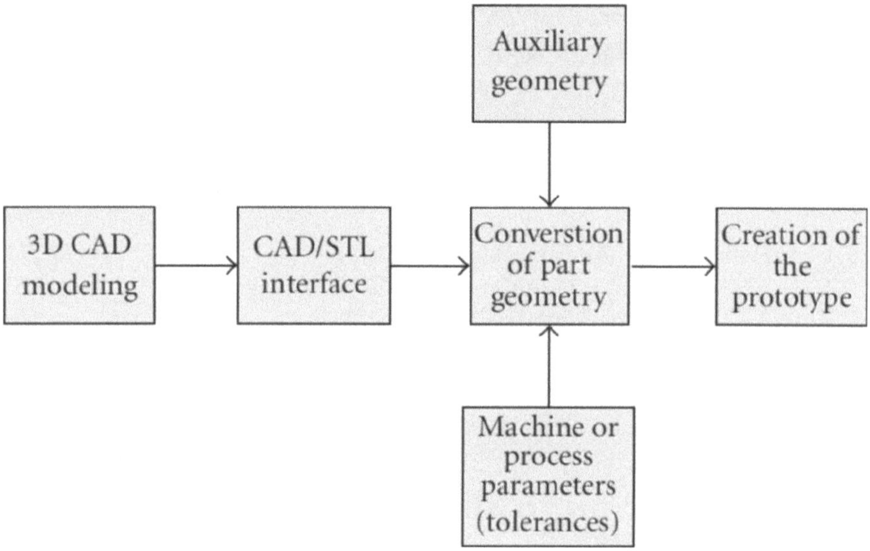

Figure 16.1 Data flow of Rapid Prototyping [1].

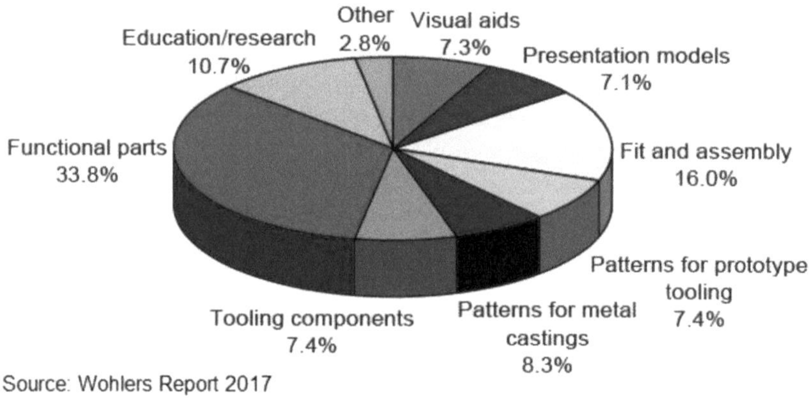

Figure 16.2 The use of additive manufacturing in the different sectors. (Source: Wohlers Report 2017.)

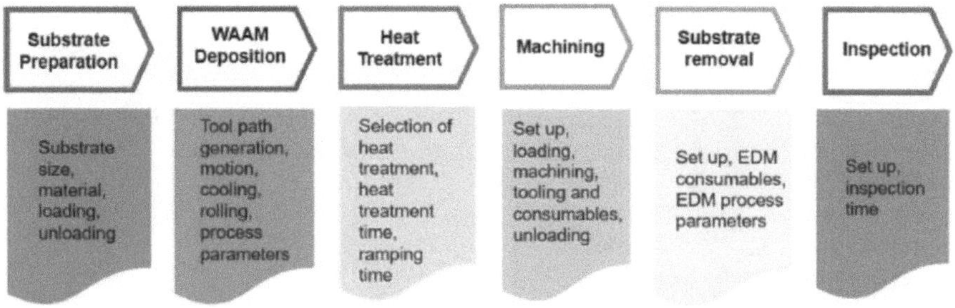

Figure 16.3 WAAM activities [6].

Metal Additive manufacturing technologies [5] are evolving in recent years for manufacturing crucial and intrusive parts for aerospace, defence, health and other sectors that include metal combinations like Titanium, Aluminium, Ferrous, Nickel, Inconel or Invar-based alloys. Wire Arc Additive Manufacturing [6,7] is a metal AM process that utilizes the metal in the form of weld rods that are held together by means of Arc Welding process and hence the name. The Wire Arc Additive Manufacturing is now also called Directed Energy Deposition-Arc (DED-arc) method. WAAM offers significant cost savings and a higher deposition rate when compared to alternative fusion sources. WAAM overcomes the shortcomings [8] of existing AM processes while also providing additional benefits such as little material wastage, improved equipment adaptability, cheap operating costs, quick production cycles and outstanding constructing perfection. A metal wire feedstock is fed in a controlled fashion into the welding arc where it is melted and gets deposited to the previous layer. The WAAM activities are as depicted in Figure 16.3.

16.2 WIRE ARC ADDITIVE MANUFACTURING (WAAM) EQUIPMENT

Path planning and modelling [9] of the WAAM robotic welding arm synergically combines them and the research is on the verge. The overall schematic diagram of a working WAAM process comprising all the major elements is as depicted in Figure 16.4. The comparison with respect to the material deposition rate and size versus its resolution and complexity is as described in Figure 16.5.

The WAAM components are as shown in Figure 16.6 that shows a propeller on the left and a turbine on the right side.

In contrast to other additive manufacturing processes, the perks of wire arc additive manufacturing, in terms of shorter lead times, less material waste, improved functionality, customized tooling for lower volume parts

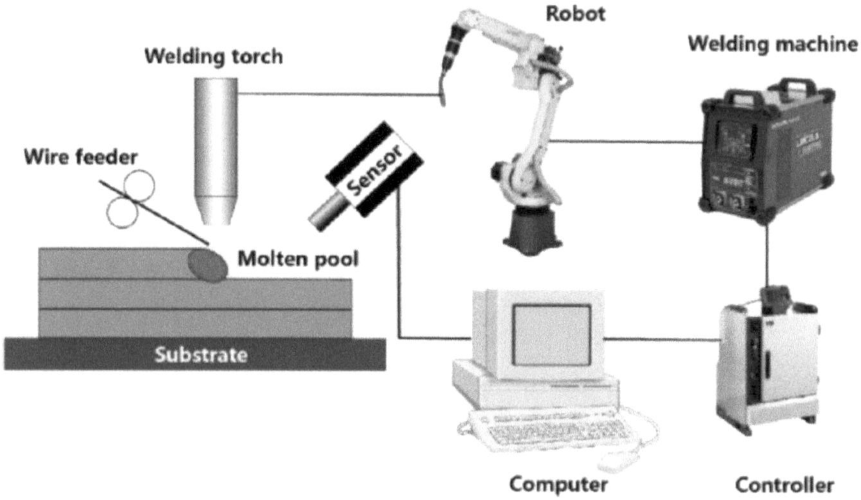

Figure 16.4 Schematic diagram of the WAAM system [8].

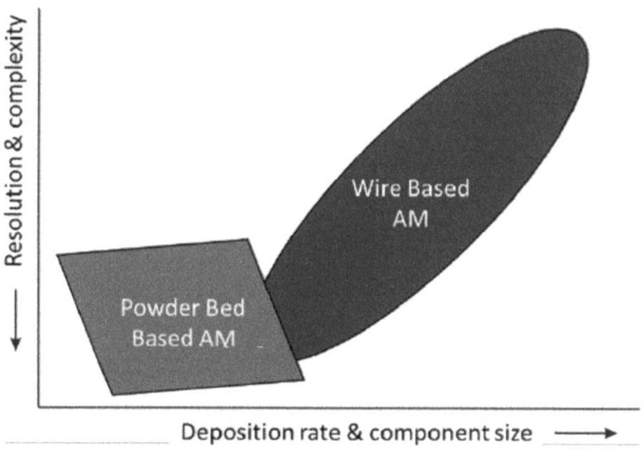

Figure 16.5 Comparison of WAAM and powder-based AM [9].

and the possibility of multi-material design, have been ascertained by Panchenko et al. and are shown in a well-organized diagram (Figure 16.7).

With the advancements in the welding systems and the emerging user-friendly software [13] that support the 3D metallic printing process to a greater extent, WAAM can be utilized not only for the creation of new parts but also for the redesign and refashioning of existing ones. WAAM can also be used to repair moulds and other items on a limited basis.

Cold Metal Transfer technology [14] has transformed the fusing of dissimilar metals and thicker structures, resulting in superior weld bead appearances and lower thermal input. CMT-laser fusion welding seems to

Figure 16.6 WAAM products – propeller [10] (left side) and turbine [11] (right side).

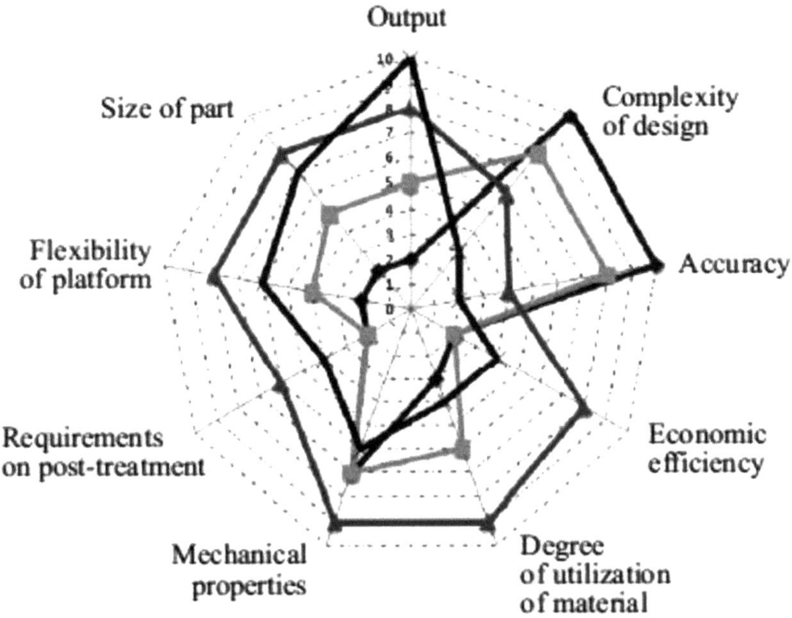

Figure 16.7 Advantages of WAAM over other AP processes [12]: (♦) bed deposition; (■) direct deposition; (×) electron beam freeform fabrication; (▲) WAAM.

be more advantageous to laser or laser hybrid welding, and as a result of its expanded qualities, it has gained use as an additive manufacturing technology in the automobile industry, defence industries and power plants. Whenever the electrode wire end touches the surface with the molten pool in the CMT process, the welding torch's servo motor is reversed via a modern electronic controller, forcing the wire to retract and promote droplet diffusion. During the metal transfer, the current tumbles to near-zero, preventing any spatter formation. When the metal transfer is complete, the arc is re-ignited, and the wire is fed forward with the predetermined welding current reflowing.

The timespan utilized to discharge a droplet of molten electrode into the weld pool is specified as a typical CMT welding electrical impulse cycle. To analyse the power distribution of various phases in the droplet transfer process, current and voltage waveform analysis is required. A CMT process is divided into three stages: peak current, background current and short-circuiting phases as elaborated in Figure 16.8. CMT technology has proved to be a viable and efficient stack for welding many similar alloy elements like Inconel 718, Aluminium 7075, Aluminium AA6061, Galvanized sheet steel, Galvannealed steel, Aluminium 5083-H116, Aluminium AA7A52, AA2219-T851, etc., and dissimilar alloys like Zinc-coated steel (Q235) and wrought aluminium (6061), Magnesium AZ31 and aluminium 1060 alloy, Magnesium AZ31 band and 6061 Al alloy, Hot dip galvanized steel and aluminium 1060 alloy, Aluminium (AA6061) and low carbon steel alloy, Magnesium AZ31 and hot-dipped galvanized mild steel, Aluminium A6061-T6 and titanium Tie6Ale4V alloys, Aluminium AA6061-T6 to galvanized steel alloy, Magnesium alloy AZ31B and pure copper T2 alloy, Pure titanium TA2 to magnesium alloy AZ31B alloy, 5083-H111 and

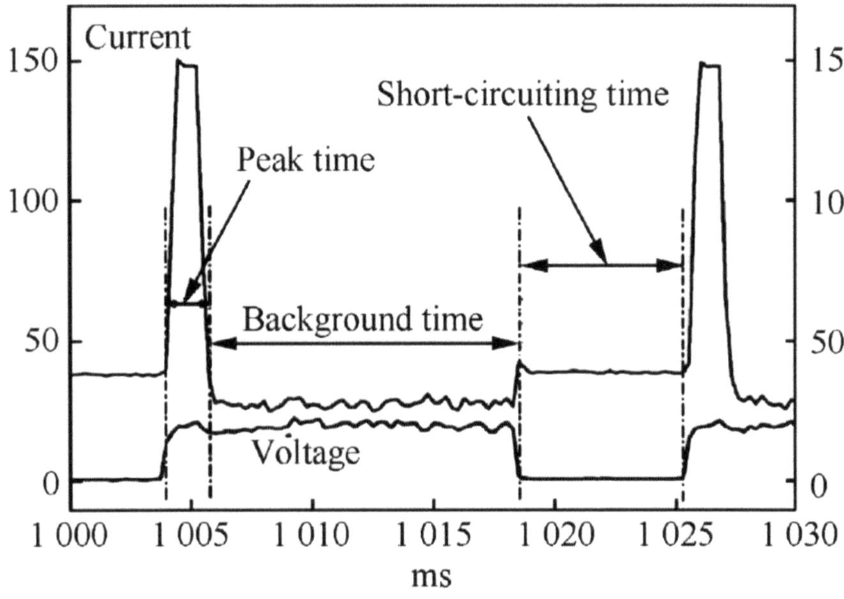

Figure 16.8 Current and voltage waveforms of CMT process [14].

6082-T651 aluminium alloys, Titanium TA2 to pure copper T2 alloy, Hot dip galvanized steel sheet and aluminium 5052 alloy, Aluminium 5A06 with pure Ni N6 plates, 5182-O and 6082-T4 aluminium alloy sheets, Titanium AMS4911L with 316L stainless steel, A6061-T6 aluminium alloy to dual phase 800 steel, AC 170 PX aluminium alloy and ST06 Z galvanized steel sheets, 304 stainless steel and 5A06 aluminium alloy sheet, etc.

Thus, the CMT process has proved to be a promising technology for the WAAM process for similar and dissimilar AM-engineered component fabrication as the spatter generation is eliminated by retracting the weld wire during the short-circuiting phase which leads to improved and enhanced mechanical properties.

The welding process, as well as its cooling or shape, can be watched and controlled in a closed loop while welding. The WAAM additive manufacturing method has been carried out using the YASKAWA 6 axis Industrial manipulator with payload capacity of 12kg inside the SVCE Welding Research Cell facility as illustrated in Figure 16.9. Figure 16.10 shows the in-process WAAM machining. Synchronizing the robot and the weld attachment is achieved with the help of the robot controller that acts as the central commanding station with which the welding end effector attachment for carrying out the WAAM process will be moved in accordance with the robot program which differs from the STL mode of 3D printing process and is carried out by using the robot programming software or via the teach pendant.

Figure 16.9 Robotic CMT-WAAM arrangement at Welding Research Cell, SVCE.

Figure 16.10 WAAM progress during operation.

The below mentioned formula in equation (16.1) is used to calculate the arc weld power [15] that is derived from the arc current and its applied potential difference and that of the total energy produced during the process of welding is as elaborated in equation (16.2).

$$P = \frac{\sum_{i=0}^{n} \int_{t_i}^{t_{i+1}} U_i I_i dt}{(t_n - t_0) * 1{,}000} (\text{kW}) \tag{16.1}$$

$$E_L = \frac{60 * P}{100 * v} (\text{kJ/cm}) \tag{16.2}$$

where

I_i is the actual welding current in Amps at time t_i,
U_i is the actual welding voltage in Volts,
$t_n - t_0$ is the duration of weld in seconds,
P is the weld power,
v is the speed of weld in m/min.

16.3 WAAM OF SHAPE MEMORY ALLOY

The shape memory effect occurs in alloys when the crystalline structure changes in response to changes in temperature and stress. Generally, shape memory alloys are intermetallic compounds having superlattice structures and metallic-ionic-covalent characteristics. Thus, they have the properties of both metals and ceramics. Shape memory alloys (SMAs) are advanced and sophisticated materials that combine temperature sensors with smart motors [16] by utilizing the shape memory effect and super-elasticity. The following are widely used shape memory alloys for engineering applications:

1. Ni–Ti alloy (Nitinol)
2. Cu–Al–Ni alloy
3. Cu–Zn–Al alloy
4. Au–Cd alloy
5. Ni–Mn–Ga and Fe-based alloys

However, the only two alloy systems that have achieved any level of commercial exploitation are Ni–Ti alloys and copper-based alloys. Properties of the two systems are quite different from each other. Some of the practical applications of shape memory alloys include microvalve/actuators, toys, medical field – blood clot filters, highly sensitive thermostats and shock absorbers.

The binary, equi-atomic intermetallic combination Ni–Ti is the foundation of the Nickel-Titanium alloy. Because it has a moderate solubility range for excess Nickel or Titanium, as well as most other metallic elements, the intermetallic compound is remarkable. Because of its solubility, the system's mechanical and transformation characteristics may be altered by alloying with a variety of elements. Excess nickel lowers the transition temperature while increasing the austenite yield strength. Contaminants like oxygen and carbon cause the transformation temperature to rise and the mechanical characteristics to deteriorate. As a result, it's also a good idea to keep the number of such pieces to a minimum. Shape Memory Alloys (SMAs) have been used in a wide range of applications owing to its ability to recover inelastic strains wider than any other metallic alloy. However, traditional fabrication of the most prevalent SMA, such as NiTi, presents significant problems, such as poor workability and possibly excessive tool wear. Except with specialized equipment, turning or milling is extremely difficult. The alloys are often difficult to weld, braze or solder.

The yield strength of Ti–Ni shape memory alloys [17–22] has been improved using a variety of procedures, including the addition of alloying elements and post-heat treatment. Ni–Ti shape memory alloys have been enriched with Mo, Co and other elements. However, if alloying elements are

Figure 16.11 WAAM wall with complex geometry. (Image courtesy: WAAM3D.)

added during the casting process, the secondary phases and intermetallic compounds will be segregated inhomogeneously. When it comes to producing excellent mechanical properties, homogeneous precipitation of alloying elements and intermetallic compounds is always desired. Furthermore, the lower martensitic transformation temperature (Ms) is a significant parameter when Ni–Ti shape memory alloys act, as it would cover a wide range of applications.

Additive manufacturing addresses these issues directly by removing the requirement for tooling and allowing the creation of complicated geometry samples straight from computer-aided designs. Shape memory alloys based on Ti–Ni are widely employed in automobiles, aeroplanes and other applications [23–25]. AM Ni–Ti-based alloys have good contact integrity, homogeneous structural morphology and excellent super-elasticity. The microstructure of CMT-based wire arc additively manufactured Ni–Ti can be altered widely, depending on process parameters and built conditions. WAAM demonstrates that in an inert atmosphere, direct energy deposition of nickel, titanium, steel and copper alloys may be used to manufacture near-net shape components with complicated geometries. Figure 16.11 illustrates the building of thin walls with complex geometry.

16.4 METHODOLOGY OF CHARACTERIZATION OF WAAM WALL

Through layer-by-layer deposition, metallic additive manufacturing (AM), a new manufacturing method defined by the repetition of local melting and solidification, may be used to build complicated geometric components directly from a digital model [26,27]. However, during the AM process, each region in the AM component undergoes fast melting and solidification, as well as repeated heating and cooling phases with directional heat dissipation. The specific heat cycle creates microstructural inhomogeneity in AM components, resulting in anisotropic mechanical characteristics in the finished product, which is a key problem that must be mitigated or avoided [28]. The columnar grain shape is significantly connected to the anisotropic characteristics of AM components. Hence to elucidate the variation of mechanical properties and the structural morphology of the built wall, the characterization is to be conducted by extracting the specimens in the horizontal and vertical directions like tensile [29] and notch-impact tests as illustrated in Figure 16.12.

L1, L2, up to L10 refer to the deposition layers of WAAM wall. HT and HI refer to the tensile and impact test specimens' direction parallel to the deposition direction. Likewise, VT and VI refer to the tensile and impact test specimens' direction perpendicular to the deposition direction. At least three specimens would be tested at room temperature to confirm the reliability of the tensile and impact testing results.

The hardness survey of the built wall both parallel and perpendicular to the deposition direction would be conducted to determine the variation in the hardness of the built wall. In a slender wall construction created using wire arc additive manufacturing, the tensile behaviour correlated with distinct microstructural distributions will also be investigated. Optical and scanning electron microscopy (SEM) are used to examine the microstructure of the specimen. The obtained micrographs are used to obtain a morphology of the built wall cross-section.

In the same way microstructure analysis is also done to find the variations in the structural morphology of built wall in both the directions by extracting the specimens from the different locations. WAAM component that is produced using the abovementioned robotic and welding apparatus is as depicted in Figure 16.13, which is cut as per ASTM standards post-WAAM process for attempting the aforesaid tensile and impact tests.

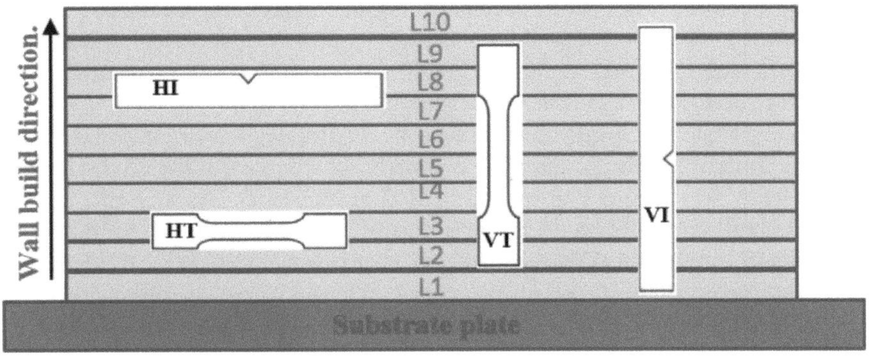

Figure 16.12 Schematic representation of the locations of specimens used for tensile and impact testing.

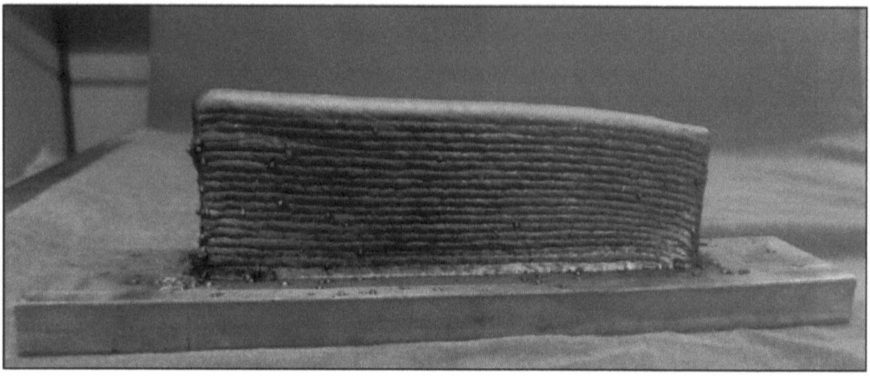

Figure 16.13 WAAM component – nickel-based alloy.

16.5 CONCLUSION

Numerous research studies for diverse application sectors have proved the immense potential of additive manufacturing technologies. It has been demonstrated that cost-effective utilization necessitates high material deposition rates and a complete production chain from element conceptualization to production. The intricate modules with uniform or regionally tailored material characteristics can also be produced using electric arc-based additive manufacturing methods. Being able to harness the new technologies, the interconnections that occur between the manufacturing process, component development, material and component qualities must be thoroughly researched.

This study helps to better understand the properties of Wire Arc Additively Manufactured components. The WAAM process using CMT could be carried out without creation of any defects as the Heat-Affected Zone (HAZ) is very narrow, which is in contrast to the conventional fusion welding methods, viz., TIG, conventional MIG, etc. The influence of arc energy on the component will have a direct impact on the mechanical and microstructure properties. Hence CMT mode of WAAM is usually preferred for the low cost and near-net shape components.

Each layer has a variable thermal history during deposition, making it difficult to control and anticipate the evolution of microstructure, which ultimately leads to variances in mechanical characteristics. As a result, future research shall concentrate on the creation of a metallurgical technique that can guide the regulation and optimization of process variables. By dynamically modifying the path planning, a closed loop control system with real-time detection can be constructed to generate 3D WAAM components.

This study reveals that in order to minimize the anisotropic mechanical behaviour in WAAM components the inhomogeneous distribution of the alloying elements and intermetallic compounds should be carefully managed. Many studies made straight wall deposition rather than complex structure and hence is a potential scope for development. The wear and corrosion behaviour of WAAM components has received minimal study. It has long been known that wear and corrosion behaviour can have a significant influence on the performance of the WAAM product.

Though WAAM has few drawbacks like anisotropic microstructures, mechanical characteristics, residual stresses, etc., it has advantages like developing intricate structures with less material wastage and readily available raw material with fewer post-processing steps in an eco-friendly manner, thus making its footprints in the defence sector, aerospace, nuclear sector in a promising way. As a result, more study is required to address the aforementioned issues by optimizing process parameters and post-deposition heat treatment.

REFERENCES

[1] K. V. Wong and A. Hernandez, "A review of additive manufacturing," *ISRN Mechanical Engineering*, vol. 2012, pp. 1–10, 2012, doi: 10.5402/2012/208760.

[2] M. Jiménez, L. Romero, I. A. Domínguez, M. D. M. Espinosa, and M. Domínguez, "Additive manufacturing technologies: An overview about 3D printing methods and future prospects," *Complexity*, vol. 2019, 9656938, 2019, doi: 10.1155/2019/9656938.

[3] O. Abdulhameed, A. Al-Ahmari, W. Ameen, and S. H. Mian, "Additive manufacturing: Challenges, trends, and applications," *Advances in Mechanical Engineering*, vol. 11, no. 2, 2019, doi: 10.1177/1687814018822880.

[4] S. Ponis, E. Aretoulaki, T. N. Maroutas, G. Plakas, and K. Dimogiorgi, "A systematic literature review on additive manufacturing in the context of circular economy," *Sustainability (Switzerland)*, vol. 13, no. 11, p. 6007, 2021, doi: 10.3390/su13116007.

[5] M. Yakout, M. A. Elbestawi, and S. C. Veldhuis, "A review of metal additive manufacturing technologies," *Solid State Phenomena*, 2018, vol. 278 SSP, pp. 1–14. doi: 10.4028/www.scientific.net/SSP.278.1.

[6] L. P. Raut and R. V. Taiwade, "Wire arc additive manufacturing: A comprehensive review and research directions," *Journal of Materials Engineering and Performance*, vol. 30, no. 7, pp. 4768–4791, 2021, doi: 10.1007/s11665-021-05871–5.

[7] K. S. Derekar, "A review of wire arc additive manufacturing and advances in wire arc additive manufacturing of aluminium," *Materials Science and Technology (United Kingdom)*, vol. 34, no. 8, pp. 895–916, 2018, doi: 10.1080/02670836.2018.1455012.

[8] Y. Li, C. Su, and J. Zhu, "Comprehensive review of wire arc additive manufacturing: Hardware system, physical process, monitoring, property characterization, application and future prospects," *Results in Engineering*, vol. 13, 100330, 2022, doi: 10.1016/j.rineng.2021.100330.

[9] K. Treutler and V. Wesling, "The current state of research of wire arc additive manufacturing (Waam): A review," *Applied Sciences (Switzerland)*, vol. 11, no. 18, p. 8619, 2021, doi: 10.3390/app11188619.
[10] M. Feldmann, R. Kühne, S. Citarelli, U. Reisgen, R. Sharma, and L. Oster, "3D-Drucken im Stahlbau mit dem automatisierten wire arc additive manufacturing," *Stahlbau*, vol. 88, no. 3, pp. 203–213, 2019, doi: 10.1002/stab.201800029.
[11] L. Gardner, P. Kyvelou, G. Herbert, and C. Buchanan, "Testing and initial verification of the world's first metal 3D printed bridge," *Journal of Constructional Steel Research*, vol. 172, 106233, 2020, doi: 10.1016/j.jcsr.2020.106233.
[12] O. V. Panchenko, L. A. Zhabrev, D. V. Kurushkin, and A. A. Popovich, "Macrostructure and mechanical properties of Al – Si, Al – Mg – Si, and Al – Mg – Mn aluminum alloys produced by electric arc additive growth," *Metal Science and Heat Treatment*, vol. 60, no. 11–12, pp. 749–754, 2019, doi: 10.1007/s11041-019-00351–z.
[13] Z. Lin, P. Liu, and X. Yu, "A literature review on the wire and arc additive manufacturing—Welding systems and software," *Science of Advanced Materials*, vol. 13, no. 8, pp. 1391–1400, 2021, doi: 10.1166/sam.2021.3971.
[14] S. Selvi, A. Vishvaksenan, and E. Rajasekar, "Cold metal transfer (CMT) technology - An overview," *Defence Technology*, vol. 14, no. 1, pp. 28–44, 2018, doi: 10.1016/j.dt.2017.08.002.
[15] Y. Ali, P. Henckell, J. Hildebrand, J. Reimann, J. P. Bergmann, and S. Barnikol-Oettler, "Wire arc additive manufacturing of hot work tool steel with CMT process," *Journal of Materials Processing Technology*, vol. 269, pp. 109–116, 2019, doi: 10.1016/j.jmatprotec.2019.01.034.
[16] L. Sun et al., "Stimulus-responsive shape memory materials: A review," *Materials and Design*, vol. 33, no. 1, pp. 577–640, 2012, doi: 10.1016/j.matdes.2011.04.065.
[17] Y. X. Tong et al., "Microstructure and martensitic transformation of an ultrafine-grained TiNiNb shape memory alloy processed by equal channel angular pressing," *Intermetallics (Barking)*, vol. 49, pp. 81–86, 2014, doi: 10.1016/j.intermet.2014.01.019.
[18] J. H. Sui, Z. Y. Gao, Y. F. Li, Z. G. Zhang, and W. Cai, "A study on NiTiNbCo shape memory alloy," *Materials Science and Engineering A*, vol. 508, no. 1–2, pp. 33–36, 2009, doi: 10.1016/j.msea.2009.01.018.
[19] V. Y. Abramov et al., "Martensitic transformations and functional properties of thermally and thermomechanically treated Ti-Ni-Nb-based alloys," *Materials Science and Engineering A*, vol. 438–440, no. SPEC. ISS., pp. 553–557, 2006, doi: 10.1016/j.msea.2006.05.170.
[20] H. Jiang, Y. Chen, D. Yang, H. Sun, J. Zeng, and L. Rong, "Effect of mo addition on the microstructure and properties of TiNiNb alloy," *Acta Metallurgica Sinica (English Letters)*, vol. 27, no. 2, pp. 217–222, 2014, doi: 10.1007/s40195-014-0039-1.
[21] B. Cui et al., "Precipitation behavior and mechanical properties of Ti-Ni-Nb-Co alloys," *Intermetallics (Barking)*, vol. 95, pp. 40–47, 2018, doi: 10.1016/j.intermet.2018.01.014.

[22] N. N. Popov, T. I. Sysoeva, E. V. Shchedrina, D. V. Presnyakov, and E. N. Grishin, "Effects of the regimes of heat treatment and of the magnitude and temperature of the inducing deformation on the characteristics of the shape-memory effect in the 43Ti-46Ni-9Nb-2Zr alloy," *Physics of Metals and Metallography*, vol. 116, no. 6, pp. 615–625, 2015, doi: 10.1134/S0031918X15060071.

[23] K. Otsuka and X. Ren, "Physical metallurgy of Ti-Ni-based shape memory alloys," *Progress in Materials Science*, vol. 50, no. 5, pp. 511–678, 2005. doi: 10.1016/j.pmatsci.2004.10.001.

[24] H. L. Tu, H. Bin Zhao, Y. Y. Fan, and Q. Z. Zhang, "Recent developments in nonferrous metals and related materials for biomedical applications in China: A review," *Rare Metals*, vol. 41, no. 5, pp. 1410–1433, 2022. doi: 10.1007/s12598-021-01905-y.

[25] D. J. Hartl and D. C. Lagoudas, "Aerospace applications of shape memory alloys," *Proceedings of the Institution of Mechanical Engineers, Part G: Journal of Aerospace Engineering*, vol. 221, no. 4, pp. 535–552, 2007, doi: 10.1243/09544100JAERO211.

[26] N. Haghdadi, M. Laleh, M. Moyle, and S. Primig, "Additive manufacturing of steels: A review of achievements and challenges," *Journal of Materials Science*, vol. 56, no. 1, pp. 64–107, 2021, doi: 10.1007/s10853-020-05109-0.

[27] P. Bajaj, A. Hariharan, A. Kini, P. Kürnsteiner, D. Raabe, and E. A. Jägle, "Steels in additive manufacturing: A review of their microstructure and properties," *Materials Science and Engineering A*, vol. 772, 138633, 2020, doi: 10.1016/j.msea.2019.138633.

[28] Y. Kok et al., "Anisotropy and heterogeneity of microstructure and mechanical properties in metal additive manufacturing: A critical review," *Materials and Design*, vol. 139, pp. 565–586, 2018, doi: 10.1016/j.matdes.2017.11.021.

[29] Z. Lyu, Y. S. Sato, S. Tokita, Y. Zhao, J. Jia, and A. Wu, "Microstructural distribution and anisotropic tensile behavior in a 2Cr13 martensitic stainless steel thin wall fabricated by wire arc additive manufacturing," *Materials Today Communications*, vol. 29, 102870, 2021, doi: 10.1016/j.mtcomm.2021.102870.

Chapter 17

Mechanical and tribological behavioral study of metal additive manufactured AlSi10Mg alloy

Kumar M, Ravikumar K, and Arunkumar P
KPR Institute of Engineering and Technology

Dhanabal A
M/s. Fabforge Innovations Pvt. Ltd.

CONTENTS

17.1 INTRODUCTION

The additive manufacturing process helps to build a 3D object from a CAD model, through a layer-by-layer material addition technique, and so it is named as Additive Manufacturing (AM). AM is a process that binds the

DOI: 10.1201/9781003430186-17

material under computer-controlled laser sintering process to produce a 3D object that is built layer by layer. The general steps involved in additive manufacturing process are modeling, printing, and finishing. Generally, there are a number of 3D printing techniques available like fused deposition modeling (FDM), stereolithography (SLA), selective laser sintering (SLS), selective laser melting (SLM), digital light processing (DLP), laminated object manufacturing (LOM), and electron beam melting (EBM), among these available techniques the current work aims on Direct Metal Laser Sintering (DMLS) process due to its reliability and efficiency. Selective Laser Sintering (SLS) principle is employed in DMLS process. DMLS is used to create a 3D metal part from its computer model and this process is an advancement of powder metallurgy.

The service life of most of the engineering components highly depends on wear and tear during its service, and in turn its noise, vibration, and energy consumption as well. In this way, AlSi-based alloys being a lightweight material offer enhanced wear and thermal performances (Aboulkhair et al., 2019). Engineering components like engine blocks, hinges, and pistons used in automobile and aircraft applications are made of AlSi alloy. Particularly, AlSi10Mg alloy with its integrated structural and functional properties has its own demand in the market. With this demand, producing highly complex, precise, and thin-walled components has become possible these days through additive manufacturing technologies (Leary et al., 2019), which decreases the conventional lengthy production cycles and costs. The common methods to fabricate AlSi10Mg alloy-based parts are Selective Laser Melting (SLM) and Selective Laser Sintering (SLS); still there are limited researches that discuss about the metallurgical imperfections formulated during SLM process. The effect of laser energy, i.e., Volumetric Energy Density (VED), on densification (Wei et al., 2017) and mechanical properties was analyzed. The results show that UTS decreases with the increase in VED due to the development of micropores like gas and keyhole-induced pores. High-performing SLM parts can be produced by meticulously understanding the laser parameters' influence (Wu et al., 2017). There are researches that suggest the keyhole mode to fabricate AlSi10Mg alloy parts by molten pool free of pores during the printing process, since this mode offers ultrafine microstructure in both the printing directions (Aboulkhair et al., 2016; Xiong et al., 2019). The Melt Pool Boundary (MPB) is the weakest section of any SLMed product, and this MPB is formed due to a sudden sharp change in state of metal from evaporation to solidification, and high cooling rate. Hence, it is recognized that lesser the MPBs higher the mechanical properties of printed parts (Hong et al., 2020). The reduced VED and lesser MPBs are the vital requirements to produce any SLMed components free of defect with ultrafine microstructure that in turn offer improved hardness (Girelli et al., 2017). It is also evident that these key process conditions would give reduced local welding failures (i.e., printing), and no sign of adhesion was noticed in the support structures. Hence, the wear and coefficient of friction were found stable and lower than as-cast AlSi10Mg parts. During the fabrication process, various grain microstructures like α-Al and Si are also observed, due to preferential growth of temperature gradient (Trevisan et al., 2017). To check the anisotropic property of printed samples, the pulse-echo ultrasonic method was carried out and the results revealed that the SLMed AlSi10Mg alloy parts exhibit anisotropy in both longitudinal and transverse directions based on the direction of printing. It's also been observed that the non-destructive testing signals of longitudinal direction confirm the state of microstructure of the printed part (Sol et al., 2018).

Microstructure refinement and improved tribological properties of AlSi10Mg alloy parts can be obtained as an effect on different melting modes in SLM. The parts made out of SLM with Keyhole Mode (KM) exhibited improved wear resistance in contrast to as-cast AlSi10Mg alloy parts due to its refined microstructure by cellular-like grain structures (Hong et al., 2020). The parts made by Laser Powder Bed Fusion (LPBF) or SLS possess improved wear resistance and stable and lower CoF since the fully densified samples do have ultrafine grain structures and high hardness. The major cause for the reduction in wear resistance is the pores present in the parts that break the bond between molten pools, in turn introducing cracks. These cracks are the source of friction that results in increased wear. Similarly, the heat treatment is the biggest thread to wear resistance (Massimo, 2019). AlSi10Mg alloy composites produced with nano-TiB_2 by Direct Metal Laser Sintering (DMLS) process offer higher wear resistance than non-reinforced and micro-reinforced AlSi10Mg alloy parts. However, in contrast DMLSed AlSi10Mg alloy exhibited higher CoF than as-cast AlSi10Mg alloy. It is the other way in the case of wear resistance due to finer microstructure fabricated through DMLS (Massimo, 2016). The other reason for the improved wear resistance is the tribo-layer formed in the contact zone that guards the materials from permanent deformation. To fabricate frictional combos, more investigations need to be done on the effect of metallurgical defect like pores on wear mechanism and lubricating performances with wet and dry conditions. There are certain researches to improve the specific wear resistance by introducing shot peening method to get hardened surfaces that reduce wear while increasing the surface roughness and in turn the CoF. This postprocessing technique also has an important role in achieving improved surface and tribological behavior of DMLSed components (Binnur et al., 2019). With this exhaustive literature experience, the present work aims to investigate the mechanical and tribological behaviors of DMLSed AlSi10Mg alloy and is being compared with as-cast AlSi10Mg alloy.

17.2 EXPERIMENTAL DETAILS

17.2.1 Selection of material

Among various aluminum alloys, AlSi10Mg alloy has a huge applications since it has imperative mechanical and tribological properties. Its most common applications are: automobile industries, aerospace, ship building, railway department, pylons and towers etc. AlSi10Mg alloy powder (Figure 17.1) was used in this process for its ease of processing, good strength, hardness and dynamic properties, and also for parts subjected to high loads etc.

The chemical composition of AlSi10Mg alloy is given in Table 17.1, and it is composed of several elements such as silicon, iron, copper, magnesium, nickel, zinc, lead, tin, and titanium identified using XRD analysis as shown in Figure 17.2.

17.2.2 Selection of additive manufacturing process

In the proposed work, Direct Metal Laser Sintering (DMLS) process is chosen for its reliability and efficacy compared to other techniques in this SLS process. DMLS process is carried out on EOS M290 DMLS machine as illustrated in Figure 17.3.

The technical specification of DMLS machine used for this work is presented in Table 17.2. This machine uses 400 W yb-fiber laser and it prints with a layer thickness of 0.03 mm using the precise F-theta laser. The dimensions of the DMLSed plate are 100×100 mm with 6 mm thickness.

Figure 17.1 AlSi10Mg alloy powder.

Table 17.1 Chemical composition of AlSi10Mg alloy

Elements	*Al*	*Si*	*Fe*	*Cu*	*Mn*	*Ni*	*Mg*	*Zn*	*Pb*	*Sn*	*Ti*
Wt %	Balance	9–11	0.55	0.05	0.45	0.05	0.2–0.45	0.10	0.05	0.05	0.15

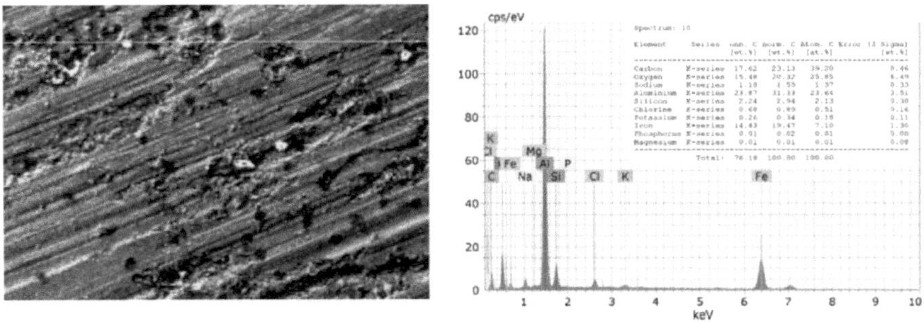

Figure 17.2 Chemical composition of AlSi10Mg.

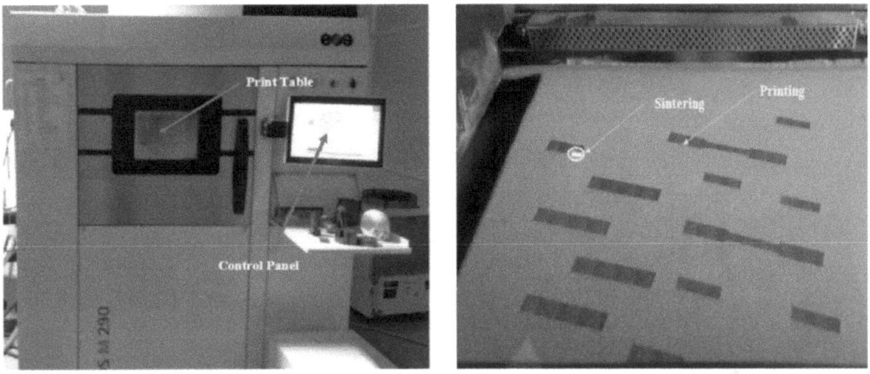

Figure 17.3 EOS M290 DMLS machine (left); Specimen Printing Table (right).

Table 17.2 DMLS machine specification

EOS M290 DMLS machine	
Laser type	yb-fiber laser
Laser capacity	400 W
Precision type	F-Theta Lens; high-speed scanner
Focus diameter	100 μm
Layer thickness	0.03 mm
Material used	AlSi10Mg powder
Fabrication method used	Powder bed fusion

17.3 EXPERIMENTAL METHODOLOGY

Next to printing, the specimens were post-processed and heated up to 300°C for 2 hours to relieve residual stress created while printing using glass beads of 70–110 μm, and is sand blasted at an operational pressure of 5 bar to achieve a good surface finish.

17.3.1 Fabrication of hardness and tensile specimens

Specimens were printed as per standards for the hardness test by DMLS process. Vickers hardness testing was employed at four different places and the mean value is taken as hardness. Specimen were printed as per ASTM E384 standard for Vickers hardness test (Figure 17.4). The tensile test specimens were also made as per the ASTM E-08 standard to test its tensile strength. The tensile test was performed at a computer-controlled universal testing machine as exhibited in Figure 17.4b.

17.3.2 Fabrication of pins for friction and wear test

The specimen to be used for the wear test is fabricated using DMLS process as per ASTM standard, and is shown in Figure 17.5. The wear test specimen is cylindrical in shape and of dimension Ø10×30 mm as per ASTM G99-05 standard.

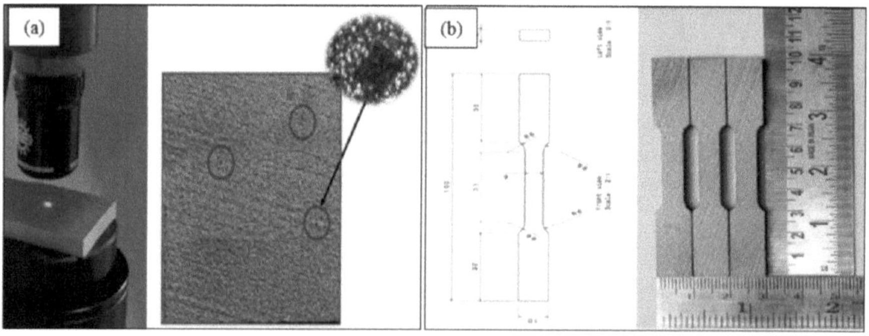

Figure 17.4 Hardness and tensile test specimens.

Figure 17.5 DMLSed pins for wear test.

17.4 EXPERIMENTATION

17.4.1 Effect of DMLS on hardness of Al10SiMg alloy

Hardness test of different pieces was conducted (Figure 17.6) and the average value is taken for discussion. The observed hardness values are listed in Table 17.3.

17.4.2 Effect of DMLS on tensile strength of Al10SiMg alloy

Tensile test of different specimens was conducted and the average value is taken for the discussion. The specimen after testing is shown in Figure 17.7. The observed tensile test results are presented in Table 17.4. The graph

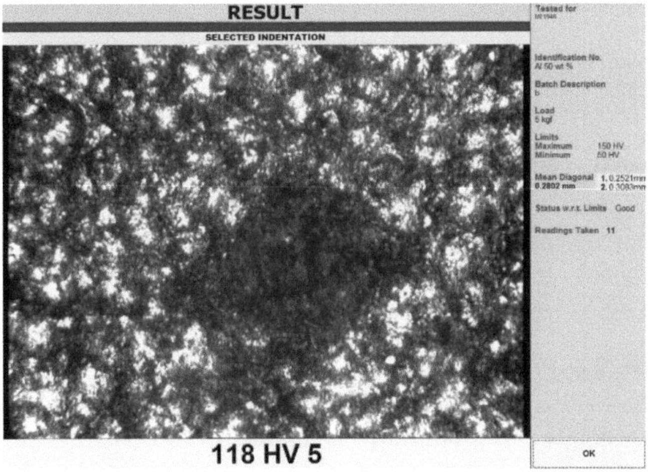

Figure 17.6 Hardness tested on specimen.

Table 17.3 Hardness test values

Sl. no.	*Specimen*	*Results*	*Vickers hardness*
1	T_1	122 120 118 126	122
2	T_2	119 124 120 123	122

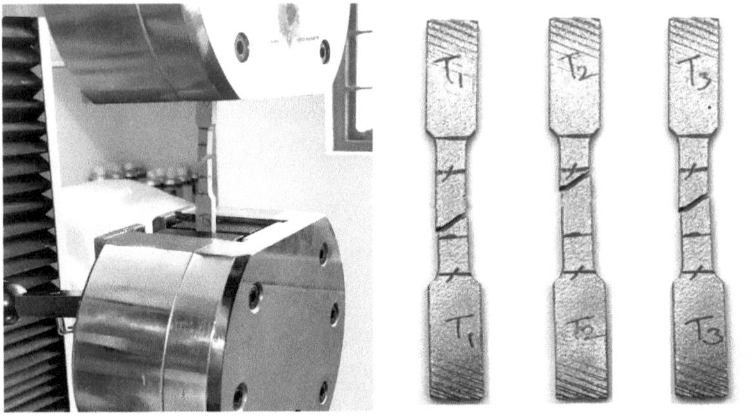

Figure 17.7 Tensile test on UTM (left); specimens after tensile test (right).

Table 17.4 Tensile test results

Sl. no.	*Specimen*	*Elongation at peak (mm)*	*Load at peak (kN)*	*UTS of DMLSed AlSi10Mg alloy (N/mm2)*	*UTS of as-cast AlSi10Mg alloy (N/mm2)*
1	T_1	7.040	10.860	352.728	
2	T_2	6.010	8.270	339.891	240–260 (Dvorakova et al., 2020)
3	T_3	8.104	12.067	365.662	

plotted in Figure 17.8 represents the load vs displacement curve of the tensile test; the load is specified in kN and displacement is specified in mm. Initial gauge length of the specimen is 25.00 mm and the final gauge length of the specimen is 26.76 mm.

17.4.3 Fractography analysis

Field Emission SEM is used to find the mechanism of fracture in the fractured specimens of each tensile test using the 1.5 nm resolution Carl Zeiss (USA) FESEM Machine with SE2 detector and the fractured area is viewed in 100, 20, 10, and 2 μm.

It is evident from Figure 17.9 that the observed fracture mode exhibits the near cup and cone formation which confirms the significant hardness with enhanced ductility that yields the increase in tensile strength (Wei et al., 2017) with reduced VED and MPB (Hong et al., 2020).

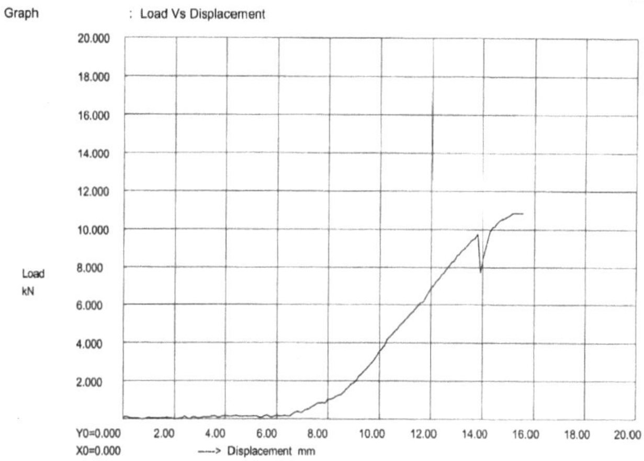

Figure 17.8 Stress strain curve of tensile test.

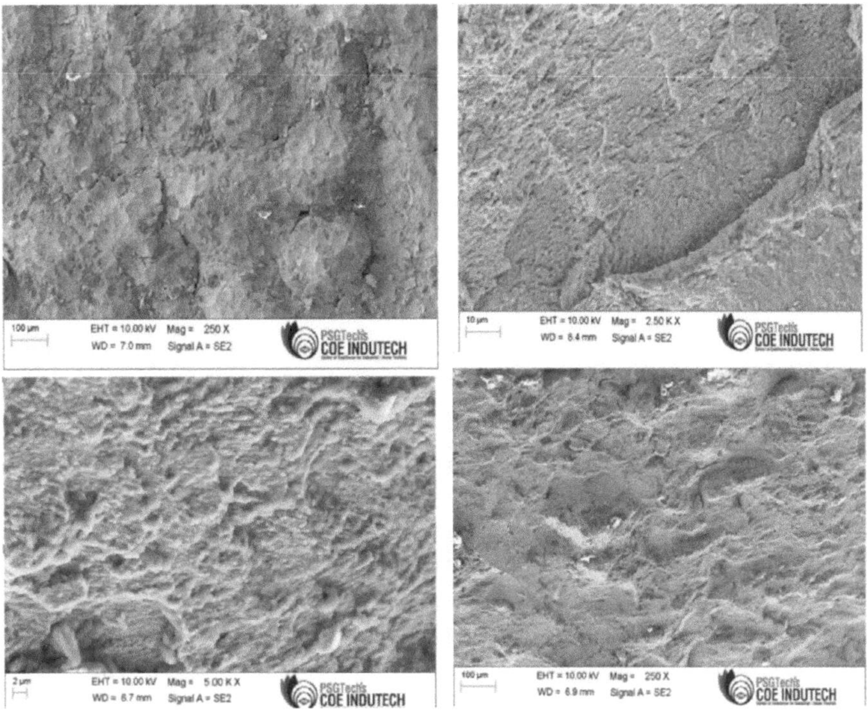

Figure 17.9 Fractography analysis of tensile tested specimen.

17.5 RESULTS AND DISCUSSION

17.5.1 Effect of DMLS process on friction and wear properties of Al10SiMg alloy

Based on Taguchi's L9 Orthogonal Array (OA), the experiments were designed, and the process parameters and their range were fixed (Table 17.5) from the pilot study and profound literature experience. The pin-on-disc apparatus used for this study is shown in Figure 17.10. As the experiment commences, the rotating disc gets into sliding contact with the surface of the pin. Hence, friction between the counterparts is produced due to which the workpiece gets heated up and softens the matrix resulting in loss of material as debris because of the intense plastic deformation. This helps to attain a fine-grained, uniform, and a defect-free structure.

The coefficient of friction and the specific wear rate are calculated using equations (17.1) and (17.2):

$$\text{Coefficient of friction (CoF)} = \text{Induced force/applied force} \tag{17.1}$$

$$\text{Specific wear rate }(\text{SWR}) = \Delta m/(\rho \times Fn \times D), \text{mm}^3/\text{Nm} \tag{17.2}$$

where

Δm = mass loss (g) = weight before wear – weight after wear,

ρ = density of material, g/mm^3,

Fn = frictional force (N),

D = sliding distance (m).

Table 17.5 Friction and wear test results

Sl. no.	*Applied load – A (N)*	*Sliding velocity – B (m/s)*	*Sliding distance – C (m)*	*Time (min)*	*Coefficient of friction*	*Specific wear rate (mm³/Nm)*
1	20	2.06	800	13.09	0.325	4.753×10^{-7}
2	20	4.18	1,600	13.15	0.205	2.376×10^{-8}
3	20	6.28	2,400	13.13	0.265	7.120×10^{-8}
4	30	2.06	1,600	25.40	0.156	9.505×10^{-8}
5	30	4.18	2,400	19.19	0.190	9.505×10^{-8}
6	30	6.28	800	4.25	0.263	3.643×10^{-7}
7	40	2.06	2,400	38.10	0.147	7.525×10^{-8}
8	40	4.18	800	6.38	0.255	2.614×10^{-7}
9	40	6.28	1,600	8.50	0.247	7.129×10^{-8}

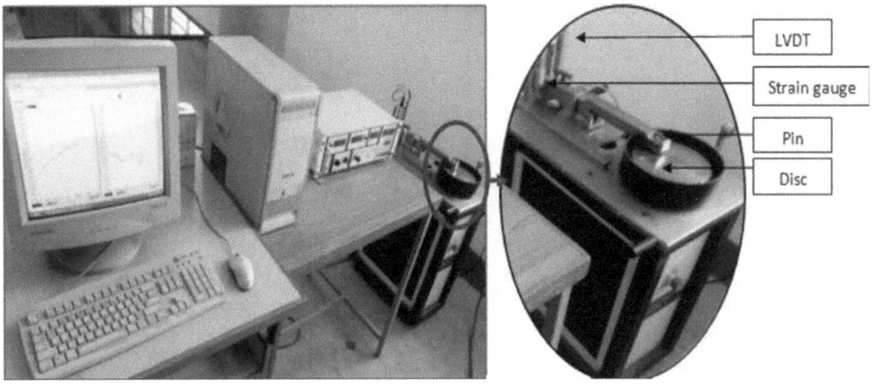

Figure 17.10 Pin-on-disc wear test apparatus.

The wear test of nine specimens was conducted and the minimum, maximum, and moderate value is taken for discussion as presented in Table 17.5. It was generally observed that with the minimum load and medium sliding distance, the observed wear is minimum. Also, when load is maximum and sliding distance is minimum, the maximum wear was recorded due to more plastic deformation. Whereas the friction is low when load is moderate and sliding distance is maximum, and friction is high when load and sliding distance are low.

From the experimental results presented in Table 17.5, it is observed from CoF results that at 40 N applied load, 2.06 m/s sliding velocity, and 2,400 m sliding distance the lowest coefficient of friction of 0.148 is observed and there exists the third-body layer, and at 20 N applied load, 2.06 m/s sliding velocity, and 800 m sliding distance the observed coefficient of friction is 0.325 which is maximum due to non-presence of third-body layer.

For 20 N applied load, 4.18 m/s sliding velocity, and 1,600 m sliding distance the minimum specific wear rate is observed as $2.376 \times 10^{-8} mm^3/Nm$, and the possible reason behind this phenomenon is the existence of tribo-layer. At 30 N applied load, 6.28 m/s sliding velocity, and 800 m sliding distance the maximum specific wear rate of $3.643 \times 10^{-7} mm^3/Nm$ is observed due to breakage of tribo-layer at that sliding velocity. From Figure 17.11a ,the influence of sliding velocity on the specific wear rate over the range of sliding distances is understood. At 2.06 m/s sliding velocity, the SWR gradually increases with increase in sliding distance and then slightly decreases with further increase in sliding distance. At 4.18 m/s sliding velocity, the SWR decreases gradually with increase in sliding distance and then slightly increases with further increase in sliding distance. At 6.28 m/s sliding velocity, the SWR decreases gradually with increase in sliding distance and then very slightly increases with further increase in sliding distance. It is evident that at medium and higher sliding velocities, the SWR is initially high due to high static friction between the pin and the disc. At medium sliding distance,

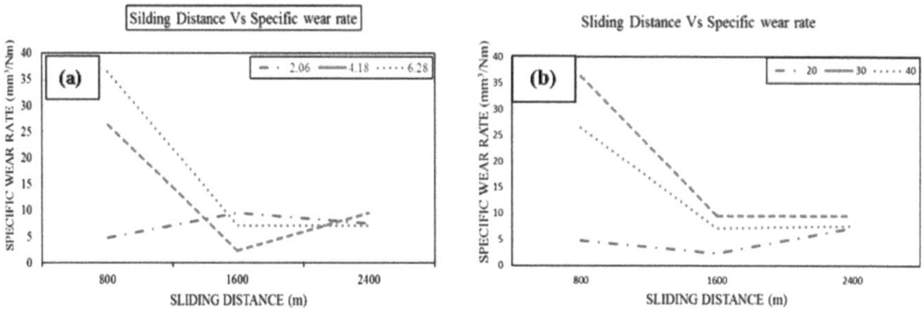

Figure 17.11 Graphical representation of specific wear rate and sliding distance at constant speed.

the SWR gradually decreases as a result of formation of a thin layer between the pin surface and the disk surface known as the "Tribo-Layer." After the formation of tribo-layer the contact area between the pin surface and the disc surface reduces which further results in less friction between surfaces and prevents wear at the pin surface. Therefore, the SWR at maximum sliding distance doesn't vary much with the SWR at medium sliding distance.

Figure 17.11b explains the interaction between the applied load and specific wear rate with respect to sliding distance. At minimum load the specific wear rate decreases with increase in sliding distance and then increases again at maximum sliding distance. The initial decrease in SWR observed for all three loading conditions is due to formation of third body, i.e., the tribo-layer between sliding surfaces, and as the sliding distance increases, the SWR starts increasing as the tribo-layer goes off due to the extended sliding distance. It is also recorded (Figure 17.11) that specimen runs at 6.28 m/s SV and 1,600 m SD exhibited improved friction and wear performance compared to all other samples.

17.5.2 SEM analysis of worn-out samples

The results of morphological analysis of worn-out samples are presented in Figure 17.12. These micrographs show the effect of sliding velocity and load over a span of sliding distance, i.e., 800–2,400 m. It is a common observation from Figure 17.12c that at maximum load and sliding velocity, the material has undergone a severe plastic deformation, hence there are a lot of patches found.

Figure 17.12d confirms that increase in sliding distance with maximum load and sliding velocity increases the specific wear rate and it is evidenced via more of adhesive patches and worn-out debris. The reason behind the low specific wear rate observed at 6.28 m/s SV and 1,600 m SD is the presence of tribo-layer during the sliding condition, and the same is confirmed with the EDAX results as shown in Figure 17.13.

Elements like Al, Si, Mg, and Fe have created their own oxide forms (Figure 17.13) during the test in the interface between sliding pairs. Hence, the tribo-layer is formed in between.

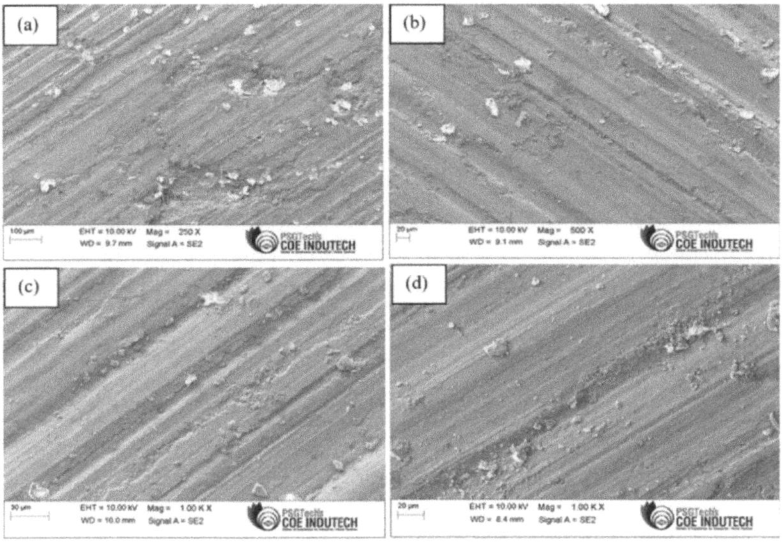

Figure 17.12 SEM image of worn-out specimen at 40 N applied load and 1,600 m sliding distance for (a) 2.06 m/s sliding velocity; (b) 4.18 m/s sliding velocity; (c) 6.28 m/s sliding velocity; and (d) 40 N applied load, 6.28 m/s sliding velocity, and 2,400 m sliding distance.

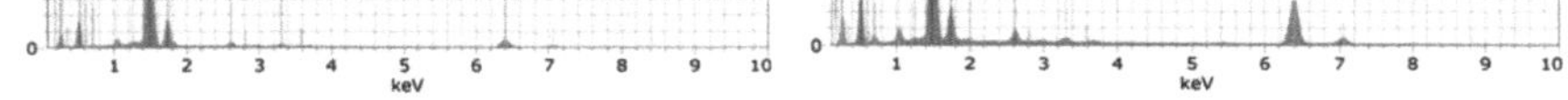

Figure 17.13 EDAX composition result of sample run for 1,600 m sliding distance at 40 N load (left); 2,400 m sliding distance at 40 N load (right).

17.6 SUMMARY

Using Direct Metal Laser Sintering Process, the AlSi10Mg alloy is fabricated, and the significant inferences obtained from mechanical and tribological tests carried out on DMLSed AlSi10Mg alloy are as follows.

17.6.1 Mechanical behavior

Tensile strength tests showed significant differences between AlSi10Mg alloy obtained by casting and DMLS technology due to improved grain size and hardness. Due to fine microstructure, DMLSed sample has higher hardness than as-cast material. Experimental results revealed that DMLSed AlSi10Mg alloy showed 9.25% higher hardness than as-cast AlSi10Mg alloy. Similarly, DMLSed AlSi10Mg alloy exhibited 76.47% higher tensile strength than as-cast AlSi10Mg alloy.

17.6.2 Tribological behavior

It was generally observed that the friction is low when load is moderate and sliding distance is maximum, and friction is high when load and sliding distance are low. With the minimum load and medium sliding distance, the observed wear rate is minimum. Also, when load is maximum and sliding distance is minimum, the maximum wear was recorded due to more plastic deformation. However, sliding for a longer time and distance causes strain hardening of specimen that even hardens the specimen and reduces the wear rate. The presence of Si in AlSi10Mg alloy improves the hardness and ultimate tensile strength of the specimen, and this is why the DMLSed specimen lasts longer even at higher loads and longer sliding distances.

REFERENCES

Aboulkhair, N. T., I. Maskery, Chris Tuck, I. Ashcroft, and N. M. Everitt. 2016. "On the Formation of AlSi10Mg Single Tracks and Layers in Selective Laser Melting: Microstructure and Nano-Mechanical Properties." *Journal of Materials Processing Technology* 230: 88–98.

Aboulkhair, N. T., M. Simonelli, L. Parry, I. Ashcroft, C. Tuck, and R. Hague. 2019. "3D Printing of Aluminium Alloys: Additive Manufacturing of Aluminium Alloys Using Selective Laser Melting." *Progress in Materials Science* 106: 100578.

Binnur, S. 2019. "Investigating the Tribological Performance of AlSi10Mg Parts Manufactured by Direct Metal Laser Sintering." *4th International Congress on 3D Printing (additive manufacturing) Technologies and Digital Industry, Antalya, TR*, 11–14 April 2019.

Dvorakova, J., Dvorak, K., & Cerny, M. 2020. Influence of DMLS method topology on mechanical properties of alloy AlSi10Mg. *MendelNet*, *27*, 445–450.

Girelli, L., M. Tocci, L. Montesano, M. Gelfi, and A. Pola. 2017. "Optimization of Heat Treatment Parameters for Additive Manufacturing and Gravity Casting AlSi10Mg Alloy." *IOP Conference Series: Materials Science and Engineering* 264: 012016.

Hong, W., Y. Ren, J. Ren, A. Cai, M. Song, Y. Liu, X. Wu, Q. Li, W. Huang, X. Wang, I. Baker. 2020. "Effect of Melting Modes on Microstructure and Tribological Properties of Selective Laser Melted AlSi10Mg Alloy." *Virtual and Physical Prototyping* 15: 570–582.

Jana, D., K. Dvorak, M. Cerny. 2020. "Influence of DMLS Method Topology on Mechanical Properties of Alloy AlSi10Mg." Mendelnet, Brno, Czech Republic, pp. 445–450.

Leary, M., T. Maconachie, A. Sarker, O. Faruque, and M. Brandt. 2019. "Mechanical and Thermal Characterisation of AlSi10Mg SLM Block Support Structures." *Materials & Design* 183: 108138.

Massimo, L., A. Aversa, D. Manfredi, F. Calignano, E. Paola Ambrosio, D. Ugues, and M. Pavese. 2016. "Tribological Behavior of Aluminum Alloy AlSi10Mg-TiB_2 Composites Produced by Direct Metal Laser Sintering (DMLS)." *Journal of Materials Engineering and Performance* 25: 3152–3160. doi: 10.1007/s11665-016-2190-5.

Massimo, L. 2019. "Tribological and Wear Behavior of Metal Alloys Produced by Laser Powder Bed Fusion (LPBF)," in Mohammad Asaduzzaman Chowdhury (ed.), *Friction, Lubrication and Wear*, IntechOpen, doi: 10.5772/intechopen.85167.

Sol, T., S. Hayun, D. Noiman, E. Tiferet, O. Yeheskel, and O. Tevet. 2018. "Nondestructive Ultrasonic Evaluation of Additively Manufactured AlSi10Mg Samples." *Additive Manufacturing* 22: 700–707.

Trevisan, F., F. Calignano, M. Lorusso, J. Pakkanen, A. Aversa, E. Ambrosio, M. Lombardi, P. Fino, and D. Manfredi. 2017. "On the Selective Laser Melting (SLM) of the AlSi10Mg Alloy: Process, Microstructure, and Mechanical Properties." *Materials* 10 (1): 76.

Wei, P., Z. Wei, Z. Chen, J. Du, Y. He, J. Li, and Y. Zhou. 2017. "The AlSi10Mg Samples Produced by Selective Laser Melting: Single Track, Densification, Microstructure and Mechanical Behavior." *Applied Surface Science* 408: 38–50.

Wu, H., J. Ren, Q. Huang, X. Zai, L. Liu, C. Chen, S. Liu, X. Yang, and R. Li. 2017. "Effect of Laser Parameters on Microstructure, Metallurgical Defects and Property of AlSi10Mg Printed by Selective Laser Melting." *Journal of Micromechanics and Molecular Physics* 2 (4): 1750017.

Xiong, Z. H., S. L. Liu, S. F. Li, Y. Shi, Y. F. Yang, and R. D. K. Misra. 2019. "Role of Melt Pool Boundary Condition in Determining the Mechanical Properties of Selective Laser Melting AlSi10Mg Alloy." *Materials Science and Engineering: A* 740: 148–156.

Chapter 18

Design and development of customized finger prosthesis with novel mechanism using additive manufacturing

Sivakumar Ganesan, Senthilmurugan Arumugam, Mohan Pushparaj, and Rajesh Ranganathan
Coimbatore Institute of Technology

CONTENTS

18.1 INTRODUCTION

Limb amputations cause severe deformity, this happens to be either by accident or is congenital. The World Health Organization (WHO) reports that about 38 million patients experience amputation in developing countries. Among these 2.4 million patients experience amputation of upper limb, namely hand, fingers and fingertips (Alturkistani et al., 2020). Patients with defects could not afford prosthetic assistance for their deformity due to the cost and effort involved in the development of prosthesis. The amputation of upper limb in the year 2005 was recorded as 1.6 million, due to various reasons, and it is expected to reach 3.6 million by the year 2050 (Ziegler-Graham et al., 2008). Among these about 6.6% of finger amputations take place due to congenital and traumatic injuries (Jang et al., 2011). The complex biomechanism of hands and fingers

DOI: 10.1201/9781003430186-18

supports the performance of highly capable grasping tasks. Tendons in the human hand help to transmit motions to the fingers by remote muscles. Finger prosthesis is broadly classified into body-powered and extracorporeal energy-powered prosthesis (Difonzo et al., 2020). The involvement of additional parts, namely motor, wire-line, used to mobilize mechanism of the prosthesis was minimized to decrease weight of the prosthesis. Therefore, body-powered finger prosthesis allows the patient to perform daily tasks by using strength of the coupled body part which decides the load-carrying capacity of prosthetics.

The design and manufacture of finger prosthesis for patients are growing due to cost and time involved with the process. Compared to conventional manufacturing methods, additive manufacturing supports customization of prosthetics to perform tasks (Varacallo, 2019). Anatomy of a finger consists of three major parts, namely ligaments, tendons and three phalanges. The very first bone adjacent to the palm is proximal phalange, the next bone is middle phalange and the smallest is distal phalange. Human finger has normally three joints, namely (i) metacarpophalangeal joint (MCP), (ii) proximal interphalangeal joint (PIP) and (iii) distal interphalangeal joint (DIP). The thumb finger does not have middle phalange as shown in Figure 18.1.

Panchik et al. (2021) investigated the process of developing a 3D printed finger prosthesis. In addition, meaningful tasks were identified to be performed using the fingers to ease the customization process. The identified meaningful tasks include assessment utilization, time management, participant responsiveness, clinical reasoning, common language and available resources (Panchik et al., 2021). Young et al. (2019) explored the potential of 3D printing in the development of finger prosthetics, in which two types of body-powered 3D-printed partial finger prosthesis is developed, and assessed its task and efficiency. This in turn helps the people in underdeveloped areas to spend less for functional finger prosthesis

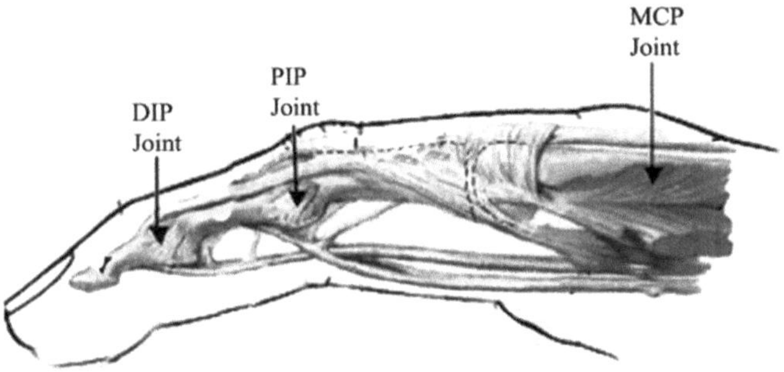

Figure 18.1 Anatomy of a finger indicating three major parts to understand its biomechanics.

(Young et al., 2019). Belvončíková et al. (2020) developed a 3D-printed finger prosthesis associated with two different flexion mechanisms, namely wire and lever mechanism and electromagnet mechanism, to achieve efficient working conditions. Based on the experimental study wire and lever flexion mechanism is the most effective one. Further, few design changes in the mechanism were suggested to improvise its performance in daily tasks (Belvončíková et al., 2020). Difonzo et al. (2020) discussed the ease of utilization of 3D printing in the development of finger prosthesis to achieve the essential mechanism mimicking the characteristics of natural movement of a finger. Furthermore, the categorization of prosthetics based on operating mechanism was reviewed along with the research trend for the past two decades. The study identified and presented Degree of Freedom (DOF), shape adaptability and linkage bar used for actuation of a developed finger prosthesis for better understanding. The scope for body-powered finger prosthesis actuated using a tendon cable was well stated (Difonzo et al., 2020).

This work involves the design and development of a novel body-powered mechanism to mimic the characteristics of a natural finger. It includes design iterations, optimization to minimize weight of the prosthetic device and number of parts in the product. The developed mechanism is divided into three functional parts, namely, base, middle and tip, of the prototype of the finger to reduce weight of the prosthetic device. The manufacturing process includes material selection, manufacture and standard testing methods. The final product was developed using ABS materials.

18.2 METHODOLOGY

The methodological approach to design and develop customized body-powered finger prosthesis was framed based on the obtained patient data as shown in Figure 18.2. The identified problem is to restore function of the hand with a passive finger for a finger amputated hand. In order to replicate the motions of a finger complete biomechanics of the finger was understood properly. Moreover, from the observation of biomechanics of a finger a novel mechanism was developed with a principle prototype. To execute the developed mechanism, a case study was adapted with a patient experiencing pinky finger amputated in an accident. The patient data was acquired to generate a completely customized prosthesis to hold and fix with MCP of the identified patient. AM manufacturing technology was used to manufacture the customized patient-specific prosthesis. Mechanism of developed finger prosthesis was validated with the support of feedback received from the patient. However, proposed methodology can be used to develop all the fingers of the hand with MCP left undamaged after the amputation.

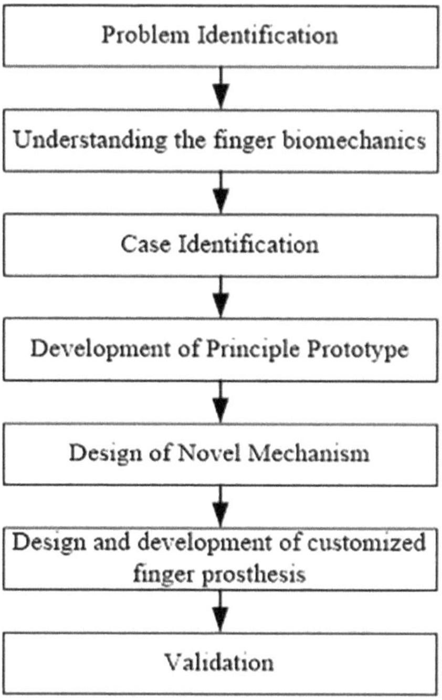

Figure 18.2 Methodology for the development of customized finger prosthesis.

18.3 METHODS

3D printing technology was used to develop a customized function of finger prosthesis. Developed novel mechanism was intended to perform operations of the finger. The mechanism consists of three major parts, namely base, middle and fore-end parts. The base part has a socket to hold the prosthesis in position on the finger of the patient. Middle part was designed to connect base and fore-end parts of the finger prosthesis. Fore-end of the prosthesis supports touching the desired objects during operations of the finger. Finally, a rigid connector part supports three major parts of the mechanism to remain connected through tendon and wrist of the patient. Design optimization was performed to decrease weight of the prosthesis. Acrylic material was used to manufacture the prototype in order to understand the outcome of desired biomechanism, and further design improvements were made with the mechanism. Finally major parts of the finger were made of ABS, which is used for socket. FDM (Fused Deposition Method) technology was used to manufacture the finger prosthesis and indulge in post-processing to achieve the finished product.

18.4 DESIGN AND DEVELOPMENT OF FINGER PROSTHESIS

18.4.1 Design process

A CAD modelling software, Solidworks 2020, was used to achieve desired design features through various design iteration. The objective of this work is to develop a functional prototype to fulfil visual and functions of the design.

The design process includes four stages in developing a desired functional design idea, namely, problem identification, data collection, analysis and develop solution through modifications in ideas. Initially, amputation of a finger leads to difficulty in performing the tasks. In order to compensate the loss of a finger an identical customized prosthesis was manufactured. Lots of nuances are involved in the customization of the product. In this work AM technology was adapted to perform complete customization based on the patient's data. The collection of patient data is significant to maintain accuracy of the developed product. Based on the information a novel mechanism was fabricated and validated. During validation process, testing was performed and changes were made through several iterations.

18.4.2 Biomechanics of finger

The three joints MCP, PIP and DIP of a finger play a predominant role in performing actions of the finger. MCP joint is the largest joint in the hand and performs the hinge operation in the hand and called as hinge joint. MCP permits flexion, extension, abduction and adduction motion. PIP joint is middle portion of the joint and acts as a middle knuckle joint. It is also called as middle knuckle joint. PIP allows motion in flexion and extension. DIP joint is the upper portion in the finger joint and acts as an upper knuckle joint. DIP permits flexion and extension motion. The soft tissues that connect muscles and bones are called tendons. The tendons stretch bone phalangeal during muscle contraction and result in finger movement. Tendons are soft tissues that connect muscles to bones. When muscles contract, tendons pull the bones causing the finger to move. Tendons are classified into two categories, namely flexor tendon and extensor tendon. The tendons on the palm side of the hand that support in bending the fingers are called flexor tendons. The tendons located on top of the hand that help in opening the fingers are called extensor tendons (Jhamtani et al., 2014). Tendons are operated based on movement of the wrist and MCP. Therefore, an MCP present in finger amputated hand can stimulate tendons through which finger motions are achieved. Hence, a finger amputated patient with MCP left out undamaged is capable of performing daily essentials with the help of prosthesis (Jhamtani et al., 2014).

18.4.3 Prototype development

Prototype of the finger was made to understand the function of the finger and its biomechanics. There are a number of prototypes that are available, namely (i) Proof of Principle Prototype, (ii) Working Prototype, (iii) Visual Prototype, (iv) User Experience Prototype, (v) Functional Prototype and (vi) Paper Prototype.

In this work proof of principle prototype is developed to test the mechanism of finger prosthesis which is to be developed as the final product. Prototype is developed with the help of acrylic material to check the working mechanism. This prototype has three parts. Acrylic material is cut into three parts and connected using nuts and bolts. A thread is used as tendon in the prototype. The mechanism is verified using this prototype. The thread is pulled, due to the pulling force, the thread moves the tip and middle part of the finger prototype. Figure 18.3 shows developed principle prototype of the mechanism using acrylic material and thread.

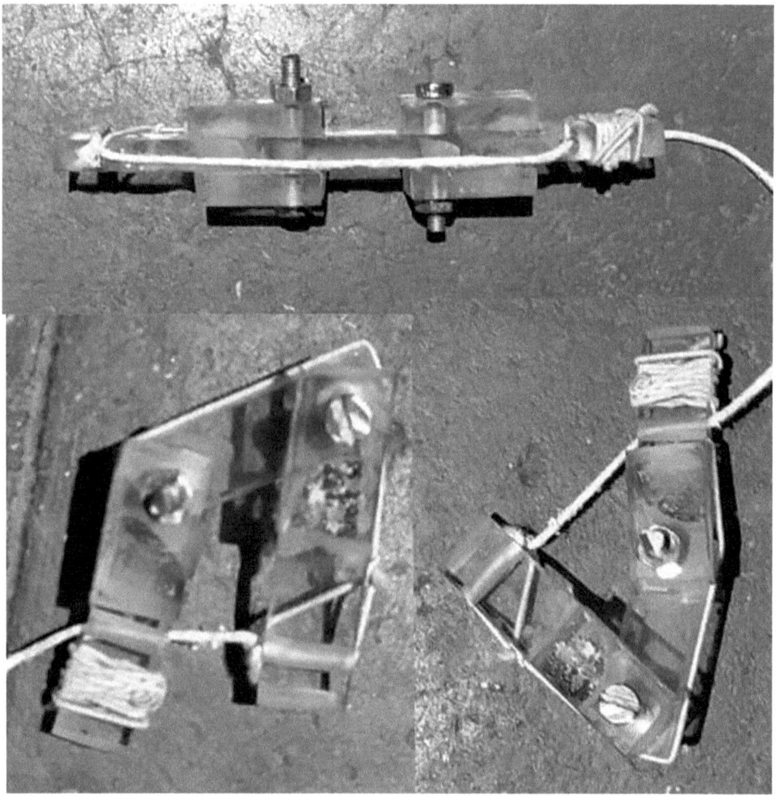

Figure 18.3 Developed principle prototype to validate functionality of the mechanism.

18.5 CASE STUDY

A 52-year-old male patient with a missing pinky finger was identified. He had a history of losing his finger in a road accident while riding a motorcycle. It is evident from the examination that there was loss of pinky finger with MCP joints left behind in the injured hand; some of the soft tissues were damaged. PIP and DIP of the finger were not present in the amputated area. There was no sign of any infection or inflammation in the injured region to plan prosthesis in place. Figure 18.4 shows the pinky finger amputated hand of the patient identified for the case study. The work focusses in rehabilitation using a prosthesis and removing the difficulties experienced with amputated finger. This helps to restore passive function of the finger by adapting the developed prosthesis. The development of prosthesis is completely based on the customization in order to provide patient-specific solution with affordable status. Moreover, the prosthesis should serve the functional purpose rather than aesthetics conditions of the finger. This supports in good retention of the product for comfortable usage by the patient.

Based on the collected patient data, prosthesis was developed as a unit finger to replace the missing one. Accordingly, a mechanism to mimic the biomechanics of the finger was validated and adapted to develop the design. A lightweight material was designed to withstand daily essential needs during the operation. As discussed, body-powered mechanism helps to execute functionality of the developed prosthesis. The process involved in the development to rehabilitate the patient was briefed to the patient and a written consent was obtained from the patient once procedure began.

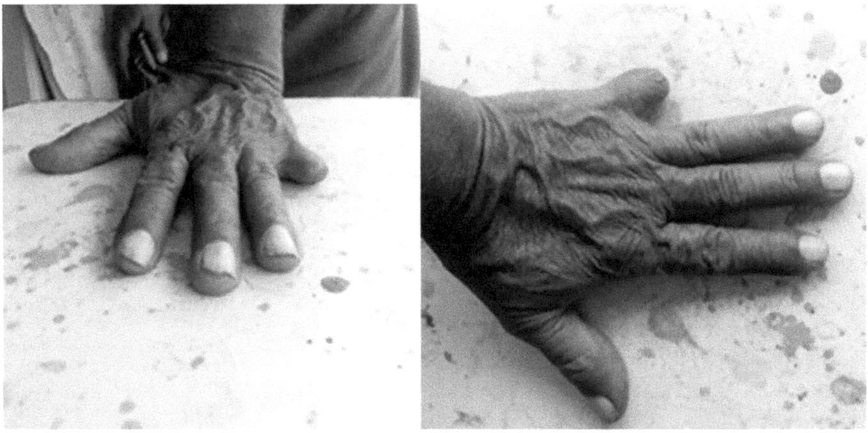

Figure 18.4 Patient with pinky finger amputated identified for case study.

18.5.1 Patient's data collection

The patient's hand profile data was obtained in order to develop a customized finger prosthesis. Here, two significant stages of the process involved are scanning and direct measurement of the hand profile. The Artec Eva 3D scanner was used to scan and obtain amputated hand profile of the patient. The accuracy of scanned model is 0.1 mm with 0.5 mm 3D resolution exported in stl format. The contralateral finger on the right hand of the patient was 3D scanned to achieve dimensions of the amputated finger. Direct measurement was taken using vernier scale, ruler on the amputated hand. Though chances for error during manual measurement are high in direct measurement when compared to 3D scanning true dimensions are necessary for design and development. The required true dimensions of the amputated hand are original finger height, amputated finger height, circumference of amputated finger and radius of finger which are 64, 23, 61 and 9.71 mm, respectively. Based on the obtained dimensions the socket to fit the MCP of the finger was developed.

From the obtained scan data of the patient's hand profile, stl format was converted into an editable format and featured in Solidworks 20 software to design a customized model of finger prosthesis into three major parts, namely base, middle and tip parts, as shown in Figure 18.5. The main characteristics of the developed model were to mimic the natural finger operation.

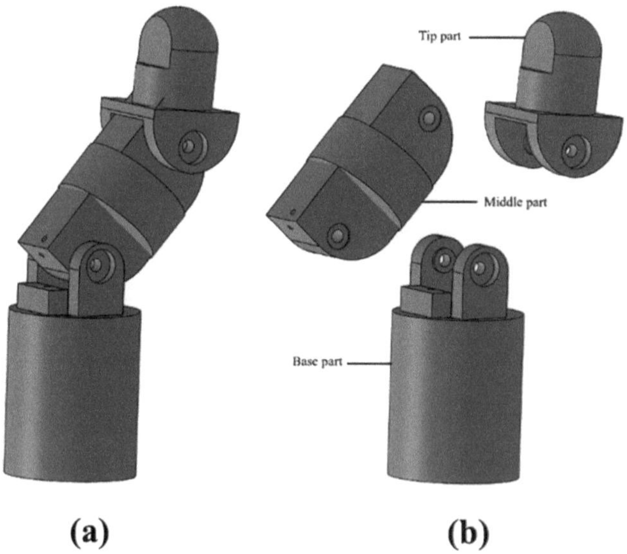

Figure 18.5 (a) Assembled parts of developed finger prosthesis and (b) exploded view of the customized finger prosthesis.

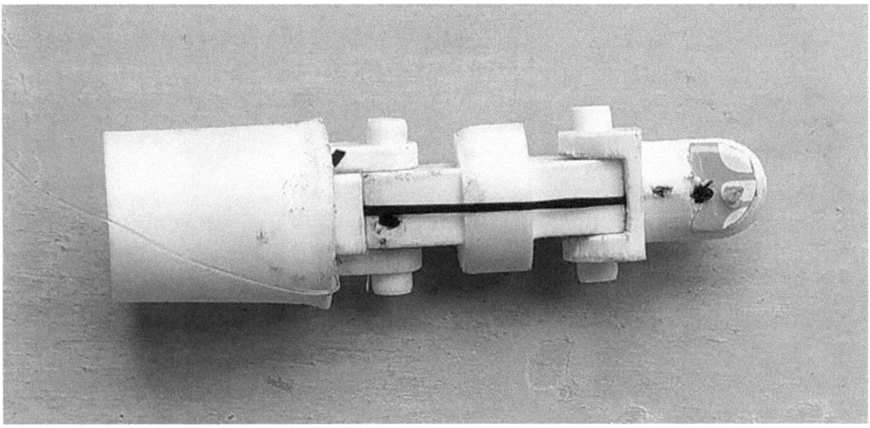

Figure 18.6 3D-printed customized finger prosthesis using a novel mechanism.

18.5.2 Manufacturing

Additive manufacturing (AM) technology is capable of creating products with complex shape and structure with appropriate material input. The advantage over conventional technique is controlled material wastage and customized products without making moulds and tools for product personalization (Kristiawan et al., 2021). Fused Deposition Modelling (FDM) is one of the popular methods in AM because of its affordable nature in product development. The fundamental concept of FDM manufacturing process is raw material being melted and formed to build a new shape and structure by layer-by-layer printing process. The raw material is the filament, placed in a roll and then forced to pass through the temperature-controlled nozzle head. The nozzle head is heated to melt the filament and converted into a semiliquid form. The semiliquid is extruded precisely in an ultrathin layer to produce layer-by-layer structure. This helps to control the contour of the layer determined by the preset program of nozzle head. Based on the developed design model from the obtained measurements customized finger prosthesis was manufactured using FDM method.

The three major parts of the design were manufactured by FDM method and post-processed for better surface finish. The post-processing, namely, filing, boring and reaming, was performed to overcome the difficulties experienced during assembly. The customized body-powered finger prosthesis was printed using FDM and is shown in Figure 18.6. The assembled finger prosthesis was intended to perform the operation of natural finger by connecting the assembled parts with fibre thread. The artificial tendon material was connected with the finger prosthesis and wrist band to keep it together. The socket designed to fix on the MCP joint supports in body-powered movement of passive finger. When MCP joint was moved, it tends to stretch the connected linkage parts. This results in flexion motion of the passive finger, and due to elastic nature of the attached tendon material extension motion was achieved once MCP joint was set free.

18.6 CONCLUSION

The design and development of a finger prosthesis as an affordable and functional part remains challenging, because each and every patient has unique anatomical structure with amputation. In this work, 3D design model and AM technologies were employed to manufacture a customized functional finger prosthesis for various amputated configurations. It is evident from the feedback of the patient that the developed prosthesis was lightweight, compact and ergonomic. Since prosthesis is a body-powered mechanism, it does not require either electrical or electronic circuits to operate. The fabricated novel mechanism could be utilized to assist all the passive fingers of the hand that hold the necessary characteristics of the real finger. The FDM method was adapted because of lower cost involved in manufacturing the product. However, surface finish of the passive finger manufactured by FDM method needs to be improved to increase aesthetic value to the finger prosthesis. Therefore, scope for employing digital 3D printing method in manufacturing the product increases quality of the product.

REFERENCES

Alturkistani, Raghad, A. Kavin, Suresh Devasahayam, Raji Thomas, Esther L. Colombini, Carlos A. Cifuentes, Shervanthi Homer-Vanniasinkam, Helge A. Wurdemann, and Mehran Moazen. 2020. "Affordable Passive 3D-Printed Prosthesis for Persons with Partial Hand Amputation." *Prosthetics and Orthotics International* 44 (2): 92–98. https://doi.org/10.1177/0309364620905220.

Belvončíková, Dominika, Lucia Bednarčíková, Monika Michalíková, Branko Štefanovič, Marianna Trebuňová, Viktória Mezencevová, and Jozef Živčák. 2020. "Development of Mechanism for Finger Prosthesis." *Acta Mechanica Slovaca* 24 (4): 6–11. https://doi.org/10.21496/ams.2020.028.

Difonzo, Erasmo, Giovanni Zappatore, Giacomo Mantriota, and Giulio Reina. 2020. "Advances in Finger and Partial Hand Prosthetic Mechanisms." *Robotics* 9 (4): 1–29. https://doi.org/10.3390/robotics9040080.

Jang, Chul Ho, Hee Seung Yang, Hea Eun Yang, Seon Yeong Lee, Ji Won Kwon, Bong Duck Yun, Jae Yung Choi, Seon Nyeo Kim, and Hae Won Jeong. 2011. "A Survey on Activities of Daily Living and Occupations of Upper Extremity Amputees." *Annals of Rehabilitation Medicine* 35 (6): 907. https://doi.org/10.5535/arm.2011.35.6.907.

Jhamtani, Rekha, A. Meenakshi, C. Thulasingam, C. Sabarigirinathan, and Bhabagrahi Sahu. 2014. "Electro-Mechanical Finger Prosthesis: A Novel Approach for Rehabilitation of Finger Amputees." *Journal of Indian Prosthodontist Society* 14 (Suppl 1): 150–154. https://doi.org/10.1007/s13191-014-0388-5.

Kristiawan, Ruben Bayu, Fitrian Imaduddin, Dody Ariawan, Ubaidillah, and Zainal Arifin. 2021. "A Review on the Fused Deposition Modeling (FDM) 3D Printing: Filament Processing, Materials, and Printing Parameters." *Open Engineering* 11 (1): 639–649. https://doi.org/10.1515/eng-2021-0063.

Panchik, Daniel, Gina C. Feeney, Angeline A. Springer, Cristina G. Vanbrocklin, and Hannah E. Winters. 2021. "Designing a 3D Printed Prosthetic to Meet Task-Specific Needs : A Case Study Designing a 3D Printed Prosthetic to Meet Task-Specific Needs : A Case Study." *Internet Journal of Allied Health Sciences and Practice* 19 (2).

Varacallo, Matthew. 2019. *Anatomy, Shoulder and Upper Limb, Metacarpophalangeal Joints Anatomy, Shoulder and Upper Limb, Metacarpophalangeal Joints*. Treasure Island (FL): StatPearls Publishing.

Young, Keaton J., James E. Pierce, and Jorge M. Zuniga. 2019. "Assessment of Body-Powered 3D Printed Partial Finger Prostheses: A Case Study." *3D Printing in Medicine* 5 (1): 1–8. https://doi.org/10.1186/s41205-019-0044-0.

Ziegler-Graham, Kathryn, Ellen J. MacKenzie, Patti L. Ephraim, Thomas G. Travison, and Ron Brookmeyer. 2008. "Estimating the Prevalence of Limb Loss in the United States: 2005 to 2050." *Archives of Physical Medicine and Rehabilitation* 89 (3): 422–429. https://doi.org/10.1016/j.apmr.2007.11.005.

Index